METAL CLUSTERS

METAL CLUSTERS

Edited by

MARTIN MOSKOVITS
Department of Chemistry
University of Toronto
Toronto, Ontario, Canada

A WILEY-INTERSCIENCE PUBLICATION
JOHN WILEY & SONS
New York · Chichester · Brisbane · Toronto · Singapore

Copyright © 1986 by John Wiley & Sons, Inc.

All rights reserved. Published simultaneously in Canada.

Library of Congress Cataloging in Publication Data:

Metal clusters.

   "A Wiley-Interscience publication."
   Includes index.
   1. Metal crystals.   I. Moskovits, Martin.
QD921.M466 1986   546'.3   86-1535
ISBN 0-471-89388-9

Printed in the United States of America

10 9 8 7 6 5 4 3 2 1

CONTRIBUTORS

DR. ROGER C. BAETZOLD, Research Laboratories, Eastman Kodak Company, Rochester, New York

DR. JOHN S. BRADLEY, Corporate Research Laboratories, Exxon Research and Engineering Co., Annandale, New Jersey

DR. ALAN BRENNER, Department of Chemistry, Wayne State University, Detroit, Michigan

DR. DAVID H. FARRAR, Department of Chemistry, University of Toronto, Toronto, Ontario, Canada

DR. PIERRE GALLEZOT, Institut de Recherches, sur la Catalyse, C.N.R.S., Villeurbanne, France

DR. BRUCE C. GATES, Department of Chemical Engineering, University of Delaware, Newark, Delaware

DR. JAMES L. GOLE, High Temperature Laboratory, Center for Atomic and Molecular Science, School of Physics, Georgia Institute of Technology, Atlanta, Georgia

DR. ROBERT J. GOUDSMIT, Department of Chemistry, University of Toronto, Toronto, Ontario, Canada

DR. MARTIN MOSKOVITS, Department of Chemistry, University of Toronto, Toronto, Ontario, Canada

DR. A. J. POË, J. Tuzo Wilson Laboratories, Erindale College, University of Toronto, Mississauga, Ontario, Canada

PREFACE

Research in atomic and molecular clusters is an interdisciplinary enterprise involving experimental and theoretical chemists and physicists. It has exploded rapidly in importance over the past 10 years, peppering scientific meetings with a barrage of symposia. It has engendered a Gordon conference and rumors (one hopes false ones) that a journal is being planned entitled *Cluster Science*.

Of all the research involving atomic and molecular clusters, the largest portion is dedicated to the study of aggregates of metal atoms. This is perhaps because this territory offers the greatest scope both in terms of chemical and physical challenge and possible utility in the field of catalysis. This book is the first attempt to cover this territory widely. Being the first of its kind, the book is in part a somewhat didactic general introduction to the field, as well as a monograph summarizing the research accomplishments realized to date.

The book travels from theory, through inorganic chemistry, past physical chemistry and chemical physics, ending in catalysis. Chapter 2 outlines the molecular orbital techniques used to predict the electronic and geometrical structure of metal clusters. It considers, in addition, the confidence with which one may carry over the results obtained with aggregates containing a few atoms to larger metal crystallites and metal crystal surfaces. In Chapter 3 the structures of stable metal cluster complexes are discussed and illustrated. The empirical rules that are currently used to account for the particular stabilities of complexes based on metal aggregates of specific shapes and nuclearity is presented. Chapter 4 discusses the kinetics of reaction and photoreaction involving metal cluster complexes, especially carbonyls. Much of its contents has not been compiled or reviewed elsewhere. The organometallic chemistry of metal clusters is reviewed in Chapter 5, including a summary of homogeneous catalysis by metal cluster compounds and a discussion of the analogy between the reactivity and structure of ligands on metal clusters versus adsorbate on metal crystal surfaces. Chapter 6 is a thorough review of the methods used to generate metal clusters in the gas phase, mainly through the use of molecular beam techniques. The results of laser-based spectroscopic techniques are presented, as well as preliminary studies involving reactions of metal clusters of various size with chlorine. Clusters frozen in noble gas matrices form the central theme of Chapter

7. Both naked metal aggregates and clusters with adsorbate attached are discussed. An introduction is also included to the very recent experiments probing the gas phase reactivity of metal clusters of various sizes toward adsorbate. Zeolites are ideal supports for very tiny metal clusters. Their molecular size pores and cavities will only admit the smallest aggregates. Chapter 8 reviews much of what is known regarding the structure of metal clusters enclosed in zeolites. The final two chapters present a review of applications of metal cluster complexes to the creation of new catalytic materials. The possibility of making catalysts of uniform cluster size distribution by denuding a cluster complex of its ligands and retaining the metal core on a catalyst support material is an exciting one. The two chapters discuss this quest and present the outcome of tests of catalysts made in this manner.

As usual, this volume rests on the efforts of many individuals, not all of whom can be acknowledged here. Thanks are due foremost to the contributors whose material forms the substance of the book; to my students and collaborators: Dr. John Hulse, Dr. Robert Lipson, William Limm, Dr. Therese Mejean, Andrew Kirkwood, Dr. Peter McBreen, Dr. Dorit Hall, Douglass Miller, Carolyn Preston, and David Zargarian, who have taught me much. Special thanks are due to my wife Linda who encouraged me to complete the task.

MARTIN MOSKOVITS

May, 1986

CONTENTS

METAL CLUSTERS

1

INTRODUCTION

MARTIN MOSKOVITS
Department of Chemistry
University of Toronto
Toronto, Ontario

The study of metal clusters is now decidedly a recognizable discipline, manifesting all the earmarks of a field of study in the process of unification, like a cosmic dust-cloud on its way to becoming a star. It has symposia and conferences organized in its name and is attended by participants whose backgrounds are so different that their presence in the same room would have been unexpected were it not for the cluster connection. Their perspectives and language have also changed. Organometallic chemists talk of the similarity between cluster complexes and molecules absorbed on metal surfaces. The same group is now aware of developments in heterogeneous catalysis and appreciate the connection between the chemistry of a metal cluster complex and reactions on large metal crystallites such as one might encounter in supported catalysts. Likewise, surface chemists using techniques such as ultraviolet photoelectron spectroscopy, electron energy loss spectroscopy, or surface infrared spectroscopy use metal cluster complexes as models on the basis of which to interpret their data. Chemical physicists studying the spectra of transition-metal diatomics produced in supersonic nozzle beams and detected by resonant two-photon ionization allude in their discussions to metal-metal multiple bonding, a concept begotten by and normally the concern of organometallic chemists. The products of organometallic synthesis are now finding their way into the hands of both chemical physicists, who wish to use them as precursers for making metal clusters in the gas phase (1) by photolysis, and catalysists, who hope to use these substances for making catalysts of uniform metal particle-size distribution by decomposing metal cluster carbonyls that have been adsorbed on oxide supports (2).

A brief historical synopsis of the study of metal clusters helps to bring it into perspective. Before 1960 metal clusters were studied mainly in the context of catalysis with little physical characterization, except for surface area studies and measurements of percent dispersion; the latter giving some idea of the cluster sizes involved. Some remarkable insights nevertheless were achieved regarding the role of multiple surface sites in catalysis which is sometimes referred to as "structure sensitivity" (3). Likewise, some commendable efforts were made in understanding the structure of the metal aggregates residing in the catalyst, as for example, in the classic electron microscopic studies of Prestridge and coworkers (4).

A major milestone in the study of clusters occurred in 1955 when Longuet-Higgins and coworkers (5) predicted the icosahedral structure of $B_{12}H_{12}^{-2}$. This prediction was confirmed experimentally as one of the many score of important chemical discoveries that were made in the fifties, sixties, and early seventies in the study of the structure and chemistry of boranes (6). Although the boranes formed the first major group of cluster compounds about which precise chemical and structural information was acquired, they will be almost totally ignored in this book. Not because they are unimportant; quite the contrary, they are so important that several excellent monographs dedicated to them already exist (6).

The 1960s also produced the first cluster carbonyls such as $Co_4(CO)_{12}$, $Fe_2(CO)_9$, and $Fe_3(CO)_{12}$, and the famous $Re_2Cl_8^{-2}$ ion which was shown by Cotton to possess the now famous metal-metal quadruple bond (7). This launched a field in its own right, culminating in the synthesis of dozens of triply and quadruply bonded bimetal complexes (8).

The 1960s, as well, saw the development of modern techniques of surface analysis for the study of molecules adsorbed on single crystal metal surfaces. In discussions on the vibrational and electronic data obtained therefrom more and more use is made of transition-metal complexes as models for molecules adsorbed on metals.

Enormous advances were made in the 1970s in the synthesis of large cluster complexes. Premier among the players here were the late Paolo Chini and coworkers (9). Compounds such as $Rh_{13}(CO)_{24}H_x^{5-x}$, $Pt_{26}(CO)_{32}^{-2}$, $Pt_{38}(CO)_{44}H_x^{-2}$, and $[Pt_3(CO)_6]^{-2}$ were reported.

The similarities between what surface scientists observe and what organometallic chemists synthesize prompted the late Earl Muetterties to formulate the "surface cluster analogy" (10) in the mid-1970s. This statement alerted both camps (surface chemists and cluster chemists) to the presence of each other. Many fundamental questions arise as a consequence. For example, Why are small naked metal clusters such good catalysts, whereas cluster carbonyls, often containing approximately the same number of metal atoms, are usually rather indifferent catalysts? How many metal atoms must a cluster contain before its properties are indistinguishable from the bulk? (This last question, it was soon realized, was really a family of questions since the answer depended dramatically on whether one was interested in a cluster's chemical, optical, electronic, or magnetic properties among others.)

Several strategies were employed in the 1970s in an attempt to address these questions. Theoreticians began to apply molecular orbital techniques (11) in improving the level of understanding of both the nature of the metal-metal bonding within clusters and the adsorbate—metal bond. Matrix isolation techniques were used to make both naked and adsorbate-covered metal clusters which could then be studied using a wide variety of spectroscopic methods (12). The related technology of metal vapor synthesis was developed (13). The new techniques, developed originally for surface science, began to be employed in the 1970s to study "real" catalysts. These methods, such as EXAFS (extended X-ray absorption, fine structure spectroscopy), revolutionize structural analysis of catalysts and catalyst materials (14).

Metal clusters were made and studied in nozzle beams first by Gole and coworkers (15) and by Schumacher and his group (16). The latter developed resonant two-photon ionization specifically for the study of alkali clusters (17). They were also able to measure the ionization potential of sodium clusters as a function of cluster size up to Na_{19} and discovered unusually stable and unusually unstable alkali cluster sizes (so-called magic numbers) among the aggregates they produced. Cluster distributions surpassing Na_{60} have now been achieved (18). Richard Smalley and his group (19) refined the nozzle-beam technique for studying just about any metal. Using an ingenious laser ablation method, researchers have now produced and studied cluster beams of a large number of metals. Two other groups—Argonne and Exxon—have developed this method further.

Most recently, the nozzle-beam technique has been turned toward the study of chemical reactions. Metal hydride, metal oxide, metal sulfide formation, and alkane dehydrogenation are among the reactions studied as a function of cluster size (20).

Enormous progress has been made in the past 20 years in metal cluster research; yet the goals now remain largely the same as then. In the fields of catalysis, chemistry, and physics the aim is to discover how a chemical process depends on cluster size; how the chemistry of a metal changes with its state of aggregation; and how structural, electronic, optical, and magnetic properties vary with cluster size, eventually reaching those of the bulk metal. Although for organometallic chemists the size of the largest soluble cluster complex and the general rule governing the sizes and structures of stable complexes remain unanswered questions, the generation of cluster complexes with unusually high catalytic activity is also still an unrealized goal.

For theory, too, the future holds its challenges. Even clusters as small as diatomics may prove difficult species to get right, as illustrated by the controversies surrounding Cr_2 (21).

REFERENCES

1. M. A. Duncan, T. G. Dietz, and R. E. Smalley, *J. Am. Chem. Soc.*, **103**, 5245 (1981); D. G. Leopold and V. Vaida, *ibid.*, **106**, 3720 (1984); D. P. Gerrity, L. J. Rothberg, and V. Vaida, *Chem. Phys. Lett.*, **74**, 1 (1980); D. A. Lichtin, R. B. Berstein, and V. Vaida, *J. Am. Chem. Soc.*, **104**, 1830 (1982).

2. X. J. Li, B. Gates, H. Knozinger, and E. Alizo, *J. Catal.*, **88**, 355 (1984); R. Vgo and R. Psaro, *J. Mole. Catal.*, **20**, 53 (1983); G. Collier, D. J. Hunt, S. D. Jackson, R. B. Moyes, I. A. Pickering, P. B. Wells, A. F. Simpson and R. Whyman, *J. Catal.*, **80**, 154 (1983); A. Brenner, *J. Mol. Catal.*, **5**, 157 (1979).

3. M. Boudart, in R. Gomer, Ed., *Interactions on Metal Surfaces,* Springer, Berlin, 1975.

4. E. B. Prestridge and D. J. C Yates, *Nature,* **234**, 345 (1971); E. B Prestridge, G. H. Via, and J. H. Sinfelt, *J. Catal.,* **50** 115 (1977).

5. H. C. Longuet-Higgins and M. de V. Roberts, *Proc. R. Soc.,* **A230**, 110 (1955).

6. E. L. Muetterties, *Boron Hydride Chemistry,* Academic, New York, 1975; W. N. Lipscomb, *Boron Hydrides,* Benjamin, New York, 1963.

7. F. A. Cotton, *Inorg. Chem.,* **4**, 334 (1965).

8. F. A. Cotton, *Acc. Chem. Res.,* **11**, 225 (1978); W. C. Troglen and H. B. Gray, *ibid.,* p. 232.

9. P. Chini, G. Longoni, and V. G. Albano, *Adv. Organometal. Chem.,* **14**, 285 (1976).

10. E. L. Muetterties, *Bull. Soc. Chim. Bdg.,* **84**, 959 (1975); *ibid.,* **85**, 451 (1976).

11. G. Blyholder and M. C. Allen, *J. Am. Chem. Soc.,* **91**, 3158 (1964); R. P. Messmer, S. K. Knudson, K. H. Johnson, J. B. Diamond, and C. Y. Yang, *Phys. Rev.,* **B13**, 1396 (1976); R. C. Baetzold, *J. Chem. Phys.,* **55**, 4363 (1971); J. C. Slata and K. H. Johnson, *Phys. Rev.,* **B5**, 844 (1972); A. B. Anderson, *J. Chem. Phys.,* **62**, 1187 (1975).

12. M. Moskovits and G. A. Ozin, Eds., *Cryochemistry,* Wiley-Interscience, New York, 1976.

13. P. S. Skell and L. D. Westcott, *J. Am. Chem. Soc.,* **85**, 1023 (1963); P. S. Skell, J. J. Havel and M. J. McGlinchey, *Acc. Chem. Res.,* **6**, 97 (1973); R. L. Timms, *ibid,* p. 118.

14. John H. Sinfet, *Bimetallic Catalysts,* Wiley, New York, 1983.

15. J. L. Gole, *Ann. Rev. Phys. Chem.,* **27**, 525 (1976).

16. A. Herrman, S. Leutwyler, E. Schumacher, and L. Woste, *Chem. Phys. Lett.,* **52**, 418 (1977).

17. A. Herrman, M. Hofmann, S. Leutwyler, E. Schumaker, and L. Woste, *Chem. Phys. Lett.,* **62**, 216 (1979).

18. E. Schumacher, M. Kappes, K. Marti, P. Radi, M. Schar, and B. Schmidhalter. *Ber. Bunsen-Ges. Phys. Chem.,* **88**, 220 (1984).

19. M. D. Morse and R. E. Smalley, *Ber. Bunsen-Ges. Phys. Chem.,* **88**, 228 (1984).

20. M. E. Geusic, M. D. Morse and R. E. Smalley, *J. Chem. Phys.,* **82**, 590 (1985); R. L. Whetten, D. M. Cox, D. J. Trevor, and A. Kaldor, *Phys. Rev. Lett.,* **54**, 1494 (1985); S. C. Richtsmeier, E. K. Parks, L. G. Pobo, and S. J. Riley, *J. Chem. Phys.,* **82**, 3659 (1985).

21. D. P. DiLella, W. Limm, R. H. Lipson, and K. V. Taylor, *J. Chem. Phys.,* **77**, 5263 (1982); V. E. Bondybey and J. H. English, *Chem. Phys. Lett.,* **94**, 434 (1983); M. M. Goodgame and W. A. Goddard, III, *J. Phys. Chem.,* **85**, 215 (1983); D. L. Michalopoulos, M. E. Geusic, S. G. Hansen, O. E. Powers, and R. E. Smalley, *J. Phys. Chem.,* **86**, 3914 (1982).

2

SOME TOPICS IN COMPUTATIONAL TREATMENT OF METAL CLUSTERS

ROGER C. BAETZOLD
Research Laboratories
Eastman Kodak Company
Rochester, New York

2.1   INTRODUCTION

The recent explosion of interest in small metal particles has been accompanied by widespread application of various theoretical methods to the problem. The variety of methods (and sometimes different results) are almost as numerous as the different groups attacking the problem. The methods that are applied have limitations even though several man-years of development may have been associated with any particular approach. These limitations often lead to differing calculated results, which have sparked numerous controversies in the field. This is characteristic of a young and growing area. I will attempt here to categorize the different theoretical approaches that have been used to study transition- and noble-metal clusters. Only some of the more obvious features of each method will be discussed since great detail is not needed for an overview of this field. Several applications of the methods to current relevant problems will be discussed, illustrating some strong and weak points associated with the various methods.

2.2   METHODS

2.2.1   Ab Initio Methods

Ab initio methods occupy a central place in theoretical studies of electronic structure. If a sufficiently flexible basis set is used with the full self-consistent-field (SCF) computational apparatus available in contemporary computer programs, excellent agreement with experiment can be obtained (1). Rarely can this limit be approached, however, for problems involving transition-metal multiple-atom clusters, and so the selection of basis sets and the use of approximation become crucial.

Ab initio methods are those in which the wavefunction is solved by accurately evaluating all of the integrals in the $2N$-electron Hamiltonian.

$$\hat{H} = \sum_{i}^{N} - \frac{\nabla_i^2}{2} + \sum_{i=1}^{N} \sum_{k=1}^{M} - \frac{z_k}{r_{ik}} + \sum_{j>i}^{N} \frac{1}{r_{ij}} + \sum_{l>k}^{M} \frac{Z_l Z_k}{R_{lk}} \qquad (2.1)$$

Here M centers containing Z electrons and separated by a distance R_{lk} interact with $2N$ electrons having nuclear-electron separation r_{ik}. At the first level of approximation the wavefunction Ψ is expressed as an antisymmetrized (2) product of spin orbitals ϕ_i in a single Slater determinant.

$$\Psi = \mathscr{A} \prod_{i=1}^{N} \phi_i \qquad (2.2)$$

Crucial at this stage is the choice of the form of the molecular orbitals (ϕ_i) since they are expanded as a linear combination of basis orbitals, χ_i. This basis set (3) may contain one function per atomic orbital (minimal), two per atomic orbital (double-zeta), or more per atomic orbital (extended). In addition, functions of a higher angular momentum quantum number (polarization) may be used to augment the description of each atomic orbital. Clearly, to account properly for changes in shape of the orbitals in a molecular or cluster environment, double-zeta or larger basis sets must be used.

The Schröedinger equation is solved (4) by minimizing the energy through the variation of the molecular orbitals. The energy is

$$
E = \sum_{i=1}^{N} 2\epsilon_i - \sum_{i,j=1}^{N} (2J_{ij} - K_{ij}) + V_N
$$

$$
\epsilon_l = \left\langle \phi_l \left| \frac{-\nabla_i^2}{2} + \sum_{k=1}^{M} \frac{-Z_k}{r_{ik}} \right| \phi_l \right\rangle + \sum_{j}(2J_{lj} - K_{lj})
$$

$$
J_{ij} = \left\langle \phi_i(1)\phi_i(1) \left| \frac{1}{r_{12}} \right| \phi_j(2)\phi_j(2) \right\rangle
$$

$$
K_{ij} = \left\langle \phi_i(1)\phi_j(1) \left| \frac{1}{r_{12}} \right| \phi_i(2)\phi_j(2) \right\rangle
$$

(2.3)

where ϵ_i is the orbital energy, J_{ij} is the coulomb integral, K_{ij} is the exchange integral, and V_N represents the nuclear repulsion. This simplest solution corresponds to two electrons per orbital and is termed *restricted Hartree-Fock*. At this level of solution the results may be easily transferred to a conceptual molecular orbital picture, but some ground-state properties often do not agree well with experiment as we will see later. Correlation effects, which allow electrons to interact instantaneously with each other rather than with the average field of the other electrons, may be introduced by the addition of a linear combination of Slater determinants. This approach, called CI (configuration interaction), improves calculated ground-state properties and permits dissociation to the proper atomic limits. Of course, this extension greatly increases the computational problem.

The application of pseudopotentials (or effective potentials) can greatly reduce the computational demands of a given problem. In this approach, a potential function is added to the Hamiltonian in equation (2.1) that will allow elimination of explicit treatment of core orbitals. The pseudopotential is usually chosen so as to lead to a computed spectrum of eigenvalues that will match the experimental spectrum of the atom or solid band structure. Goddard et al. (5) have pointed out several potential problems with this approach, including one of the most damaging for problems of interest here. That problem is whether choosing the correct eigenvalue spectrum guarantees that the correct shape and sizes of valence orbitals will be obtained. He advocates an ab initio effective-potential method that fits the effective potential to an all-electron ab initio calculation and therefore gives the correct shape of the valence orbitals.

One of the major problems with restricted Hartree-Fock calculations is that at the dissociation limit ionic states are formed. This problem is exemplified by the textbook example for H_2, where at infinite separation the wavefunction contains equal ionic and covalent contributions, whereas only covalent contributions are desired. One solution to this problem is the *unrestricted Hartree-Fock* theory in which different spatial functions are used for the pairs of up- and down-spin orbitals. Alternatively, valence-bond methods that contain only covalent contributions at large distances may be used.

The generalized valence-bond method (GVB) developed by Goddard and co-workers (1,5) has the correct dissociation limit and solves for the optimal orbitals as in Hartree-Fock theory. For H_2, the Hartree-Fock wavefunction

$$\Psi^{HF} = \mathscr{A}[\phi_1\phi_2(\alpha\beta)], \qquad \langle\phi_1|\phi_2\rangle = 0 \tag{2.4}$$

is replaced by a wavefunction in which $\langle\phi_1|\phi_2\rangle \neq 0$. By use of this prescription the singlet electron pairs are allowed to correlate

$$\Psi^{GVB} = \mathscr{A}[\phi_1'\phi_2'(\alpha\beta - \beta\alpha)], \qquad \langle\phi_1'|\phi_2'\rangle \neq 0 \tag{2.5}$$

This modifies slightly the form of the variational equations to be solved but automatically includes some of the correlation effects in the solution. In addition, further configuration interaction calculations may be carried out to obtain more of the correlation energy.

2.2.2 X_α Methods

There are several variations of calculations carried out in the X_α scheme. The starting point is an equation derived by Slater (6) in which the exchange terms in the Hartree-Fock equations are replaced by the X_α potential

$$V_{X_\alpha} = -6\alpha\left(\frac{3}{4\pi}\rho\right)^{1/3} \tag{2.6}$$

derived from the free-electron theory. In this equation ρ is the electron density and α is a parameter for each atom. The one-electron Hartree-Fock-Slater equation is

$$\{-\nabla_1^2 + V_c(1) + V_{X_\alpha}(1)\}\phi_i(1) = \epsilon_i\phi_i(1) \tag{2.7}$$

where ϵ_i and ϕ_i are the eigenvalue and the molecular spin orbital, respectively, ∇_1^2 represents the kinetic energy operator, and $V_c(1)$ is the coulomb term (6). Note the use of the local energy-independent X_α potential to replace the nonlocal energy-dependent exchange potential of the Hartree-Fock equations in this procedure. The procedure considerably reduces the computational complexity of the problem to be solved compared to the Hartree-Fock method. All electrons are treated in this method, and apparently some electron-correlation contribution is provided by the X_α potential. The eigenvalues computed in X_α theory do not have the same meaning as eigenvalues computed in Hartree-Fock theory, where in the Koopmans' theorem approximation their negative value gives the ionization potential. Of course, this approximation neglects relaxation processes inherit in the ionization process. In the X_α methods a transition-state method is used to compute the ionization potentials

$$-\text{IP} = \frac{\partial \langle E_{X_\alpha} \rangle}{\partial n_i} \qquad (2.8)$$

where $\langle E_{X_\alpha} \rangle$ is the statistical total energy and n_i is the occupation number of the ith orbital. This procedure accounts for relaxation effects inherent in the ionization (7).

The most commonly used form of the theory (SCF-X_α-SW) makes muffin-tin approximations in which spherical potentials are used to partition space around a given molecule or cluster. This approximation has given good eigenvalues and wavefunctions but does not presently permit computation of an accurate energy. This is a limitation for many applications of the theory. Also, the spherical potentials of the muffin tin make it more appropriate for high-symmetry problems. Some work has been done using the overlapping spheres method in which more asymmetric spatial potentials are treated to divide space. This introduces an arbitrary choice of the degree of overlap. Other versions of the X_α theory have to permit computation of the total energy. These include the LCAO-X_α (8) and discrete variational methods (9).

2.2.3 Semiempirical Methods

Semiempirical methods of quantum chemistry are generally aimed at understanding the trends within a series of like structures after the parameters are fixed by comparison with experiment for some model species. Thus, the numerous integrals in equation (2.2) are replaced by their effective corresponding experimental counterparts. These methods are much less sophisticated than the procedures we have described before but they remain an important tool in quantum chemistry. Conceptual models are most easily derived from them, and the essential symmetry properties of wavefunctions are retained at this level.

The CNDO (complete neglect of differential overlap) (10) and MINDO (modified intermediate neglect of differential overlap) (11) methods exemplify one line of development of the theory. The central approximations of this method involve

$$\left\langle \chi_i(1)\chi_l(1) \left| \frac{1}{r_{12}} \right| \chi_j(2)\chi_k(2) \right\rangle = \langle il|jk \rangle = \langle ii|jj \rangle \delta_{il} \cdot \delta_{jk} \qquad (2.9)$$

$$S_{ij} = \langle \chi_i|\chi_j \rangle = \delta_{ij}$$
$$\delta_{ij} = \text{Krönicker delta}$$

This approximation of neglecting differential overlap is based on the knowledge that $\chi_i(1)\chi_j(1)$ is usually small unless $i = j$. In addition to this approximation, atomic ionization potential and electron affinity terms are identified with some of the core integrals in the Fock matrix. The resulting equations to be solved for a closed-shell case are

$$\det(F - E) = 0$$

$$F_{\mu\mu} = -\tfrac{1}{2}(IP + EA) + \gamma_{dd}^{A,A}(\tfrac{1}{2} - P_{\mu\mu})$$
$$+ \sum_B [(P_{dd}^B - M_B)\gamma_{dd}^{AB} + (P_{ss}^B - N_B)\gamma_{sd}^{AB}] \qquad (2.10)$$

$$F_{\mu\nu} = \beta_{AB}^0 S_{\mu\nu} - P_{\mu\nu}\gamma_{\mu\nu}^{AB}$$

where IP = atomic ionization potential
EA = atomic electron affinity
$\gamma_{dd}^{A,A}$ = one center electron-repulsion integral between d orbitals
$P_{\mu\nu}$ = bond density matrix element
$P_{d,d}^B$ = d-electron population on atom B
M_B = number of d electrons on atom B
P_{ss}^B = s,p-electron population on atom B
N_B = number of s,p electrons on atom B
$\gamma_{i,j}^{AB}$ = electron repulsion integral between orbitals of i,j type on atoms A,B
$\beta_{A,B}^0$ = empirical parameter for atoms A,B
$S_{\mu\nu}$ = overlap integral

The electron-repulsion and overlap integrals are calculated, and the equations are solved self-consistently. One major problem with the procedure has been finding appropriate parameters for simultaneously computing a wide range of different properties. It has been particularly difficult to fit bond lengths and bond energy with the same parameter set. A promising development that alleviates this problem is the MINDO procedure as modified for transition elements by Blyholder et al. (12,13). The off-diagonal term in equation (2.10) is modified by dividing by R, the internuclear separation. Favorable results for FeO clusters have been obtained.

The extended Hückel method (EHM) of Hoffmann (14) is probably the simplest procedure available for the treatment of the electronic structure of transition-metal atoms. It does not have its origins in a mathematical derivation starting from some elementary principle, but rather, it rests on the experience gained over many years of use starting with the Hückel theory employed in organic chemistry. It is useful to test models and explore concepts, but almost all workers would not advocate its use in quantitative work.

The extended Hückel theory begins with a secular equation similar in form to that which arises in ab initio theory.

$$\det(H - ES) = 0 \qquad (2.11)$$

In this method the effective Hamiltonian matrix elements are given by

$$H_{ii} = -IP_i$$

$$H_{ij} = \tfrac{1}{2}KS_{ij}(H_{ii} + H_{jj})$$

(2.12)

where IP_i is the experimental atomic ionization potential for the orbital i, S_{ij} is the computed overlap element between the i,j orbitals, and K is the Wolfsberg-Helmholtz constant, usually taken as 1.75. Slater orbitals with exponents determined by atomic Hartree-Fock theory are used. Energies are computed as a sum of the occupied energy levels (ϵ_i)

$$E = \sum_i n_i\epsilon_i$$

(2.13)

where n_i is the occupancy. The method avoids many of the issues complicating the ab initio theory, such as correlation energy, by appropriate choice of one-electron parameters. It is easily transferred into a conceptual picture of bonding.

The simplest version of the extended Hückel theory presented earlier has many limitations. One of these involves proper accounting for repulsive effects and therefore determination of bond lengths. Structural data are usually needed to fix the bond length. Of course, repulsive potentials of various types may be added to the Hamiltonian but they usually require some additional parameter. With this matter settled, the IP_i parameters are to be chosen for the appropriate valence-state configuration and charge. Tables of such parameters are available (15), and self-consistent procedures may be used to arrive at the final parameter set. We note that the parameters for a transition-metal atom in a high-oxidation-state cluster would be quite different from the appropriate parameters for the same atom in a metal. One main strength of the method is its ability to show how the frontier orbitals, which dominate many interactions, change as a function of the chemical variables at hand. This will be illustrated later.

2.3 RESULTS

2.3.1 Diatomic Species

Diatomic gas-phase metal fragments are one of the few cluster species for which fairly reliable experimental information exists for the noble and transition metals. It thus becomes possible to show how the various levels of computational approximation (particularly in ab initio theory) give results that approach the experimental limit. This has been done well for Ni_2 and Cu_2, and these results will be reviewed. In X_α and semiempirical theories, diatomic species often form the starting point for parameter selection, so the reader should be careful in making judgments of the quality of theory for these problems.

One of the most careful investigations of the effect of various levels of approximation on computed properties of Ni_2 has been presented by Noell et al. (16). They performed calculations using an effective potential for the argon core of the Ni atom at the Hartree-Fock, GVB, and configuration-interaction levels. The effective-potential method was first compared with all-electron calculations, giving good agreement in justification of the potential chosen. The basis sets employed Gaussian functions optimized for the $3d^9 4s^1$, ^3D state of the Ni atom, followed by an optimization to the lowest energy of a Ni_2 ground state. This ground state of Ni_2 was shown earlier (17) to be a triplet arising from $3d\delta$ hole states, whereas the $4s\sigma_g$ molecular orbital provides the major source of bonding.

The molecular wavefunctions were chosen at various levels of sophistication. First, a single-configuration, restricted Hartree-Fock (RHF) wavefunction was chosen. Second, a two-configuration wavefunction including correlation effects associated with $4s\sigma_g^2 \rightarrow 4s\sigma_u^2$ excitation was included at the GVB level. The next level of sophistication involved inclusion of configuration-interaction calculations, treating all distributions of the valence electrons with treatment of angular correlation effects between the bonding pair of electrons (POLSDCI). The most elaborate wavefunction involved all configurations having six or fewer open shells (SDCI), and thus represents the most complete CI calculation. The results are shown in Table 2.1.

Let us first review the experimental data. The computed dissociation energy reflects the degree of accuracy of the Ni_2 wavefunction. The experimental value of this quantity is determined from the third-law analysis of mass spectroscopic data, which requires information on the electronic partition functions and molecular data. Since these data were not available when the experiments (18) were originally performed, estimates placed D_e at 54.0 ± 5 kcal/mol^{-1}. Using the recently computed electronic-state information, Noell et al. (16) recomputed this value to be 45.5 ± 5 kcal/mol^{-1}. Good experimental values for the bond length and the vibration frequency are not available.

Table 2.1 clearly shows that a substantial amount of configuration interaction must be included for proper computation of the dissociation energy. Restricted Hartree-Fock calculations give only about a fourth of the dissociation energy even though the computed bond length is not nearly so dependent on the wavefunction. As an illustration of the sensitivity to choice of basis sets, a line is included in Table 2.1 showing GVB level calculations (17) used with a different basis set in separate

TABLE 2.1   Calculations of Ni_2 Properties at Various Levels

Type of Calculation	Reference	R_e (bohr)	D_e (kcal/mol^{-1})	W_e (cm^{-1})
RHF	16	4.50	11.0	199
GVB	16	4.60	17.8	169
POLSDCI	16	4.52	30.0	190
SDCI	16	4.27	43.4	—
GVB	17	4.3	31.9	287
Experiment	16,18	—	46 ± 5	—

work where the $3d^8 4s^2$ state of the Ni atom is considered. Clearly, the two results at the GVB level differ significantly, indicating the importance of the choice of a basis set. Finally, we note that X_α calculations for Ni_2 (19) differ significantly from these even in the qualitative bonding picture. Those studies predict no δ holes in the ground state. Overall, Table 2.1 shows that very sophisticated wavefunctions including many correlation effects are needed to compute bond energy accurately.

Diatomic copper clusters have been investigated in a number of recent papers. The popularity of this species for study stems from its good experimental characterization and the fact that qualitatively the bonding should be dominated by the $4s$ electrons, with the $3d$ electrons playing a leser role. Pelissier (20) has investigated Cu_2 by a set of effective-potential calculations using various levels of configuration interaction and compared results with previous calculations and experiment. Some of his comparisons are shown in Table 2.2. At the SCF level only about half of the experimental dissociation energy is calculated. This is true at either the all-electron or the effective-potential level of calculation. Bond lengths are much less sensitive to correlation effects, but they generally shorten as CI is added to the calculation. The extended Hückel (EHT) and X_α-SW calculations do a good job of reproducing the experimental values. In each procedure the $4s\sigma_g$ molecular orbital provides the dominant source of bonding, although in the X_α-SW procedure (25) s-d hybridization is much larger than the other procedures.

The dominant $4\sigma_g$ bonding orbital discussed earlier for Ni_2 and Cu_2 is also found for Fe_2 (27), but not for all other transition-metal diatomic clusters. Calculations for Cr_2 by Goodgame and Goddard (28) at the GVB-CI level predict an antiferromagnetic dimer as the ground state. A computed value of the dissociation energy of 0.3 eV is reported versus the experimental value of 1.0 ± 0.3 eV. Recent CI calculations (29) for Sc_2 predict a weakly bound ground state of 3 to 5 kcal/mol^{-1}. Recomputation of the mass spectrophotometric dissociation energy using the electronic-state information suggests that the experimental value may well be this low.

Some major conclusions from the study of these dimers are:

1. The Hartree-Fock procedure with appropriate degree of CI yields good agreement with experiment.

TABLE 2.2 Calculations of Properties of Cu_2

Type of Calculation	Reference	r_e (Å)	W_e (cm^{-1})	D_e (eV)
All-electron-SCF + CI	21	2.25	—	1.75
All-electron-SCF	22	2.34	247	0.84
Valence-electron-SCF	23	2.25	—	0.50
Valence-electron-SCF + CI	23	2.22	—	1.84
Valence-electron-SCF	20	2.40	195	0.55
Valence-electron-SCF + CI	20	2.25	265	1.93
EHT	24	2.25	390	2.11
X_α-SW	25	2.17	272	2.8
Experiment	26	2.22	266	2.05

2. Without configuration interaction the bond energy is poorly described. Since the amount of computation necessary to include CI in larger-cluster calculations is enormous, other approximations will be needed in this area.

3. The dominant bonding due to $4\sigma_g$ molecular orbitals of Fe_2, Ni_2, and Cu_2 need not be true in all transition-metal dimers.

2.3.2 Multicenter Clusters

The intrinsic properties of multicenter metal clusters are important in such areas as photography and heterogeneous catalysis. Thus, there has been a growing interest in the area. One of the principal concerns has been an elucidation of the size at which a given cluster property becomes bulklike. The question now seems settled for many properties as developed through a combined effort in theory and experiment, although the early work in this field produced rather diverse viewpoints.

Extended Hückel theory (30–32) and CNDO (31,33) were the first versions of theory applied to the problem of metal clusters. These results showed cluster properties quite different from those of the bulk. For example, small Ag_N clusters preferred a linear versus two- or three-dimensional structure (31). In addition, the binding energy per atom for clusters up to 55 atoms reached only about one third to one half the bulk cohesive energy (34). The ionization potential decreased with cluster size nonmonotonically, showing odd–even effects (31,35). Later, X_α-SW calculations were published (36) that criticized many aspects of the semiempirical work but emphasized the similarity of computed orbital-energy spectra for 4- to 13-atom clusters with the computed bulk spectra. One point in common in both the semiempirical and X_α approaches was that the s energy levels of clusters such as Cu, Pd, or Ag overlapped the d energy levels at a small cluster size [e.g., 2 atoms for Pd (31), and fewer than 13 atoms for Cu (37)]. Further EHT calculations were employed to compare the orbital spectra with computed bulk spectra for clusters such as Ag (35) and Pd (38), using identical parameters for the bulk and the clusters. The spread of the d-orbital energy levels increased with cluster size, and the largest cluster, a 79-atom cluster with face-centered cubic (fcc) geometry, had only 86% of the bulk width. Concurrently, photoemission experiments in many laboratories were being conducted on supported clusters of Pd, Ag, and Au. Photoemission measures the orbital spectrum modulated by transition-matrix elements, and several workers (38–42) have concluded that 100- to 200-atom clusters are required to get nearly bulk photoemission properties. These studies have included UPS measurements of Pd on C (41) or SiO_2 (41), Au on C (39), and Ag on C (39), and XPS (X-ray photoelectron spectroscopy) measurements of Pd, Pt on C (38), and Au on C (39), Al_2O_3 (40), or NaCl (42). The criteria used in comparison to bulk behavior included the width and shape of d states as well as photon-energy-dependent effects characteristic of bulk behavior.

Examples of the size-dependent effects observed in photoemission are shown in Figure 2.1, where UPS spectra for Pt, Ag, and Cu clusters of different mean sizes are shown. The clusters were prepared by vacuum deposition onto carbon supports

CLUSTER DIFFERENCE SPECTRA

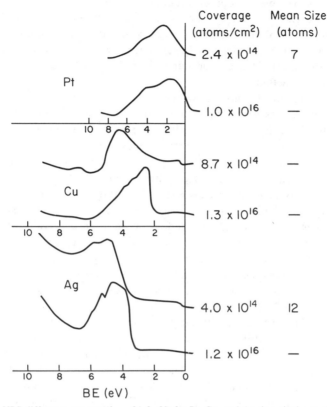

Figure 2.1 UPS difference spectra ($h = 21.2$ eV) for Pt, Cu, and Ag deposited onto C support. The low-coverage spectra corresponding to particles of a mean number of atoms (see text) are compared with spectra for continuous films. Energies are measured relative to the Fermi energy. The intense peaks at (0–6 eV) Pt, (2–5 eV) Cu, and (4–8 eV) Ag are due to emission from d electrons.

and were characterized by electron microscopy as described later. As is common practice in UPS work, difference spectra are shown in Figure 2.1, corresponding to the subtraction of the photoemission intensity of the clean carbon support from the corresponding intensity after metal deposition. Two spectra are shown for each metal, corresponding to small cluster or nearly continuous films. In each metal, intensity increased at the top of the d states as cluster size increased. The most intense region of these spectra corresponds to emission from d electrons since the s-electron emission has a much smaller cross section. Considerably more detailed study than is presented here would be needed to identify the threshold size for bulk behavior. This has been done in many examples. One particularly convincing set of experiments is the storage-ring work (39) for gold and silver clusters. Here bulk behavior was noted by the presence of photon-energy-dependent modulations in the spectra characteristic of the development of periodic behavior. In this experiment the energy of photons was varied in the study of clusters of various sizes. When the $\bar{k}$ wavevector becomes a good quantum number in clusters that are sufficiently large, the $\bar{k}$ conservation rule in photoemission gives spectral modulation with photon energy. Clusters of more than 100 atoms were required to produce this effect. The support material in these studies, carbon, interacts weakly with the cluster (43), showing that the threshold size is an intrinsic property of the clusters and not the cluster-support interaction.

Recent ab initio calculations for Ni (44) and Cu (45) clusters confirm the idea that these small clusters have properties far different from those of the bulk. The calculations for Ni employed an effective-potential method in the Hartree-Fock scheme with limited CI. The general picture that emerged was of the clusters being an open-shell species of Ni atoms in d^9 configurations bound together by $4s$ electrons. The $4p$ orbital increased in importance in the wavefunction as cluster size or dimensionality increased. At small cluster size, Ni_3 prefers linear $(D_{\infty h})$ to triangular (D_{3h}) geometry. The binding energy per atom increases with cluster size up to 6 atoms, the largest cluster considered, and is far from the bulk value. In ab initio calculations, Demuynck et al. (45) considered Cu clusters up to 13 atoms with an effective potential and a basis set of double-zeta quality for the $3d$ and $4s$ orbitals that are important in bonding. They found that the binding energy per atom increases with the number of atoms and does not reach a plateau for 13 atoms, appearing rather different from the bulk value. Favored geometries such as linear for Cu_3 and icosahedral for Cu_{13} differed significantly from the bulk geometry.

One of the apparent differences between the ab initio calculations for Ni and Cu and the earlier X_α-SW calculations (36) concerned whether the s orbitals overlap the d orbitals at a small cluster size. The X_α-SW calculations of orbital energies showed considerable overlap of a broad s set of levels with the d levels at 8 or 13 atoms, whereas the ab initio calculations for Cu_8 showed the $3d,4s$ levels well separated and that they just begin to overlap at Cu_{13}. The ab initio calculations for Ni clusters also showed the s,p orbitals above the d orbitals with no overlap up to the largest size considered, 6 atoms. This apparent discrepancy has been resolved (46–48) by breaking symmetry in the ion state where the hole left upon ionization is localized on one center rather than forced to be delocalized as imposed by symmetry considerations in the earlier calculations.

Post and Baerends (46) performed LCAO-X_α calculations for Cu clusters up to 9 atoms. Contrary to the X_α-SW calculations discussed earlier, the energy of the cluster could be calculated accurately; it increased regularly with size but reached only 53% of the bulk value at 9 atoms, in good agreement with the ab initio results. The density-of-states picture showed overlap of the s and d orbitals at even the smallest cluster, Cu_2. To examine the origin of the overlap discrepancy with ab initio results, Post and Baerends considered Cu_2 and computed the ion state with a localized d hole; this led to raising the energy of d levels as can be traced to an atomic-like relaxation energy and permitted overlap of s,d orbitals. Cox et al. (47) performed broken-symmetry Hartree-Fock calculations for Cu clusters and found that the relaxation energy of d orbitals in the cluster remains at nearly the atomic value (0.2 au). This raises the d-orbital energies (Koopmans's ionization levels) relative to the s-orbital energies and permits overlap of s and d manifolds at clusters as small as 5 atoms. Newton (48) reexamined clusters of Ni_2 and Ni_4 within this approximation, finding considerable s,d overlap at 4 atoms. He presented interesting data for Ni_2 ionization, showing that the s,p electrons readjust in the presence of the local d hole so that Ni atom with the d hole has a charge of $+0.30$ compared to the charge of $+0.70$ on the other atom of Ni_2^+. This type of screening becomes even more pronounced in larger Ni clusters. Finally, Messmer et al. (49) presented results for Cu_4, comparing the X_α-SW orbital energies and transition-state energies. The distribution of energy levels is qualitatively similar with the overlap of the s,d levels. There now seems to be agreement that an overlap of the s,d orbitals occurs at small Cu, Ni cluster sizes, but that the electronic structure remains far from that of the bulk.

The electronic structure of small Cu clusters has also been investigated by all-electron SCF calculations (50). Rather complex Gaussian basis sets were used with the inclusion of polarization functions, but CI calculations were not undertaken. Many features of this calculation agree well with some of the earlier effective-potential calculations for Cu_N. For example, the binding energy per atom increases linearly from Cu_2 to Cu_6 as reported in the earlier calculations. One point of difference concerns the geometry of Cu_3, which is computed to be more stable for triangular rather than for linear geometry. Bandwidths, defined as the difference in energy between occupied orbitals of a given type, increase with size but remain smaller than the bulk experimental value. The position of the d orbitals is 8 eV below the highest occupied s molecular orbital. A reorganization energy shift of 6 eV is discussed for repositioning these levels only 2 eV below the highest occupied level.

One paper of significant interest concerns the ab initio calculations (51) with a relativistic core potential and CI of Ag_1, Ag_2, and Ag_3 properties. There is good agreement between this and earlier semiempirical CNDO calculations of Ag_N. The geometry of the Ag_3 species in its various charged forms is: cation, equilateral triangle; neutral, slightly bent; anion, linear. These structures are quite understandable from simple molecular orbital theory based on dominant bonding through the $5s$ orbitals. A sawtooth behavior of ionization potential and electron affinity is found in which greater stability is associated with the closed-shell species. Similar effects were found in CNDO calculations (35).

The finite cluster size leads to geometric inequalities in the local atom environment, giving to a variety of "surface" properties. One example is the distribution of charge between the surface and the interior of a cluster (52). The shape and the width of the local density of states reflects the surrounding environment for a particular cluster atom (53). As the number of nearest neighbors increases, the width of the local density of states increases, reflecting the larger number of covalent interactions. This effect leads to a surface-bulk charge redistribution, and thus causes a potential to be set up between the surface and the interior of the cluster. This effect may be illustrated for the 19-atom Pd and Cu fcc clusters sketched in Figure 2.2. Table 2.3 shows the self-consistent potential for d electrons centered on various atoms of the clusters. These calculations (54) were performed by EHT, by the procedure of Basch, Viste, and Gray (15), to arrive at self-consistent results. The

M_{19} FCC STRUCTURE

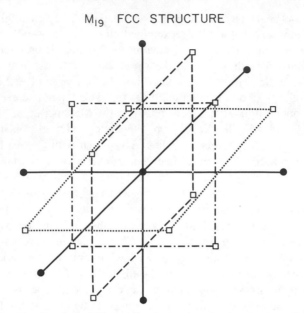

Figure 2.2 Structure of a spherical 19-atom fcc cluster. A center atom is surrounded by a shell of 12 nearest neighbors and another shell of 6 next nearest neighbors. The 16-atom cluster discussed in the text results from removing 3 atoms in the outermost shell.

potential on outer atoms of the 16- (formed by removing 3 outermost atoms in Fig. 2.2) or 19-atom cluster is shifted closer to vacuum relative to interior atoms with more nearest neighbors. The total potential shift between center and outer atoms is greater for the 16-atom cluster with an irregularly shaped surface versus the 19-atom spherical cluster. This shift of surface d potential in clusters is thought to be related to similar surface core shifts, as observed recently in photoemission experiments on ordered W (55), Ir (56), Au (57), and Pt (58) crystal surfaces.

TABLE 2.3 Self-consistent Potential of d Electrons

Cluster	Atom Location	d Potential (eV)
Cu_{19}	Center	− 12.45
	Middle	− 11.67
	Outer	− 11.39
Pd_{19}	Center	− 10.12
	Middle	− 9.77
	Outer	− 9.59
Cu_{16}	Center	− 12.67
	Middle	− 11.51
	Outer	− 11.15
Pd_{16}	Center	− 10.33
	Middle	− 9.85
	Outer	− 9.76

2.3.3 Clusters as Models for Chemisorption and Reaction

Models of chemisorption or reaction require treatment of the interaction between a molecular species and the underlying metal or ionic support. Typically, experimental data are measured on single-crystal, film, or large-particle (up to several hundred Ångstroms in diameter) systems. Thus, it becomes necessary to model the continuous infinite support by a cluster or a finite-thickness film. When clusters are used in this approximation, local interactions with the adsorbed species are assumed to dominate and long-range interactions to play a small role. In view of the fact that ionization potential and electron affinity have not merged to the common value (the work function) in small clusters, this approximation may be very severe. On the other hand, two-dimensional film systems of finite thickness have a well-defined Fermi level but suffer from the limitation that a rather small repeating unit cell pattern must be used. In this section several examples are given where these various approximations are used.

Goddard and coworkers have treated the chemisorption of various species, including CO, H, and hydrocarbon fragments on Ni clusters using a variety of ab initio techniques. An early treatment (59) of CH_2 and CH_3 chemisorbed to one Ni atom was examined by GVB calculations with CI. The Ni core was replaced by an argon-effective potential, reducing Ni to a 10-electron atom. The bonding of the $3d^9 4s^1$ Ni atom involved principally the $4s$ orbital of Ni with a σ molecular orbital of the hydrocarbon fragment. In each case electrons are transferred to the hydrocarbon ($CH_3 : 0.42e^-$; $CH_2 : 0.56e^-$). The bond energies were 65 and 60 kcal/mol^{-1} for $NiCH_2$ and $NiCH_3$, respectively, with the corresponding metal carbon bond lengths of 1.78 and 1.87 Å. A small amount of π bonding was observed for Ni—CH_2. These fragments are thought to be important in heterogeneous catalysis involving hydrocarbons. Few experimental data are available about their properties. Some recent analyses (60) of experimental data have estimated the metal-alkyl bond strength at 30 kcal/mol^{-1}.

Chemisorption on larger Ni clusters has been formulated under different approximations by Upton and Goddard (61). In these Hartree-Fock calculations an effective potential was used that included the argon core and an average $3d^9$ configuration. This reduced the problem to one $4s$ electron per Ni atom, not permitting explicit participation of the $3d$ orbitals in bonding with an adsorbate. Calculations comparing this treatment on all atoms of an Ni_{20} cluster interacting with H were compared with those for a similar cluster, permitting four Ni atoms closest to H to have full variational freedom of the $3d^9$ electrons. Bond energies, vibration frequencies, and bond lengths agree well between the two calculations, prompting the authors to conclude adequacy of the $3d^9$ average potential for describing Ni—H bonding. This effect agrees with the fact that the $4s$ orbitals are much more diffuse than the $3d$ orbitals and would be expected to form the dominant bond from Ni to an H adsorbate, as earlier calculations had found (62). There does remain some question as to how good this approximation would be in calculations on semi-infinite systems. Cohesion energies of the bulk transition metals are related to the number of d electrons (63), which dominate the effect of s electrons. In general, adsorbate bonding to transition-metal semi-infinite systems is ascribed to interactions with the d electrons of the metal (64). Despite these possible reservations, the cluster-calculated results agree well with experimental measurements of H adsorbed to the (100) Ni surface. The H is predicted to adsorb in a fourfold site with 3.0 eV bonding energy, compared to the experimental value of 2.74 eV. Likewise, the electron transfer to H, vibrational frequency of Ni—H, and orbital spectrum agree well with experiments, making the technique very promising.

The treatment of CO chemisorbed to Ni_{14} has been considered with the GVB method (65), using an effective potential including the $3d^9$ electrons like that described earlier. The CO molecule adsorbs on top of one Ni atom, having the bond axis perpendicular to the surface with the C end closest to Ni. Good agreement with experimental values is found for bond strength, vibrational frequencies, and photoemission spectrum. It was concluded that the π system of CO is unaffected by bonding to Ni_{14}. The calculated dipole moment $\mu = +0.069D$ with the negative end pointing toward the cluster. Experimental values of the work-function change for CO adsorption on Ni (100) indicate a dipole pointing in the opposite direction. Allison and Goddard conclude that the Ni_{14} cluster poorly describes this effect.

Comprehensive calculations of the NiCO system have been reported (66) using a SCF method with CI. These calculations included all of the Ni electrons, avoiding the use of effective potentials. These results show the importance of correlation effects in some of the π^* molecular orbitals of CO and give a picture of Ni—CO bonding different from that of earlier treatments that did not take into account this effect. There is net electron transfer to CO of 0.2 electrons following the σ CO donation of $0.4e^-$ and π-$3d$ backbonding of $0.6e^-$. This more comprehensive calculation of Ni—CO bonding arrives at a picture known for some time as the Blyholder model (67), as derived from arguments based on organometallic compounds and semiempirical calculations. In this model the σ donation from CO takes place to the s orbitals of the metal, and back donation from the d-metal orbitals to π^* molecular orbitals of CO dominates. In addition, numerous other calculations (68,69) have supported this picture.

The dissociation of H_2 had been examined some time ago within the extended Hückel scheme (70). Although the method is inadequate to describe the dissociation limit of H_2, it is useful to investigate the factors important in the initial stages of bond activation. The H_2 molecule was placed in a di-σ arrangement over a bond

on a 9-atom metal atom (M) cluster. The H—H bond is stretched as the M—H bond length remains constant. Before the H—H bond is stretched, the molecule is a net electron acceptor from an Ni_9 cluster as a result of mixing σ_u levels of H_2 into the occupied molecular orbitals. This population of σ_u orbitals weakens the H—H bond and promotes elongation. As the H—H bond is stretched, the σ_u level moves closer to the metal energy levels, causing a larger σ_u population and further weakening

H—H. Opposing these forces is the destabilization of the $\sigma_g H_2$ molecular orbital caused by H—H elongation. The net effect sketched in Figure 2.3 is a small activation barrier for dissociation, which is in accord with experiment. The utility of this calculation is aimed at the picture of bond activation rather than quantitative aspects.

The usefulness of small clusters to quantitatively model chemisorption on semi-infinite surfaces is difficult to assess. There are few experiments that examine properties of chemisorbed species as a function of cluster size. Some experiments, however, that do bear on this question involve the use of ultraviolet photoemission spectroscopy (UPS). In these experiments (71) metal is evaporated onto a weakly interacting support (carbon) to produce a distribution of cluster sizes that can be

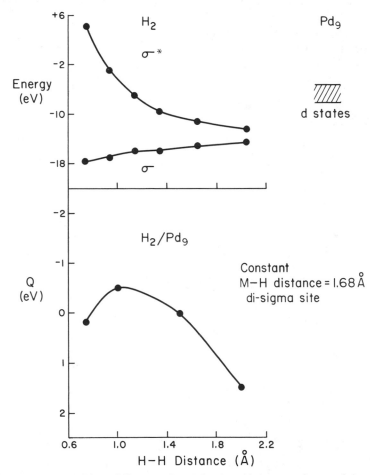

Figure 2.3 Relative positions of H_2, σ_g, and σ_u molecular orbitals versus the occupied energy levels for a Pd_9 cluster. The computed (EHM) energy barrier for dissociation of H_2 is shown for the reaction coordinate described in the text.

studied by electron microscopy (72). The cluster deposits are characterized by a coverage of metal deposited and mean cluster size. The clusters are kept in ultra high vacuum and exposed to a chemisorbing gas (Cl_2) to achieve saturation coverage. The chemisorbing gas does not adsorb to the carbon substrate under conditions of the experiment. The UPS spectrum is recorded before and after gas exposure and subtracted to give the chemisorption-induced changes in the spectra. Figure 2.4 shows examples for Cl_2 on Ag and Pt clusters. Difference spectra at small mean cluster size and on continuous films are shown.

The spectra of Figure 2.4 show significant cluster-size-dependent effects. For Ag clusters, the two-peak spectra invert in intensity ratio as the mean cluster size is changed. Mean cluster sizes of 40 to 50 atoms are required to obtain ''bulk-like'' spectra (71). These changes have been related to the ionization potential of Ag clusters, which decreases with size and thus facilitates ease of electron transfer to Cl. Significant changes in the region of the Ag *d* orbitals are also noted, indicating the role of halogen-*p*/silver-*d*-orbital interactions.

The case of Cl_2 adsorption on Pt clusters shown in Figure 2.4 is more complex than for Ag clusters and not so easily interpreted. Clearly, there is a broadening of

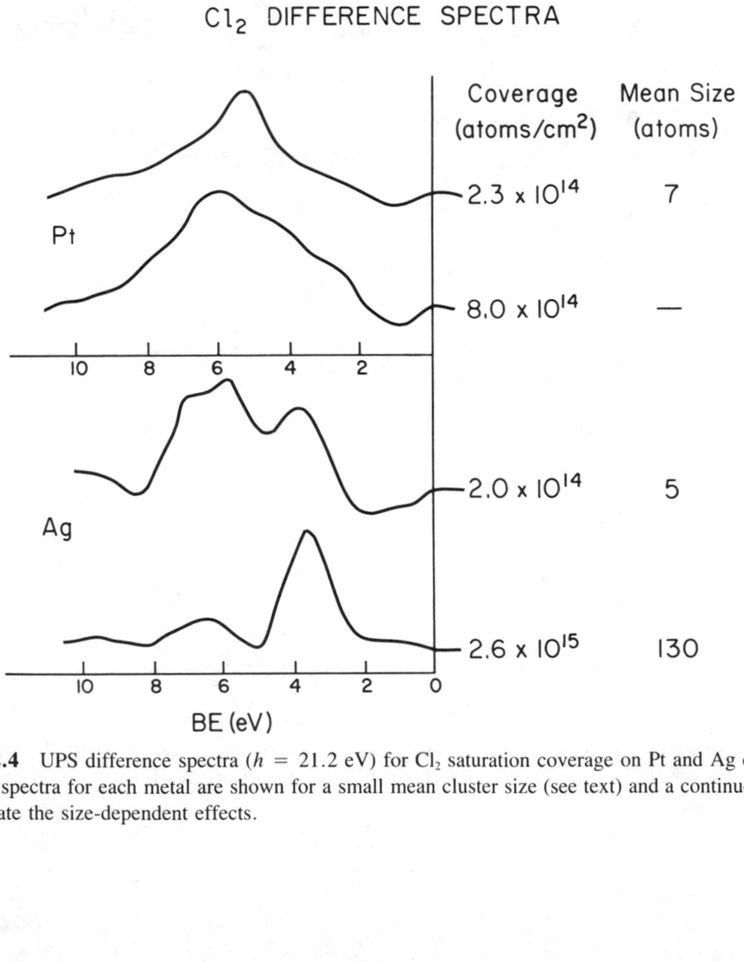

Figure 2.4 UPS difference spectra ($h = 21.2$ eV) for Cl_2 saturation coverage on Pt and Ag clusters. The two spectra for each metal are shown for a small mean cluster size (see text) and a continuous film to illustrate the size-dependent effects.

the Cl-induced spectrum, which increases with cluster size. The cause of this effect is unknown, but rather large Pt clusters of ~50 atoms are required to attain a bulk-like spectrum.

2.3.4 Organometallic Complexes

Organometallic cluster complexes have been a focal point for theoretical activity in metal clusters because of their well-defined structure. Thus, unlike the bare metal clusters, these species have a starting point for characterization. Numerous physical studies focusing on the electronic structure and reactivity patterns have been reported. In addition, an active area of research (73) has involved probing the cluster-surface analogy for insights into chemisorption and catalysis.

The ability of various levels of theory to provide structural information of cluster complexes is of interest. Although the Hartree-Fock theory gives good predictions of the equilibrium molecular geometry (74) for first- and second-row elements, the situation is less clear for transition metals. Good reviews of the area exist (75), and there are early indications (76) that bond lengths computed for organometallic compounds are too long. A recent investigation (77) of LCAO-SCF calculations on ferrocene and decamethylferrocene has been directed toward this question.

Calculations using basis sets ranging from minimal to better than triple-zeta were used in this calculation for ferrocene. The carbon ring-metal distance is computed to be 1.88 Å with the best basis set, and the authors conclude that this is near the Hartree-Fock limit. The experimental value is 1.65 Å. Similar results and conclusions were reached regarding decamethylferrocene. Despite the difficulty in computing metal ring distances, ligand geometries were computed to be close to experiment. The authors conclude the need for CI calculations as a possible remedy for this problem. Despite this rather depressing result for the Hartree-Fock theory, continued systematic investigations of this type are needed to provide ground rules for interpretation of the calculations. Other theories, such as X_α or semiempirical, usually avoid this problem altogether by taking the experimental geometry as a starting point for the calculation.

An interesting example of a study of organometallic reactivity concerns the reductive elimination reaction of d^8 complexes:

$$\begin{array}{c} L \\[-2pt] \diagdown \quad \diagup L \\ M \\ \diagup \quad \diagdown \\ R \qquad R' \end{array} \longrightarrow ML_2 + R\!-\!R'$$

where R = H, alkyl
$\quad L$ = ligand
$\quad M$ = Ni^{+2}, Pd^{+2}, Pt^{+2}

Tatsumi et al. (78) investigated this reaction using a wavefunction (correlation) analysis within the framework of extended Hückel calculations. They concluded that the better the σ-donating capability of the leaving groups (R) is, and the weaker the donor properties of the trans ligands (L) is, the more readily the elimination reaction proceeds. Potential energy surfaces were constructed from the EHT calculations to arrive at these conclusions. Interestingly, angular variations of the R and L groups on the metal atom reveal the essential features of the elimination reaction without the need for evaluating effects of the M—R distance. In this treatment, preparation of the organometallic compound for elimination is the primary basis for energy barriers leading to various reaction pathways.

SCF-X_α-SW calculations appeared later (79) examining the same reductive elimination reaction studied earlier. Despite the vast difference in the extended Hückel and X_α-SW methodologies, Balazs et al. (79) conclude behavior consistent with the potential surfaces of Tatsumi (78). This conclusion comes despite the frequently raised criticism of EHT to compute potential surfaces. The reader is also directed to the supplementary information in Ref. (79) for a good comparison of the relative merits of different computational techniques for this type of problem.

Shustorovich (80) recently analyzed the bonding patterns of various ligands with metal surfaces and clusters. He showed that the antibonding levels of the ligand play a much greater role than was previously ascribed to them. This effect is traced to the coefficients in the antibonding and bonding ligand orbitals. This may be shown for H_2 where

$$\Psi^* = \frac{1}{\sqrt{2(1-S)}} (\chi_1 - \chi_2)$$

$$\Psi = \frac{1}{\sqrt{2(1+S)}} (\chi_1 + \chi_2)$$

where $\chi_i = 1s$ orbital
$\quad\quad S = $ overlap integral

Because the S overlap is large (0.68), the coefficients in the antibonding molecular orbital become quite large, enhancing the matrix element

$$\langle \Psi^* | H | d \rangle$$

d metal d orbital

versus its bonding counterpart and leading to significant backbonding to the ligand antibonding molecular orbitals. This effect is general for ligands other than H_2. This result may be thought of as arising from the greater spatial extent of the antibonding versus bonding molecular orbitals. Because of this, the acceptor role of ligand orbitals needs attention in evaluating metal-ligand interactions.

An example of the effect of antibonding ligands has been shown for the chemisorption of CH_4 to transition-metal films (81). Figure 2.5 shows the relative energy levels of the CH_4 frontier orbitals (σ^*_{C-H} and σ_{CH}) versus the d levels of the metal film. Note that because the energy separation between the σ^*_{C-H} and the metal d-

RELATIVE ENERGY LEVELS

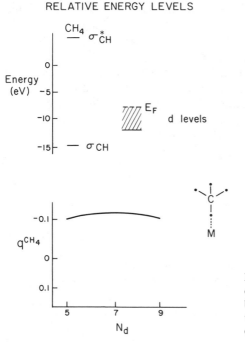

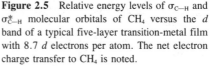

Figure 2.5 Relative energy levels of σ_{C-H} and σ^*_{C-H} molecular orbitals of CH_4 versus the d band of a typical five-layer transition-metal film with 8.7 d electrons per atom. The net electron charge transfer to CH_4 is noted.

orbital energy is quite large, this interaction is often considered negligible. However, because of the size of the coefficients in the σ^*_{C-H} molecular orbital, this interaction dominates and CH_4 is an electron acceptor. The charge on CH_4 computed by Mulliken analysis is for the perferred geometry computed for the transition elements with more than six d electrons per atom.

2.4   SUMMARY

Some of the recent developments in theoretical studies of bare and organometallic metal clusters were reviewed. The major computational methods were discussed along with their strengths and weaknesses. Although there is much competition between the proponents of each particular method, there seems to be a place for each method. Ab initio methods have shown very good results for bare diatomic molecules when a sufficient degree of correlation is included in the problem. It is now well documented experimentally and theoretically that small clusters containing 10 to 12 atoms are not bulk-like, as manifested in their orbital density of states, ionization potential, bond energy, and other properties. Nevertheless, such small clusters are often used as models for infinite surface regimes where experiments are usually performed. Often good representations for the infinite surface are found with the cluster models. Successful calculations have been carried out to model reaction sequences such as reductive elimination, which predict many features in agreement with experiment. An important new conceptual finding concerns the role of antibonding molecular orbitals in ligand-metal bonding, which were generally previously neglected because of their energy position.

REFERENCES

1. H. F. Schaefer, III, Ed., *Methods of Electronic Structure Theory*, Plenum Press, New York, 1977.
2. The antisymmetrizer operator $\mathscr{A}$ gives a sum over all permutations where the sign of each term is + or − depending on whether there is an even or odd number of permutations.
3. H. F. Schaefer, III, Ed. *The Electronic Structure of Atoms and Molecules*, Addison-Wesley, Reading, Mass., 1972.
4. An excellent review of ab initio concepts is J. L. Whitten, *Acc. Chem. Res.*, **6**, 238 (1972).
5. W. A. Goddard, III, J. J. Barton, A. Redondo, and T. C. McGill, *J. Vac. Sci. Technol.*, **15**, 1274 (1978).
6. J. C. Slater, *The Self-Consistent Field for Molecules and Solids*, McGraw-Hill, New York, 1974.
7. D. R. Salahub, A. E. Forti, and V. H. Smith, Jr., *J. Am. Chem. Soc.*, **78**, 7863 (1978).
8. H. Sambe and R. H. Felton, *J. Chem. Phys.*, **62**, 1122 (1975).
9. E. J. Baerends, D. E. Ellis, and P. Ros, *Chem. Phys.*, **2**, 41 (1973).
10. J. A. Pople, D. P. Santry, and G. A. Segal, *J. Chem. Phys.*, **43**, 5129 (1965).
11. R. C. Bingham, M. J. S. Dewar, and O. H. Lo, *J. Am. Chem. Soc.*, **97**, 1285 (1975).
12. G. Blyholder, J. Head, and F. Ruette, *Theor. Chim. Acta*, **60**, 429 (1982).
13. G. Blyholder, J. Head, and F. Ruette, *Inorg. Chem.*, **21**, 1539 (1982).
14. R. Hoffmann, *J. Chem. Phys.*, **39**, 137 (1963).
15. H. Basch, A. Viste, and H. B. Gray, *J. Chem. Phys.*, **44**, 10 (1966).
16. J. O. Noell, M. D. Newton, P. J. Hay, R. L. Martin, and F. W. Bobrowicz, *J. Chem. Phys.*, **73**, 2360 (1980).
17. T. H. Upton and W. A. Goddard, III, *J. Am. Chem. Soc.*, **100**, 5659 (1978).
18. A. Kant, *J. Chem. Phys.*, **41**, 1872 (1964).
19. N. Rosch and T. N. Rhodin, *Phys. Rev. Lett.*, **32**, 1189 (1974).
20. M. Pelissier, *J. Chem. Phys.*, **75**, 775 (1981).
21. P. Joyes and M. Leleyter, *J. Phys. B*, **6**, 150 (1973).
22. C. Bachman, J. Demuynck, and A. Veillard, *Gazz. Chim. Ital.*, **108**, 389 (1978).

23. R. N. Dixon and I. Robertson, *Mol. Phys.,* **36**, 1099 (1978).

24. A. B. Anderson, *J. Chem. Phys.,* **68**, 1744 (1978).

25. G. A. Ozin, H. Hubert, D. McIntosh, S. Mitchell, J. G. Norman, and L. Noodleman, *J. Am. Chem. Soc.,* **101**, 3504 (1979).

26. J. Lochet, *J. Phys. B,* **11**, 55 (1978).

27. I. Shim and K. A. Gingerich, *J. Chem. Phys.,* **77**, 2490 (1982).

28. M. M. Goodgame and W. A. Goodard, III, *J. Phys. Chem.,* **85**, 215 (1981).

29. G. Das, *Chem. Phys. Lett.,* **86**, 482 (1982).

30. C. R. Hare, T. P. Sleight, W. Cooper, and G. A. Clarke, *Inorg. Chem.,* **7**, 669 (1968).

31. R. C. Baetzold, *J. Chem. Phys.,* **55**, 4363 (1971).

32. A. B. Anderson and R. Hoffmann, *J. Chem. Phys.,* **61**, 4545 (1974).

33. G. Blyholder, *Surf. Sci.,* **42**, 249 (1974).

34. R. C. Baetzold and R. E. Mack, *J. Chem. Phys.,* **62**, 1513 (1975).

35. R. C. Baetzold, *J. Chem. Phys.,* **68**, 555 (1978).

36. R. P. Messmer, S. K. Knudson, K. H. Johnson, J. B. Diamond, and C. Y. Yang, *Phys. Rev. B,* **13**, 1396 (1976).

37. R. C. Baetzold, *J. Phys. Chem.,* **82**, 738 (1978).

38. R. C. Baetzold, M. G. Mason, and J. F. Hamilton, *J. Chem. Phys.,* **72**, 366 (1980).

39. M. G. Mason, S.-T. Lee, and G. Apai, *Chem. Phys. Lett.,* **76**, 51 (1978); G. Apai, S.-T. Lee, and M. G. Mason, *Solid State Commun.,* **37**, 213 (1981); S.-T. Lee, G. Apai, M. G. Mason, R. Benbow, and Z. Hurych, *Phys. Rev. B,* **23**, 505 (1981).

40. K. S. Liang, W. R. Salaneck, and I. A. Aksay, *Solid State Commun.,* **19**, 329 (1976).

41. R. Unwin and A. M. Bradshaw, *Chem. Phys. Lett.,* **58**, (1978); Y. Takasu, R. Unwin, B. Tesche, and A. M. Bradshaw, *Surf. Sci.,* **77**, 219 (1978).

42. H. Roulet, J. M. Mariot, G. Dufour, and C. F. Hague, *J. Phys. F,* **10**, 1025 (1980).

43. M. G. Mason, *Phys. Rev. B.,* **27**, 748 (1983).

44. H. Basch, M. D. Newton, and J. W. Moskowitz, *J. Chem. Phys.,* **73**, 4492 (1980).

45. C. Bachmann, J. Demuynck and A Veillard, *Faraday Symp. Chem. Soc.,* **14**, 170 (1980); J. Demuynck, M.-M. Rohmer, A. Strich, and A. Veillard, *J. Chem. Phys.,* **75**, 3443 (1981).

46. D. Post and E. J. Baerends, *Chem. Phys. Lett.,* **86**, 176 (1982).

47. P. A. Cox, M. Benard, and A. Veillard, *Chem. Phys. Lett.,* **87**, 159 (1982).

48. M. D. Newton, *Chem. Phys. Lett.,* **90**, 291 (1982).

49. R. P. Messmer, R. C. Caves, and C. M. Kao, *Chem. Phys. Lett.,* **90**, 296 (1982).

50. H. Tatewaki, E. Miyoshi, and T. Nakamura, *J. Chem. Phys.,* **76**, 5073 (1982).

51. H. Basch, *J. Am. Chem. Soc.,* **103**, 4657 (1981).

52. E. Shustorovich and R. C. Baetzold, *J. Am. Chem. Soc.,* **102**, 5989 (1980).

53. F. Cyrot-Lackman, *Adv. Phys.,* **16**, 393 (1967).

54. R. C. Baetzold, *J. Phys. Chem.,* **82**, 738 (1978).

55. T. M. Duc, C. Guillot, Y. Lassailly, J. Lecante, Y. Jugnet, and J. C. Vedrine, *Phys. Rev. Lett.,* **43**, 789 (1979).

56. J. F. VanderVeen, F. J. Himpsel, and D. E. Eastman, *Phys. Rev. Lett.,* **44**, 189 (1980).

57. P. H. Citrin, G. K. Wertheim, and Y. Bayer, *Phys. Rev. Lett.,* **41**, 1425 (1978).

58. R. C. Baetzold, G. Apai, E. Shustorovich, and R. Jaeger, *Phys. Rev. B*, **26**, 4022 (1982).

59. A. K. Rappé and W. A. Goddard, III, *J. Am. Chem. Soc.*, **99**, 3966 (1977).

60. J. Halpern, *Acc. Chem. Res.*, **15**, 238 (1983).

61. T. H. Upton and W. A. Goddard, III, *Phys. Rev. Lett.*, **42**, 472 (1979); C. F. Melius, T. H. Upton, and W. A. Goddard, III, *Solid State Commun.*, **28**, 501 (1978).

62. D. J. M. Fassaert and A. van der Avoird, *Surf. Sci.*, **55**, 313 (1976).

63. D. G. Pettifor, *Phys. Rev. Lett.*, **42**, 846 (1979).

64. J. R. Smith, Ed., *Theory of Chemisorption*, Springer, Berlin, 1980.

65. J. N. Allison and W. A. Goddard, III, *Surf. Sci.*, **115**, 553 (1982).

66. P. S. Bagus and B. O. Ross, *J. Chem. Phys.*, **75**, 5961 (1981).

67. G. Blyholder, *J. Vac. Sci. Technol.*, **11**, 865 (1974); *J. Phys. Chem.*, **79**, 756 (1975).

68. H. Itoh and G. Ertl, *Z. Naturforsch. Teil A*, **37**, 346 (1982).

69. G. Doyen and G. Ertl, *Surf. Sci.*, **43**, 197 (1974).

70. R. C. Baetzold, *Surf. Sci.*, **51**, 1 (1975).

71. R. C. Baetzold, *J. Am. Chem. Soc.*, **103**, 6116 (1981); *Inorg. Chem.*, **21**, 2189 (1982).

72. J. F. Hamilton, D. R. Preuss, and G. R. Apai, *Surf. Sci.*, **106**, 146 (1981).

73. E. L. Muetterties, *Science*, **196**, 839 (1977).

74. W. A. Latham, L. A. Curtiss, W. H. Hehre, J. B. Lisle, and J. A. Pople, *Prog. Phys. Org. Chem.*, **11**, 175 (1974).

75. H. F. Schaefer, III, *J. Mol. Struct. Thermochem.*, **76**, 117 (1981).

76. H. P. Lüthi, J. Ammeter, J. Almlöf, and K. Korsell, *Chem. Phys. Lett.*, **69**, 540 (1980).

77. H. P. Lüthi, J. Ammeter, J. Almlöf, and K. Faegri, Jr., *J. Chem. Phys.*, **77**, 2002 (1982).

78. J. Tatsumi, R. Hoffmann, A. Yamamoto, and J. K. Stille, *Bull. Chem. Soc. Jpn.*, **54**, 1857 (1981).

79. A. C. Balazs, K. H. Johnson, and G. M. Whitesides, *Inorg. Chem.*, **21**, 2162 (1982).

80. E. Shustorovich, *J. Phys. Chem.*, **87**, 14 (1983).

81. R. C. Baetzold, *J. Am. Chem. Soc.*, **105**, 4271 (1983).

3

STRUCTURE OF METAL CLUSTER COMPLEXES

DAVID H. FARRAR AND ROBERT J. GOUDSMIT
Department of Chemistry
University of Toronto
Toronto, Ontario

3.1 INTRODUCTION

3.1.1 Preamble

A metal cluster may be defined as a group of two or more metal atoms in which there are substantial and direct bonds between the metal atoms (1). At the time of this writing there are over a thousand metal cluster structures in the literature,

TABLE 3.1

Sc	Ti	V	Cr⁺	Mn	Fe	Co	Ni	Cu	Zn⁺
Y	Zr	Nb	Mo	Tc	Ru	Rh	Pd	Ag	Cd⁺
La	Hf	Ta	W	Re	Os	Ir	Pt	Au	Hg

⁺ = mixed metal

comprising metal halides, neutral carbonyls, carboxylate anions, carbonyl hydrides, isocyanides, phosphites, phosphines, and many mixed, metal-type molecules. These may be conveniently categorized into low-oxidation state metal halides and metal carbonyls. The former are common in the lower halide chemistry of titanium, niobium, tantalum, molybdenum, and tungsten, whereas the latter are prevalent in the iron, cobalt, nickel, and copper subgroups. Table 3.1 shows the d block metals, with those metals for which clusters have been characterized indicated by boldface type.

With few exceptions, to be discussed later, the metal atom framework in metal clusters is a fragment of hexagonal or cubic close packing, or of body-centered cubic (bcc) packing. Also like bulk metals, metal clusters are able to form interstitial complexes containing elements such as hydrogen, carbon, nitrogen, sulfur, phosphorus, and arsenic in the interstices. An example of an encapsulated hydrogen atom is the hexanuclear compound $[HRu_6(CO)_{16}]^-$ (2) shown in Figure 3.1(a). The hydride has been detected, by a neutron diffraction study, at the center of the metal octahedron. It is worth noting how often the five regular platonic solids (the tetrahedron, the octahedron, the cube, the icosahedron, and the dodecahedron) occur in cluster chemistry. Examples of these species will be mentioned throughout this chapter. In addition to these polyhedra, other highly regular geometries are found. The trigonal prism represents a commonly found basis for a metal skeleton and the cube-octahedron is known for both metal core and ligand envelope geometries.

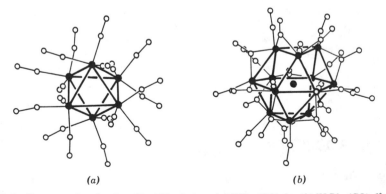

(a)   (b)

Figure 3.1   Two examples of carbonyl hydride clusters. (a) $[HRu_6(CO)_{16}]^-$; (b) $[H_3Rh_{13}(CO)_{24}]^{2-}$ (only surface Rh atoms have been connected). Throughout this chapter metal atoms will be represented by solid circles and ligand atoms by open ones. The metal atom core is connected by thicker lines.

Another hydrido-cluster is the centered-cube-octahedral $[H_3Rh_{13}(CO)_{24}]^{2-}$ (3), shown in Figure 3.1(b), where the hydrides are rapidly migrating inside the metal framework. The arrangement of this core can be regarded as a fragment of a hexagonal close-packed lattice.

In this chapter we present a variety of metal clusters, stressing the geometry of the metal framework. The d block will be examined group by group, highlighting salient examples and discussing the types of complexes that are prevalent in that part of the periodic table. We will conclude by attempting to show how many of these cluster geometries may be rationalized in terms of a few, relatively simple theories. Particular emphasis will be placed on the Wade rules (4) and on their recent extensions.

3.1.2 Metal Clusters

We begin with a brief examination of some clusters containing encapsulated carbon atoms, as this will serve to exemplify the diversity of known polyhedral geometries. This will not only indicate the elegance of the field, but will also show the kinds of unusual coordination that clusters can impose on main group atoms. Some examples of lower nuclearity carbido clusters are shown in Figure 3.2.

The tetranuclear carbido iron clusters have been extensively investigated by the

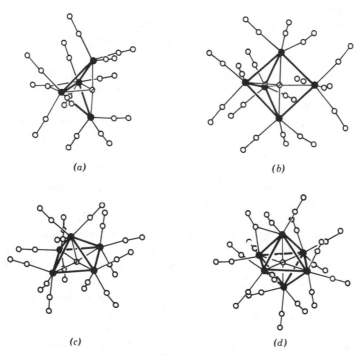

(a)   (b)

(c)   (d)

Figure 3.2   Some examples of carbido clusters of the iron triad. (a) [Fe$_4$(CO)$_{13}$C]; (b) [Os$_5$(CO)$_{16}$C]; (c) [Ru$_5$(CO)$_{15}$C]; (d) [Ru$_6$(CO)$_{17}$C]. ⌀ = carbido, C atom.

research groups of J. S. Bradley (5), D. F. Shriver (6), and the late E. L. Muetterties (7). The parent carbonyl compound $[Fe_4(CO)_{13}C]$ was synthesized and structurally characterized by Bradley and coworkers (8) and found to consist of a cradle of iron atoms (commonly referred to as a butterfly arrangement) supporting a carbon atom (Fig. 3.2a). The carbon atom sits in the middle of the cluster, approximately colinear with the "wing-tip" iron atoms, and can be regarded as bonded to all four metal atoms. (Throughout this chapter cluster geometries will be described in terms of the polyhedron formed by the metal vertices. However, this does not assign a formal metal–metal bond along polyhedral edges. The limitations of such an approach will be dealt with in Section 3.3. Similarly, although the carbon atom in $[Fe_4(CO)_{13}C]$ has close contacts with each of the metal atoms, it is not appropriate to assign *formal* metal–carbon bonds. This will be demonstrated further when the interstitial carbon atom becomes six-coordinate, showing the need to examine cluster orbitals as opposed to atomic ones.) The carbido carbon atom in $[Fe_4(CO)_{13}C]$ is exposed and can be attacked readily under ambient conditions. The similarity of this cluster with a carbon atom exposed on a metal surface is marked and, as such, the reactions of simple molecules with $[Fe_4(CO)_{13}C]$ have proved an excellent model for metal surface reactivity.

There are two main geometries known for the pentanuclear carbido clusters of the iron subgroup. The closely related geometries of $[M_5(CO)_{15}C]$ (9) and $[M_5(CO)_{16}C]$ (10) (Fig. 3.2b and c), M = Ru or Os, studied by J. Lewis, B. F. G. Johnson, and J. N. Nicholls, can be readily interconverted by the breaking or reforming of a single polyhedral edge. The structure of $[M_5(CO)_{15}C]$ is best described as a square-based pyramid of metal atoms, with the carbon atom lying approximately in the middle of the square base. Facile addition of CO to this cluster gives $[M_5(CO)_{16}C]$ which may be regarded as a butterfly of four metal atoms with a fifth metal atom bridging the wing tips (usually referred to as a "bridged butterfly" arrangement). The carbon atom lies in the center of the cluster and, as in $[M_5(CO)_{15}C]$, may be regarded as five-coordinate. Cleavage of a polyhedral edge linking the unique metal atom with the basal plane, followed by a minor rearrangement will interconvert the molecular geometries of the $[M_5(CO)_{15}C]$ with that of $[M_5(CO)_{16}C]$. Some rationalizations for such structural changes will be stated in Section 3.3.

In the hexanuclear carbido species $[Ru_6(CO)_{17}C]$ (11) the metal core geometry is an octahedron, where the carbon atom is totally enveloped in the metal framework. The carbide is six-coordinate with all the metal–carbon distances equal. Six-coordinate carbides are also found in the cobalt subgroup, but the carbon is commonly situated at the center of a trigonal prism of metal atoms, as in $[Rh_6(CO)_{15}C]^{2-}$ (12) (Fig. 3.3a). A similar arrangement is found for the octanuclear species $[Rh_8(CO)_{19}C]$ (13) (Fig. 3.3b); one of the additional metal atoms has capped a rectangular face of the prism, giving a pyramidal arrangement, and the other has bridged an edge. However, in the isoelectronic cobalt cluster $[Co_8(CO)_{18}C]^{2-}$ (14) (Fig. 3.3c), the eight metal atoms adopt a tetragonal antiprismatic arrangement, where all the cobalt atoms interact with an off-centered carbide. Four of the Co—C bonds average 1.99 Å, whereas the remaining four average 2.15 Å. The reasons for this major difference between the metal skeletons of these clusters are not fully understood, but two

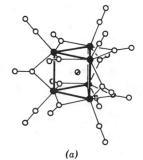

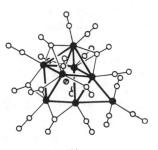

(a)   *(b)*

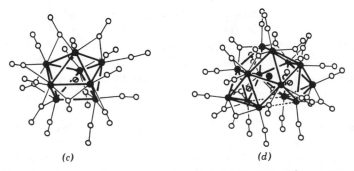

(c)   *(d)*

Figure 3.3 Examples of carbido clusters of the cobalt triad. (*a*) $[Rh_6(CO)_{15}C]^{2-}$; (*b*) $[Rh_8(CO)_{19}C]$; (*c*) $[Co_8(CO)_{18}C]^{2-}$; (*d*) $[Rh_{15}(CO)_{28}C_2]^{-}$.

explanations are presently favored. The first examines the size of the cavity inside the metal polyhedron where the carbide is accommodated. It is believed that the space inside a trigonal prismatic geometry of cobalt atoms is insufficient to take a carbon atom. Therefore, the metal polyhedron is forced to expand to a tetragonal antiprismatic structure. With rhodium the cavity would be large enough and no radical rearrangement is required. The second rationalization considers the steric hindrance caused by the additional carbonyl ligand in the rhodium cluster and states that the metal core geometry must alter to relieve the crowding. In order for the cluster to remain electronically satisfied, major rearrangements of the metal skeleton may be required. An elegant approach to the interaction between the steric require-ments of the ligands and the electronic needs of the metal core has been developed by Johnson (15). He proposed that simple packing arguments, concerning the ligand polyhedron, could explain the distribution of the ligands about the metal core. Johnson extended this approach to rationalize certain fluxional processes and ligand mobility in clusters (16).

Some examples of the possible coordination modes of a carbon atom with a transition metal cluster have been mentioned. The range of structures involving the bonding of a metal core to a carbon atom that is part of an organic moiety is vast and beyond the scope of this chapter (17). However, there are instances where two carbons can be regarded as carbides within the same cluster. The simplest situation is exhibited by the decanuclear cluster $[Ru_{10}(CO)_{24}C_2]^{2-}$ (18) where the structure may be described as two octahedra sharing an edge, with carbon atoms in both octahedral cavities. There is no formal C—C bond. The complicated metal skeleton of the dicarbide $[Rh_{15}(CO)_{28}C_2]^-$ (19) has precise C_2 symmetry (Fig. 3.3d). It may be regarded as a centered and tetracapped pentagonal prism in which the central atom is 12-coordinate, or as two fused octahedra sharing one vertex, plus four extra atoms forming tetrahedra fused with the octahedra and themselves. The two carbide atoms occupy octahedral cavities and are again not formally linked. (Pentagonal symmetry, although not so common in packing arrangements for bulk metals as hexagonal or cubic configurations, is known in metal whiskers.)

It is clear that there exists a great diversity in cluster geometries and in the variety of ligands that can be accommodated either on the "surface" of the metal polyhedron or as an interstitial unit. The comparison between the bonding of organic substrates on bulk metals with the coordination of ligands in metal clusters will not be dealt with in this chapter. It is sufficient to note that the metal-surface analogy has provided models for possible chemisorption states found on a metal surface (20).

3.2   STRUCTURAL TRENDS FOR TRANSITION-METAL CLUSTERS

In spite of the range of metals involved in cluster formation, it is possible to classify the d block broadly in terms of the type of compound that each group will exhibit. This is partly a function of the electronic demands of the metals concerned. This section will not be a comprehensive survey of the literature, but is intended merely to give an indication of some typical samples. More extensive accounts may be found elsewhere (21,22).

The cluster chemistry on the left of the d block is dominated by the low-valency halide clusters where the metals have formal oxidation numbers of $+2$ to $+3$. The general formulas for these species are $[M_6X_8]$, $[M_6X_{12}]$ (Fig. 3.4) and $[M_3X_9]$. There is generally little distortion in the metal octahedron for $[M_6X_{12}]^{n+}$ (M = Nb, Ta) (23), with the 12 halogen atoms, X, acting as μ_2-bridging groups located over the centers of the 12 edges of the octahedron. Such units usually act as the core for larger species, $[(M_6X_{12}X_6]^{n-}$, the additional six halogen atoms coordinating terminally to the six metals (24).

The basic cluster unit for molybdenum and tungsten is $[M_6X_8]^{n+}$, where the six

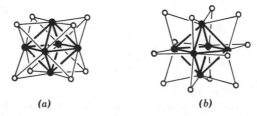

(a)   (b)

Figure 3.4   Halide complexes of Nb and Ta. (a) $[M_6X_8]$; (b) $[M_6X_{12}]$.

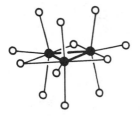

Figure 3.5 $[Re_3Cl_{12}]^{3-}$.

metal atoms again define an octahedron but the eight halogen atoms now act as μ_3-face-capping ligands on the eight triangular faces of the metal polyhedron (25). Six additional halogen atoms may also be added to this basic unit as terminal ligands to the metal atoms.

The trinuclear cluster $[Re_3Cl_{12}]^{3-}$ (Fig. 3.5) has a triangular metal skeleton, with three chloride atoms acting as μ_2-bridges along the edges. There are three terminal chlorines bound terminally to each metal, giving approximately octahedral coordination around each rhenium (26).

These three groups of clusters account for a number of complexes formed by the metals on the left of the d block. Trinuclear, triangular niobium clusters have been reported (21); however, most clusters of the vanadium subgroup have hexametal octahedral cores. Some distortion in these octahedra is found, especially on oxidation or substitution of the ligands. Infinite chains of M_6 units linked by halide bridges are also known.

For the chromium subgroup, the variation in the metal core on oxidation and substitution of ligands is again present, as is the ability to form chain structures. The variety of nuclearities, geometries, and ligands increase from the left of the d block. With molybdenum, tri-, tetra-, and pentanuclear complexes exist. The compounds $[M_3O_2(O_2CCH_3)_6(H_2O)_3]^{3+}$ (M = Mo, W (Fig. 3.6)) (27), where the metal triangle has two μ_3-face-capping oxygen atoms and acetate bridges along the edges, are typical of a range of trinuclear clusters. The tetranuclear compounds $[(\eta^5-$

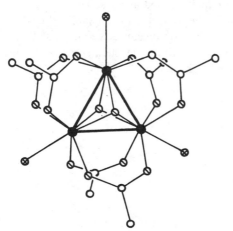

Figure 3.6 $[W_3O_2(O_2CH_3)_6(H_2O)_3]^{2+}$; $\bigcirc$ = and O atom; $\otimes$ = H_2O ligand.

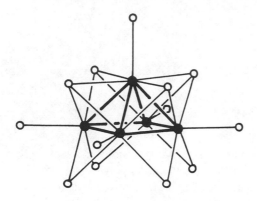

Figure 3.7 $[Mo_5Cl_{13}]^{2-}$

$(C_5H_5)MS]_4$ (where M = Cr or Mo) are of the cubane type, where the four metals combine with the four sulfur atoms to give a cubic arrangement (28). The dianion $[Mo_5Cl_{13}]^{2-}$ exhibits a square-based pyramidal metal skeleton (29) (Fig. 3.7).

The number of carbonyl clusters also increases toward the center of the d block. They are known for the vanadium and chromium subgroups. $[(\eta^5\text{-}C_5H_5)_3Nb_3(CO)_7]$ (30) (Fig. 3.8) exhibits an unusual carbonyl ligand bound in both a σ and π fashion: $[Mo_4(\mu_3\text{-}OH)_4(NO)_4(CO)_8]$ (31) (Fig. 3.9) is another cubane-like structure.] There is an extensive manganese subgroup carbonyl cluster chemistry. $[Mn_4(\mu_3\text{-}OR)_4(CO)_{12}]$ (32) (Fig. 3.10) is cubane-like. $[H_3Mn_3(CO)_{12}]$ (33) consists of an equilateral triangle of metals with hydrogen atoms bridging the edges; the carbonyls are all terminally bound. This compound can be readily converted to the linear species $[Mn_3(CO)_{14}]^-$ (34) by treatment with an alcoholic base (Fig. 3.11).

The tetranuclear rhenium clusters, presented in Figure 3.12, show a variety of metal core geometries: $[H_4Re_4(CO)_{13}]^{2-}$ (35) has a tetrahedral arrangement of metal atoms with the hydride atoms postulated as edge bridging; in $[Re_4(CO)_{16}]^{2-}$ (36) the metal atoms define a parallelogram; the structure of $[H_4Re_4(CO)_{15}]^{2-}$ (36) may be derived from $[Re_4(CO)_{16}]^{2-}$ by the breaking of a Re—Re bond to leave a triangle of rhenium atoms with the other metal terminally bound as a "spike."

The polynuclear complexes at the right of the d block will be examined next, as the iron subgroup will be used in the rationalization of cluster geometries. It will be clear working back to this group that these carbonyl clusters show nearly the whole range of ligands and structures found elsewhere in the d block.

Figure 3.8 $[(\eta^5\text{-}C_5H_5)_3Nb_3(CO)_7]$; —○—○ are terminal —C≡O ligands; —⊘—⊗ is a σ and π bound C≡O ligand.

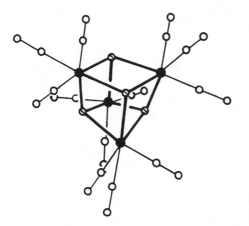

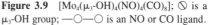

Figure 3.9 $[Mo_4(\mu_3\text{-}OH)_4(NO)_4(CO)_8]$; $\lozenge$ is a μ_3-OH group; —O—O is an NO or CO ligand.

There are no carbonyl clusters in the copper subgroup. Copper itself forms complexes with three to eight metal atoms (Fig. 3.13), ranging from $[Cu_3Cl_2(Ph_2PCH_2PPh_2)_3]^+$ (the silver analog has been structurally characterized (38)) where a triangle of copper atoms is capped asymmetrically by μ_3-chlorine and bridged by diphosphine ligands, to the square antiprismatic geometry of $[Cu_8(C_7H_8O)_8]$ (39). The intermediate compounds display tetrahedral, butterfly, and distorted octahedral arrangements stabilized by donor ligands coordinated through nitrogen, sulfur, or phosphorus. The square planar geometry of $[Cu_4(CH_2SiMe_3)_4]$ (40) is unusual (Fig. 3.13a).

Gold, unlike silver, is beginning to show an extensive cluster chemistry. Most of the larger complexes can be regarded as being derived from the centered icosahedral $[Au_{13}(Ph_2PCH_2PPh_2)_6]^{4+}$ (41). However, $[Au_6(P(C_7H_7)_3)_6]^{2+}$ (42) contains a distorted octahedron of metal atoms, whereas $[Au_6(PPh_3)_6]^{2+}$ (43) has an edge-shared bi-tetrahedral geometry (Fig. 3.14). Clearly, the factors affecting the structure of the polyhedral skeleton are subtle.

The nickel subgroup represents one of the major areas studied in transition-metal

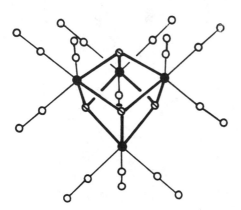

Figure 3.10 $[Mn_4(\mu_3\text{-}OH)_4(CO)_{12}]$.

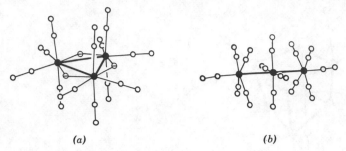

(a)   (b)

Figure 3.11. (a) [H$_3$Mn$_3$(CO)$_{12}$]; (b) [Mn$_3$(CO)$_{14}$]$^-$.

cluster chemistry. The coordination can vary from the linear [Pt(CN)$_4$]$^{2-}$ (44), where planar [Pt(CN)$_4$] units are stacked to form a one-dimensional solid, to the tetracapped octahedron of [Pd$_{10}$(CO)$_{12}$(P^{n}Bu$_3$)$_6$] (45) (Fig. 3.15). Although more large polyhedral clusters are being reported, most are either triangular or built up from triangular units. Noteworthy are the [Pt$_3$(CO)$_6$]$_n^{2-}$ stacks (46), where the dimer, trimer, and the like (Fig. 3.16) consist of prismatic units linked through the triangular faces. Some square planar [Pd$_4$(CO)$_4$(CH$_3$CO)$_4$] (47), [Pt$_4$(CH$_3$CO$_2$)$_8$]) (48), and cubic [Ni$_8$(CO)$_8$(PPh)$_6$] (49) structures are known (Fig. 3.17).

Two very large platinum clusters show a great similarity to bulk metals: [Pt$_{26}$(CO)$_{32}$]$^{2-}$

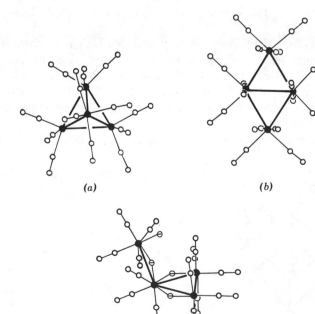

(a)   (b)

(c)

Figure 3.12 (a) [H$_4$Re$_4$(CO)$_{13}$]$^{2-}$; (b) [Re$_4$(CO)$_{16}$]$^{2-}$; (c) [H$_4$Re$_4$(CO)$_{15}$]$^{2-}$.

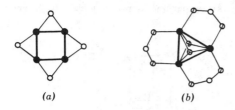

(a)   (b)

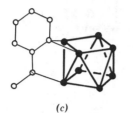

(c)

Figure 3.13 Examples of copper clusters. (a) [Cu$_4$(μ_2-CH$_2$SiMe$_3$)$_4$], ○ = μ_2-CH$_2$SiMe$_3$ group; (b) [Cu$_3$Cl$_3$(Ph$_2$PCH$_2$PPh$_2$)$_3$]$^{+3}$, ⊗ = Cl, ⊘ = PPh$_2$; (c) [Cu$_8$(C$_7$H$_8$O)$_8$]; only one anisole ligand is shown for clarity.

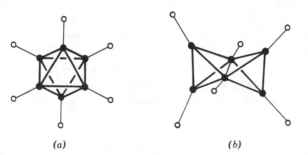

(a)   (b)

Figure 3.14 Gold clusters. ○ represents the phosphine ligands. (a) [Au$_6$(P(C$_7$H$_7$)$_3$)$_6$]$^{2+}$; (b) [Au$_6$(P-(C$_6$H$_5$)$_3$)$_6$]$^{2+}$.

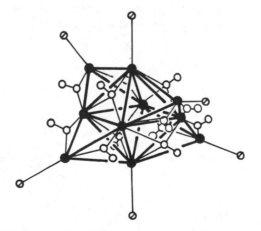

Figure 3.15 [Pd$_{10}$(CO)$_{12}$(P^{n}Bu$_3$)$_6$]; ⊘ = P^{n}Bu$_3$; the complex contains four μ_3-CO and eight μ_2-CO ligands.

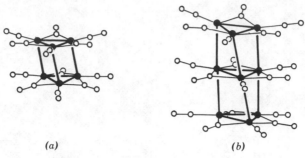

(a)   (b)

Figure 3.16   (a) $[Pt_3(CO)_6]_2^{2-}$; (b) $[Pt_3(CO)_6]_3^{2-}$.

(50) has a hexagonal close-packing arrangement, whereas $[Pt_{38}(CO)_{44}H_x]^{2-}$ exhibits a cubic close packing (50,51).

The clusters in the cobalt subgroup range from trinuclear, triangular geometries to some of the largest clusters yet characterized. Like their platinum analogs the higher rhodium clusters represent fragments of close packing. For instance, $[H_3Rh_{13}(CO)_{24}]^{2-}$ (Fig. 3.1b) has a metal skeleton consistent with the hexagonal close packing, whereas the structures of $[Rh_{14}(CO)_{25}]^{4-}$ and $[Rh_{15}(CO)_{27}]^{3-}$ (52,53) show a stepwise body-centered cubic/hexagonal close-packed interconversion (53) (Fig. 3.18). Tetrahedra, octahedra, trigonal prisms, and tetragonal antiprisms are known for their smaller clusters (21).

The iron subgroup shows the most diverse set of metal geometries. There is a complete range of nuclearities from 3 to 11 and most of the polyhedra encountered elsewhere in the section have examples in this subgroup. For these reasons and by virtue of the large number of structures published by the Lewis and Johnson group at Cambridge, osmium will be used as the model for the rationalization of cluster geometries in Section 3.3.

Mixed metal clusters have not been discussed as their structures are usually related to those containing only one type of metal (22).

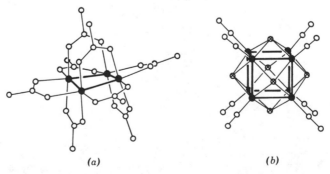

(a)   (b)

Figure 3.17   (a) $[Pt_4(CH_3CO_2)_8]$; (b) $[Ni_8(CO)_8(PPh)_6]$, $\otimes = \mu_4$-PPh.

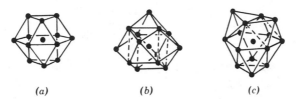

(a)   *(b)*   *(c)*

Figure 3.18 Examples of the metal geometries found in large rhodium clusters. (*a*) $[H_2Rh_{13}(CO)_{24}]^{3-}$; (*b*) $[Rh_{14}(CO)_{25}]^{4-}$; (*c*) $[Rh_{15}(CO)_{27}]^{3-}$; only surface metal atoms have been connected.

3.3 RATIONALIZATION OF CLUSTER GEOMETRIES

3.3.1 The 18-Electron Rule (54)

The 18-electron rule can be applied to mono- and polynuclear complexes. With the trinuclear cluster $[Os_3(CO)_{12}]$ (55), shown in Figure 3.19(*a*), each metal center has four carbonyl ligands associated with it and, as such, may be regarded as having 16 valence electrons. The presence of two metal–metal bonds allows each osmium to obtain the stable 18-electron configuration. In this case, the metal–metal bonds are regarded as two-electron/two-center interactions and are expected to lie along the axis joining the metals.

Similarly, the molecular geometry of the hydrido cluster $[H_2Os_3(CO)_{10}]$ (56), (Fig. 3.19*b*), which has two hydrides bridging one edge of the triangle described by the three metal atoms, is explained by invoking a metal–metal double bond. The cluster as a whole has 46 valence electrons [i.e., $(8 \times 3) + (2 \times 10) + (1 \times 2)$] compared to 48 valence electrons [$(8 \times 3) + (2 \times 12)$] for $[Os_3(CO)_{12}]$ and, as such, requires four metal–metal bonds. Therefore, one of the triangle's edges (the short bond bridged by the two hydrides) is considered to be a double bond. This is consistent with the tendency of $[H_2Os_3(CO)_{10}]$ to react with two-electron donors such as CO or PPh_3 (57). A more precise description of the bonding in this cluster was proposed by M. B. Hall (58).

3.3.2 The Effective Atomic Number Rule (59)

The effective atomic number (EAN) rule is again based on the assumption that the metal atoms attain a noble gas configuration by forming the required number of metal–metal bonds. Although polyhedral edges still represent two-electron/two-center bonds, it is not necessary that each metal itself be electronically correct, but merely that the cluster as a whole be satisfied. For example, $[Os_5(CO)_{16}]$ (60), shown in Figure 3.20, has 72 [$(8 \times 5) + (2 \times 16)$] cluster electrons. However, for all five osmium atoms to attain a closed shell, 90 (18×5) electrons are required. The missing nine electron pairs may be obtained by the formation of nine metal–metal bonds, as is found in the trigonal bipyramidal metal framework. This cluster could not be rationalized by the 18-electron rule since the two osmium atoms in the apical position are formally 17-electron species, whereas the osmium in the equitorial plane

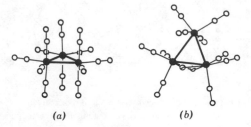

(a)   (b)

Figure 3.19 (a) $[Os_3(CO)_{12}]$; (b) $[H_2Os_3(CO)_{10}]$, $\oslash$ = bridging hydride ligand.

with four carbonyl ligands is formally a 20-electron species (incipient bringing CO ligands help to remove the electron imbalance).

The EAN rule breaks down for many larger clusters precisely because polyhedral edges can no longer represent formal two-electron/two-center bonds. The 84-valence electron cluster $[Os_6(CO)_{18}]$ (61) would obey the 18-electron rule at each individual osmium atom if it were to have octahedral (O_h) symmetry. Instead, the structure was found to consist of a bicapped tetrahedron of metals (Fig. 3.21), each with three terminal carbonyls. This is still a deltahedron having 12 edges and 8 faces, but one of C_{2v} symmetry, and, as such, it obeys the EAN rule only as a whole. Dative bonds are required to achieve the postulated inert gas configuration.

3.3.3 Wade's Rules (4)

The term "Wade's rules" is applied to a set of analogies between transition-metal and boron hydride clusters developed independently by K. Wade (4a) and D. M. P. Mingos (4c). The stoichiometries of the boranes did not allow simple bonding schemes as found for the hydrocarbons. There were insufficient electrons to allow classical two-electron/two-center bonds between adjacent pairs of atoms. Indeed,

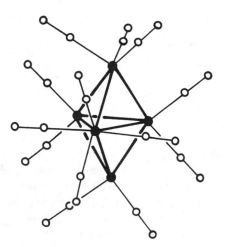

Figure 3.20 $[Os_5(CO)_{16}]$.

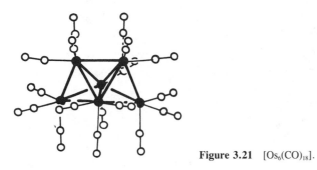

Figure 3.21 $[Os_6(CO)_{18}]$.

the high connectivity of many atoms in these structures precluded such formalisms. These compounds were labeled electron-deficient (62) and in order to rationalize them "multicenter bonding," where one pair of electrons may bind three or more atoms, was invoked. With transition-metal clusters the situation is further complicated with some complexes being electron-precise, whereas others are "electron-deficient." Wade considered the molecular orbitals of the cluster as whole, as opposed to those of the individual metal atoms. By analogy with the boron hydride clusters, the following rules and assumptions were stated (4a):

1. Only three hybrid orbitals are used by each metal carbonyl fragment for cluster bonding. The remaining six orbitals are involved with ligand bonding and nonbonding electrons.

2. These six orbitals are filled first, requiring $6n$ electron pairs for the cluster, where n is the number of metal atoms.

3. The remaining three orbitals are involved in skeletal bonding. The fundamental metal deltahedron can then be established from the number of skeletal electron pairs (S) remaining, where $S = \frac{1}{2}(N - 12n)$ and N is the number of valence electrons in the cluster. For an n-vertex polyhedron, $S = n + 1$; for example, the pentanuclear cluster $[Os_5(CO)_{16}]$ (60) (Fig. 3.20) has $N = 72$ $[(8 \times 5) + (2 \times 16)]$, $n = 5$, and $S = 6$, and this is in agreement with the observed trigonal bipyramidal structure. For clusters where $S > n + 1$, Wade has used the terminology of the boron hydrides; that is,

$S = n + 1$ defines a closo structure, where the n metal atoms define the n vertices of the appropriate polyhedron

$S = n + 2$ defines a nido structure, where one vertex of the parent polyhedron is missing

$S = n + 3$ defines an arachno structure where two vertices are missing from the basic polyhedron

The tetranuclear nitrido cluster $[Os_4(CO)_{12}N]^-$ (63) (Fig. 3.22) has $N = 62$ $[(8 \times 4) + (2 \times 12) + (1 \times 1) + (5 \times 1)]$, $n = 4$, and $S = 7$. The predicted structure of an arachno octahedron is in agreement with the observed "butterfly"

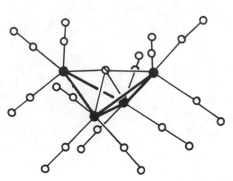

Figure 3.22 $[Os_4(CO)_{12}N]^-$; $\ominus$ = μ_4-N ligand.

geometry. The pentanuclear carbido clusters $[Os_5(CO)_{15}C]$ (64) and $[Os_5(CO)_{16}C]$ (60) (Fig. 3.20) are readily interconverted and possess related molecular frameworks. The addition of a carbonyl ligand converts $[Os_5(CO)_{15}C]$ from a nido-octahedron ($N = 74$, $n = 5$, and $S = 7$) to the arachno pentagonal bipyramidal $[Os_5(CO)_{16}C]$ ($N = 76$, $n = 5$, and $S = 8$). The theory does not state which vertices are excised, but it may be noted that the transformation of the square-based pyramid to the bridged butterfly may be regarded as the result of the breaking of a single metal–metal bond. It will be shown that the addition of two-electron donors tend to be concomitant with bond cleavage (65).

An extension of Wade's rules for $S < n + 1$ considers the fundamental polyhedron derived from S, and for $S = n$ caps it once, for $S = n - 1$ caps it twice, and so on (66) [e.g., the decanuclear carbido dianion $[Os_{10}(CO)_{24}C]^{-2}$ (67), shown in Figure 3.23, has $N = 134$, $n = 10$, and $S = 7$; these values are in agreement with the tetracapped octahedral structure observed].

This theory allows structural changes involving addition or loss of electron pairs to be predicted. For instance, the monocapped trigonal bipyramidal structure $[Os_6(CO)_{18}]$ (61), where $S = 6$, changes on reduction to an octahedral arrangement in $[Os_6(CO)_{18}]^{2-}$ (68), where $S = 7$. It must be noted that this predictive power is limited since it does not differentiate between a *closo* structure and a *nido*-capped polyhedral ar-

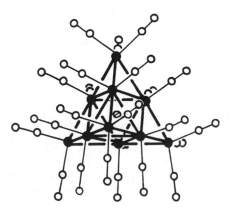

Figure 3.23 $[Os_{10}(CO)_{24}C]^{2-}$; $\oslash$ = carbido C-atom.

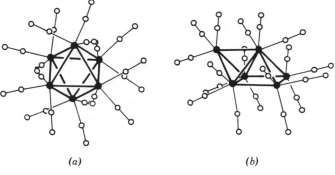

(a)   *(b)*

Figure 3.24   *(a)* $[Os_6(CO)_{18}]^{2-}$; *(b)* $[H_2Os_6(CO)_{18}]$.

rangement: for example, $[Os_6(CO)_{18}]^{2-}$ and $[H_2Os_6(CO)_{18}]$ (68) both have $S = 7$ but the former is a regular octahedron, whereas the latter has a monocapped square pyramidal metal geometry (Fig. 3.24). Wade theory predicts the thermodynamically most stable arrangement, but easily accessible geometries may be stabilized by other factors.

It has been suggested that the addition of a pair of electrons to a cluster is concomitant with the breaking of metal–metal bonds (65) as would be expected from both EAN and Wade rules. The "bow-tie" structure of $[Os_5(CO)_{19}]$ (69,70) can be rationalized (71) by considering the effect of adding three additional electron pairs to the trigonal bipyramidal structure of $[Os_5(CO)_{16}]$ (60) (Fig. 3.25).

Recent theoretical studies by D. M. P. Mingos (72) have extended the basic Wade approach by considering the type of units that make up the cluster as a whole. Since the structure must be known in order to apply these extended rules, there is a limit to their predictive usefulness. They have, however, proved to be particularly effective for rationalizing a wide range of cluster geometries. The approach (72,73) of considering a cluster to be comprised of ML_4, ML_3, and ML_2 fragments allows a ready explanation of the unusual planar structures of the $[Os_6(CO)_{21}]$ derivatives, $[Os_6(CO)_{21-n}L_n]$ (74) (Fig. 3.26). The metals in this complex do obey the 18-electron rule and it is inappropriate to consider the cluster as an arachno-tetragonal antiprism

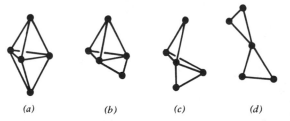

(a)   *(b)*   *(c)*   *(d)*

Figure 3.25   A pathway for the conversion of $[Os_5C(CO)_{16}]$ to $[Os_5(CO)_{19}]$, showing the proposed metal frameworks (72). *(a)* $[Os_5(CO)_{16}]$: $N = 72$, $S = 6$; *(b)* "$[Os_5(CO)_{17}]$": $N = 74$, $S = 7$; *(c)* "$[Os_5(CO)_{18}]$": $N = 76$, $S = 8$; *(d)* $[Os_5(CO)_{19}]$: $N = 78$, $S = 9$.

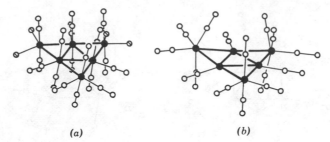

(a)   (b)

Figure 3.26 (a) $[Os_6(CO)_{17}(P(OMe)_3)]_4$; $\ominus$ = $P(OMe)_3$; (b) $[Fe_3Pt_3(CO)_{15}]^{2-}$, central triangle of "Pt(CO)" moieties "bridged" by "Fe(CO)$_4$."

as the basic Wade rules would predict. The molecular orbital calculations also correctly proposed that these systems will take part in a two-electron reduction (72,75).

Mingos has employed Hoffman's (76) isolobal analogies to rationalize among others, the closely related mixed metal cluster $[Pt_3Fe_3(CO)_{15}]^{2-}$ (77). This complex, depicted in Figure 3.26(b), also has six metal atoms in a plane and can be regarded as the analog of $[Os_6(CO)_{21}]^{2-}$. The basic approach of the isolobal connection considers the frontier orbitals of different metal fragments. If the orbitals have similar symmetries, extents in space and energies, and the fragments donate the same number of electrons to the cluster, then they may be considered as "isolobal" (78).

Various cluster geometries, especially those of mixed metal complexes, may be readily rationalized by employing this concept in conjunction with the Wade rules. Synthetic approaches have also been based on this approach but these are discussed in considerable detail elsewhere (76,79).

The largest osmium carbonyl cluster yet isolated, $[Os_{11}(CO)_{27}C]^{2-}$ (80) (Fig. 3.27), has a metal core geometry that may be considered as a bicapped square pyramid sharing its face with a trigonal prism, which is also bicapped on its triangular faces. [This is similar to the metal skeleton of $[Rh_8(CO)_{19}C]$ (13) Fig. 3.3b]. Mingos (72) regards such species as condensed polyhedra and, as such, can be rationalized by subtracting the electron count characteristic to the atoms common to both polyhedra from the sum of the characteristic electron counts for the parent polyhedra. For example, the square pyramidal and prismatic units in $[Os_{11}(CO)_{27}]^{2-}$ provide

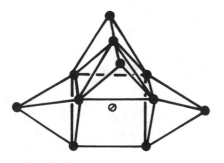

Figure 3.27 The metal core of $[Os_{11}(CO)_{27}C]^{2-}$.

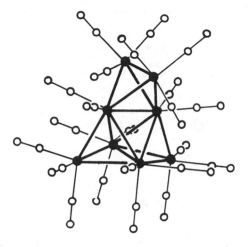

Figure 3.28 $[HOs_8(CO)_{22}]^-$.

98 and 114 cluster valence electrons, respectively. After subtracting 64 electrons from their sum to account for the fused square face, one can obtain the correct total electron count of 148 cluster valence electrons.

The hydrido monoanion $[HOs_8(CO)_{22}]^-$ (81) (Fig. 3.28) has a structure that is readily rationalized by a similar approach. The metal geometry obeys the EAN rule, as did $[Os_{11}(CO)_{27}C]^{2-}$ (80), but on deprotonation the dianion is formed, which has a bicapped octahedral arrangement as predicted by Wade. This again emphasizes the subtle factors determing cluster geometry. The structure of $[HOs_8(CO)_{22}]^-$ is considered to be a result of the synthetic route employed (81).

It is clear that the extended Wade rules represent a useful guide to the rationalization of cluster geometries. However, for a number of the types of polynuclear complexes discussed in this chapter, the simple polyhedral skeletal electron pair theory is inappropriate.

For the classical $[M_4S_4]$ cubane-like clusters, the bonding may be considered in terms of localized molecular orbitals. [The related species $[(Cp)_2Mo_2Fe_2(CO)_8S_2]$ (82) may be rationalized in terms of the Wade approach]. Delocalized molecular orbitals are invoked to rationalize the early transition-metal halide clusters (83). Calculations have shown that d-orbital interactions are major contributors to the stability of these species.

3.4   CONCLUSION

This chapter has been concerned with presenting a number of characteristic clusters of the d-block metals. We have attempted to show how these may be classified by type and how, in turn, each type may be rationalized by either qualitative or quantitative molecular orbital approaches. Where both methods have been applied it is clear that any simple bonding description should only be used with caution. However, like the valence-shell electron-pair repulsion theory (VSEPR) (84) for main-group complexes, the Wade rules are not only very useful, but they also provide a base for the rigorous theoretical arguments that are beginning to encompass them.

REFERENCES

1. F. A. Cotton and G. Wilkinson, *Advanced Inorganic Chemistry*, 4th ed., Wiley, New York, 1980, p. 1081.

2. C. R. Eady, B. F. G. Johnson, J. Lewis, M. C. Malatesta, P. Machin, and M. McPartlin, *J. Chem. Soc. Chem. Commun.*, 945 (1976).

3. S. Martinengo, G. Ciani, A. Sironi, and P. Chini, *J. Am. Chem. Soc.*, **100**, 7096 (1978).

4. The Wade rules is the term applied to the concepts developed by K. Wade. (a) K. Wade, *J. Chem. Soc. Chem. Commun.*, 792 (1971); (b) K. Wade, *Adv. Inorg. Chem. Radiochem.*, **18**, 1 (1976); (c) R. E. Williams, *Adv. Inorg. Chem. Radiochem.*, **18**, 67 (1976); (d) D. M. P. Mingos, *Adv. Organometal. Chem.*, **15**, 1 (1977).

5. J. S. Bradley, *Phil. Trans. R. Soc. London*, **A308**, 103 (1982).

6. E. M. Holt, K. Whitmire, and D. F. Shriver, *J. Chem. Soc. Chem. Commun.*, 778 (1980).

7. M. Tachikawa and E. L. Muetterties, *Prog. Inorg. Chem.*, **28**, 203 (1981).

8. J. S. Bradley, G. B. Ansell, M. E. Leonowicz, and E. W. Hill, *J. Am. Chem. Soc.*, **103**, 4968 (1981).

9. B. F. G. Johnson, J. Lewis, J. N. Nicholls, J. Puga, P. R. Raithby, M. J. Rosales, M. McPartlin, and W. Clegg, *J. Chem. Soc. Dalton*, 277 (1983).

10. B. F. G. Johnson, J. Lewis, W. J. H. Nelson, J. N. Nicholls, J. Puga, P. R. Raithby; M. J. Rosales, M. Schroder, and M. D. Vargas, *J. Chem. Soc. Dalton*, 2447 (1983).

11. A. Sirigu, M. Bianchi, and E. Benedetti, *J. Chem. Soc. Chem. Commun.*, 596 (1969).

12. V. G. Albano, M. Sansoni, P. Chini, and S. Martinengo, *J. Chem. Soc. Dalton*, 651 (1973).

13. V. G. Albano, P. Chini, S. Martinengo, M. Sansoni, and D. Strumolo, *J. Chem. Soc. Chem. Commun.*, 299 (1974).

14. V. G. Albano, P. Chini, G. Ciani, S. Martinengo, and M. Sansoni, *J. Chem. Soc. Dalton*, 463 (1978).

15. B. F. G. Johnson, *J. Chem. Soc. Chem. Commun.*, 211 (1976).

16. B. F. G. Johnson, and R. E. Benfield, "Ligand Mobility in Clusters," in B. F. G. Johnson, Ed., *Transition Metal Clusters*, Wiley, New York, 1980, p. 471.

17. A. J. Deeming, "Some Reactions of Metal Clusters," in B. F. G. Johnson, Ed., *Transition Metal Clusters*, Wiley, New York, 1980, p. 391.

18. C. T. Hayward, J. R. Shapley, M. R. Churchill, C. Bueno, and A. L. Rheingold, *J. Am. Chem. Soc.*, **104**, 7347 (1982).

19. V. G. Albano, M. Sansoni, P. Chini, S. Martinengo, and D. Strumolo, *J. Chem. Soc. Dalton*, 970 (1976).

20. E. L. Muetterties, *Angew. Chem. Int. Ed. Engl.*, **17**, 545 (1978).

21. P. R. Raithby, "The Structure of Metal Cluster Compounds," in B. F. G. Johnson, Ed., *Transition Metal Clusters*, Wiley, New York, 1980, p. 5; M. I. Bruce, "Index of

Structures Determined by Diffraction Methods," in G. Wilkinson, F. G. A. Stone, and E. W. Abel, Eds., *Comprehensive Organometallic Chemistry*, Vol. 9, Pergamon, New York, 1982, p. 1209.

22. M. I. Bruce, *J. Organometal. Chem.*, **257,** 417 (1983).

23. F. A. Cotton and G. Wilkinson, *Advanced Inorganic Chemistry*, 4th ed., Wiley, New York, 1980, p. 1095.

24. R. A. Field, D. L. Kepert, B. W. Robinson, and A. H. White, *J. Chem. Soc. Dalton*, 1858 (1973).

25. P. C. Healy, D. L. Kepert, D. Taylor, and A. H. White, *J. Chem. Soc. Dalton*, 646 (1973).

26. J. A. Bertrand, F. A. Cotton, and W. A. Dollase, *Inorg. Chem.*, **2,** 1166 (1963).

27. A. Bino, F. A. Cotton, Z. Dori, S. Koch, H. Kuppers, M. Millar, and J. C. Sekutowski, *Inorg. Chem.*, **17,** 3245 (1978).

28. H. Vahrenkamp, *Angew. Chem. Int. Ed. Engl.*, **14,** 322 (1975).

29. K. Jodden, H. G. von Schnering, and Harold Schafer, *Angew. Chem. Int. Ed. Engl.*, **14,** 570 (1975).

30. W. A. Herrmann, M. L. Ziegler, K. Wiedenhammer, and H. Biersack, *Angew. Chem. Int. Ed. Engl.*, **18,** 960 (1979).

31. V. Albano, P. Bellon, G. Ciani, and M. Manassero, *J. Chem. Soc. Chem. Commun.*, 1242 (1969).

32. E. Horn, M. R. Snow, and P. C. Zeleny, *Aust. J. Chem.*, **33,** 1659 (1980).

33. S. W. Kirtley, J. P. Olsen, and R. Bau, *J. Am. Chem. Soc.*, **95,** 4532 (1973).

34. R. Bau, S. W. Kirtley, T. N. Sorrell, and S. Winarko, *J. Am. Chem. Soc.*, **96,** 988 (1974).

35. A. Bertolucci, G. Ciani, M. Freni, P. Romiti, V. G. Albano, and A. Albinati, *J. Organometal. Chem.*, **117,** C37 (1976).

36. M. R. Churchill and R. Bau, *Inorg. Chem.*, **7,** 2606 (1968).

37. V. G. Albano, G. Ciani, M. Freni, and P. Romiti, *J. Organometal. Chem.*, **96,** 259 (1975).

38. V. U. Schubert, D. Neugebauer, and A. A. M. Aly, *Z. Anorg. Allg. Chem.*, **464,** 217 (1980).

39. A. Camus, N. Marsich, G. Nardin, and L. Randaccio, *J. Organometal. Chem.*, **174,** 121 (1979).

40. J. A. J. Jarvis, B. T. Kilbourn, R. Pearce, and M. F. Lappert, *J. Chem. Soc. Chem. Commun.*, 475 (1973).

41. J. W. A. van der Velden, F. A. Vollenbroek, J. J. Bour, P. T. Beurskens, J. M. M. Smits, and W. P. Bosman, *Recl. Trav. Chim. Pays-Bas*, **100,** 148 (1981).

42. P. Bellon, M. Manassero, and M. Sansoni, *J. Chem. Soc. Dalton*, 2423 (1973).

43. C. E. Briant, K. P. Hall and D. M. P. Mingos, *J. Organometal. Chem.*, **254,** C18 (1983).

44. For examples of one-dimensional inorganic complexes see J. S. Miller and A. J. Epstein, *Prog. Inorg. Chem.*, **20,** 1 (1976).

45. E. G. Mednikov, N. K. Eremenko, V. A. Mikhailov, S. P. Gubin, Y. L. Slovokhotov, and Y. T. Struchkov, *J. Chem. Soc. Chem. Commun.*, 989 (1981).

46. J. C. Calabrese, L. F. Dahl, P. Chini, G. Longoni, and S. Martinengo, *J. Am. Chem. Soc.*, **96,** 2614 (1974).

47. I. I. Moiseev, T. A. Stromnova, M. N. Vargaftig, G. J. Mazo, L. G. Kuz'mina, and Y. T. Struchkov, *J. Chem. Soc. Chem. Commun.*, 27 (1978).

48. M.A.A.F. de C. T. Carrondo and A. C. Skapski, *Acta Cryst.*, **B34,** 1857 (1978).

49. L. D. Lower and L. F. Dahl, *J. Am. Chem. Soc.*, **98,** 5046 (1976).

50. A. Ceriotti, D. Washecheck, L. F. Dahl, G. Longoni, and P. Chini, as cited in P. Chini, *J. Organometal. Chem.*, **200,** 37 (1980).

51. E. L. Muetterties, *Am. Chem. Soc. Chem. Eng. News*, **60,**(35), 28 (1982).

52. S. Martinengo, G. Ciani, A. Sironi, and P. Chini, *J. Am. Chem. Soc.*, **100,** 7096 (1978).

53. P. Chini, *J. Organometal. Chem.*, **200,** 37 (1980).

54. K. Wade, "Some Bonding Considerations," in B. F. G. Johnson, Ed., *Transition Metal Clusters*, Wiley, New York, 1980, p. 193 and references therein.

55. M. R. Churchill and B. G. DeBoer, *Inorg. Chem.*, **16,** 878 (1977).

56. A. G. Orpen, A. V. Rivera, E. G. Bryan, D. Pippard, G. M. Sheldrick, and K. D. Rouse, *J. Chem. Soc. Chem. Commun.*, 723 (1978).

57. A. J. Deeming and S. Hasso, *J. Organometal. Chem.*, **114,** 313 (1976).

58. D. E. Sherwood and M. B. Hall, *Inorg. Chem.*, **21,** 3458 (1982).

59. B. F. G. Johnson and J. Lewis, *Adv. Inorg. Chem. Radiochem.*, **24,** 225 (1981).

60. B. E. Reichert and G. M. Sheldrick, *Acta Cryst.*, **B33,** 173 (1977).

61. R. Mason, K. M. Thomas, and D. M. P. Mingos, *J. Am. Chem. Soc.*, **95,** 3802 (1973).

62. K. Wade, *Electron Deficient Compounds*, Nelson, London, 1971.

63. M. A. Collins, B. F. G. Johnson, J. Lewis, J. M. Mace, J. Morris, M. McPartlin, W. J. H. Nelson, J. Puga, and P. R. Raithby, *J. Chem. Soc. Chem. Commun.*, 689 (1983).

64. P. F. Jackson, B. F. G. Johnson, J. Lewis, and J. N. Nicholls, *J. Chem. Soc. Chem. Commun.*, 564 (1980).

65. R. Mason and D. M. P. Mingos, *J. Organometal. Chem.*, **50,** 53 (1973).

66. D. M. P. Mingos and M. I. Forsyth, *J. Chem. Soc. Dalton*, 610 (1977).

67. P. F. Jackson, B. F. G. Johnson, J. Lewis, W. J. H. Nelson, and M. McPartlin, *J. Chem. Soc. Dalton*, 2099 (1982).

68. M. McPartlin, C. R. Eady, B. F. G. Johnson, and J. Lewis, *J. Chem. Soc. Chem. Commun.*, 883 (1976).

69. D. H. Farrar, B. F. G. Johnson, J. Lewis, J. N. Nicholls, P. R. Raithby, and M. J. Rosales, *J. Chem. Soc. Chem. Commun.*, 273 (1981).

70. B. F. G. Johnson, J. Lewis, P. R. Raithby, and M. J. Rosales, *J. Chem. Soc. Dalton*, 2645 (1983).

71. D. H. Farrar, B. F. G. Johnson, J. Lewis, P. R. Raithby, and M. J. Rosales, *J. Chem. Soc. Dalton*, 2051 (1982).

72. D. G. Evans and D. M. P. Mingos, *Organometallics*, **2,** 435 (1983).

73. D. M. P. Mingos, *J. Chem. Soc. Chem. Commun.*, 706 (1983); D. M. P. Mingos and D. G. Evans, *J. Organometal. Chem.*, **251,** C13 (1983).

74. R. J. Goudsmit, B. F. G. Johnson, J. Lewis, P. R. Raithby, and K. H. Whitmire, *J. Chem. Soc. Chem. Commun.*, 640 (1982).

75. R. J. Goudsmit, B. F. G. Johnson, J. Lewis, and M. Schroder, unpublished results.

76. R. Hoffmann, *Angew. Chem. Int. Ed. Engl.*, **21**, 711 (1982).

77. G. Longoni, M. Manassero, and M. Sansoni, *J. Am. Chem. Soc.*, **102**, 7973 (1980).

78. M. Elian, M. M. L. Chen, D. M. P. Mingos, and R. Hoffmann, *Inorg. Chem.*, **15**, 1148 (1976).

79. F. G. A. Stone, *Acc. Chem. Res.*, **14**, 318 (1981).

80. D. Braga, K. Henrick, B. F. G. Johnson, J. Lewis, M. McPartlin, W. J. H. Nelson, A. Sironi, and M. D. Vargas, *J. Chem. Soc. Chem. Commun.*, 1131 (1983).

81. D. Braga, K. Henrick, B. F. G. Johnson, J. Lewis, M. McPartlin, W. J. H. Nelson, and M. D. Vargas, *J. Chem. Soc. Chem. Commun.*, 419 (1982).

82. T. P. Fehlner, C. E. Housecroft, and K. Wade, *Organometallics*, **2**, 1426 (1983).

83. F. A. Cotton and T. E. Haas, *Inorg. Chem.*, **3**, 10 (1964).

84. R. J. Gillespie and R. S. Nyholm, *Quart. Rev. Chem. Soc.*, **11**, 339 (1957).

4

KINETICS OF REACTIONS OF METAL CARBONYL CLUSTERS

A. J. POË
J. Tuzo Wilson Laboratories
Erindale College
University of Toronto
Mississauga, Ontario

4.1   INTRODUCTION

Growth of interest in metal carbonyl clusters over the past decade or so has been evidenced mainly by synthetic and structural studies (1–3), and the volume of kinetic work has been minute by comparison (3–5). Even the putative value of clusters as catalysts has received relatively little basic mechanistic examination. Parshall (6) was led to comment on homogeneous catalysis by transition-metal complexes in general, that "the review of mechanism" (in his book) "is often distressingly brief because so little is reported on the subject" and Johnson (1) referred to the study of cluster reactivity as still being "in its infancy" in 1980.

Kinetic studies in this area are of considerable importance in providing (1) evidence for the existence of multiple pathways for apparently simple reactions, (2) evidence for the existence of various intermediates in a given reaction path, and (3) precise quantitative measurements of the reactivity of the cluster with respect to their various reaction paths. Thus, it is found, not infrequently, that a reaction can proceed to the same product by a variety of reaction paths so that there is no simple answer to the question: "What is *the* mechanism of the reaction?" On the other hand, different paths of reaction of a cluster with the same reactant can lead to different products and kinetic studies can show unambiguously which path leads to which product. In addition, quite closely related complexes that differ, for example, only in the detailed nature of a substituent can react by quite different paths or at widely different rates so that generalizations about a given class of complexes must not too readily be based on results for one or two members of the class.

Although the existence of various intermediates of defined composition is often demonstrable without reasonable doubt on the basis of the stoichiometric rate equation, the nature of the intermediates and the function of their various components usually remains a matter for informed speculation. This problem can occasionally be resolved if the intermediates can be synthesised by different methods and characterized under much milder conditions than those involved in the kinetic study. More frequently, support for the proposed nature of an intermediate can be obtained by synthesis and characterization of very closely analogous species that have, however, valid reasons for being more stable than the proposed intermediate. Finally, the systematic quantification of the reactivity of clusters with respect to the various reaction paths available to them provides a sound basis for an understanding of the factors governing the relative importance of these paths and could, in principle, lead to rational tailoring of potential catalysts to suit specific purposes.

4.1.1 Nonbridged or Weakly Bridged Dimetal Carbonyls

The major reactions undergone by mononuclear carbonyls—substitution, oxidative addition or elimination, reductive elimination, and insertion-migration—have been the subject of extensive kinetic study for many years, and their mechanistic pathways are now well established and much is known about their energetics (3–5,7,8).

Some systematic effort has also been devoted to the kinetic study of dinuclear metal carbonyls containing either an unsupported metal–metal bond or a metal–metal bond that is weakly bridged, usually by CO ligands (9,10). An important and distinctive pathway that might be expected for thermal reactions of such complexes is homolysis of the metal–metal bond to form two separate 17-electron radical species, a process that is known to be of great importance in photolytic reactions (11). This initial step can lead to either mononuclear products, when suitable radical scavengers are present, or substituted dinuclear products, the latter being formed by rapid substitution at the 17-electron metal-centered radicals followed by dimerization involving substituted radicals. Whether mono or di substituted products are formed depends on the relative steady-state concentrations of substituted and unsubstituted radicals which in turn depends on the relative rates of substitution and dimerization. Homolysis was believed (9) to have been established for the decacarbonyls $M_2(CO)_{10}$ (M_2 = Mn_2, Tc_2, Re_2, and MnRe) on the basis of thermal decomposition kinetics of a type thought to be peculiar to reversible homolysis followed by radical scavenging. This conclusion has had to be modified in the light of the failure to observe the crossover reactions, expected if homolysis is occurring, and involving chemically or isotopically distinct decacarbonyls (12). The crossover reaction between $Mn_2(CO)_{10}$ and $Re_2(CO)_{10}$ has been shown to proceed, under forcing conditions, via aggregation processes rather than via fragmentation (13). The kinetic behavior found for the thermal decompositions is not, therefore, unique to the homolysis mechanism, and thus is only a necessary and not a sufficient condition for that mechanism to be established.

On the other hand, quite a large number of disubstituted dimanganese carbonyls are still believed to undergo homolysis on the basis of their kinetic behavior supported by the observation of appropriate crossover reactions (9c). This conclusion allows the activation enthalpies to be taken as good estimates of the strengths of the Mn—Mn bonds, and these appear to decrease substantially with increasing size of the substituents, electronic effects being quite small. The nature of the reactions of the decacarbonyls is still uncertain but their complicated kinetics argue strongly (14) against the proposed (12,15) simple CO dissociative process, and some form of reversible homolysis induced by the radical scavenger may be occurring (9c).

Another distinguishing feature of dinuclear carbonyls is the possibility that what seem to be coordinatively unsaturated intermediates, formed by CO dissociation, may be stabilized by a form of CO bridging in which the CO acts as a 4-electron donor (16). A CO on one metal donates two of its π electrons to the other metal from which a CO has been dissociated. Continued adherence to the 18-electron rule following CO dissociation can also be maintained by formation of CO-bridged metal–metal multiple bonds (17). Both these possibilities are indicated in equation (4.1).

$$
\begin{array}{cc}
\underset{M}{\overset{O}{\underset{|}{C}}} \underset{M}{\overset{O}{\underset{|}{C}}} & \longrightarrow \quad M\overset{C\!\equiv\!O}{-\!\!-\!\!-}M \quad \text{or} \quad M\overset{\overset{O}{C}}{=\!\!=\!\!=}M
\end{array} \qquad (4.1)
$$

In addition, reversible dissociation of *two* ligands, one from each metal, can occur due to the stabilization of the product by formation of metal–metal triple bonds (9). This has been demonstrated for reactions of $Co_2(CO)_6(P\text{-}n\text{-}Bu_3)_2$ (18) and $Cp_2Mo_2(CO)_6$ $(Cp = \eta^5\text{-}C_5H_5)$ (19) with alkynes.

Yet another distinctive process that could occur in dinuclear carbonyls is metal migration in which a metal atom, with its dependent ligands, migrates from the other metal to a C atom on a CO attached to that metal (20). This results in loss of the metal–metal bond and formation of a CO-bridged dimer in which one of the metals is coordinatively unsaturated and subject to nucleophilic attack. The process has been invoked for reactions involving insertion of $SnCl_2$ into metal–metal bonds (21,22), as exemplified in equations (4.2) to (4.4) $(L = P\text{-}n\text{-}Bu_3)$ (21), and for some reactions of alkynes with dinuclear carbonyls (18,19).

$$L(OC)_3Co\text{—}Co(CO)_3L \longrightarrow L(OC)_3Co(\mu\text{-}CO)Co(CO)_2L \qquad (4.2)$$

$$L(OC)_3Co(\mu\text{-}CO)Co(CO)_2L \xrightarrow{\text{SnCl}_2} L(OC)_3Co(\mu\text{-}CO)Co(CO)_2L(SnCl_2) \qquad (4.3)$$

$$L(OC)_3Co(\mu\text{-}CO)Co(CO)_2L(SnCl_2) \longrightarrow L(OC)_3Co(\mu\text{-}SnCl_2)Co(CO)_3L \qquad (4.4)$$

Just as substitution reactions of a given mononuclear carbonyl can occur by both associative and dissociative paths (7,8), so it is true that more than one of the paths previously outlined can be followed in reactions of a given dinuclear carbonyl. In addition, a given dinuclear carbonyl could follow one or more of the paths peculiar to such carbonyls and also one of the sort found for mononuclear carbonyls.

The great mechanistic diversity possible for dinuclear carbonyls must, in principle, also be available for metal-carbonyl clusters and, of course, the processes outlined earlier could occur through photochemical as well as through thermal activation. For the purposes of this review, carbonyl clusters have been defined as carbonyl complexes containing at least three atoms all bound firmly together, of which at least two are transition metals but one or more can be nonmetallic, metalloid, or nontransition-metal atoms. This definition includes the basic feature of a cluster that distinguishes it from a dimetal carbonyl, namely, simple breaking of one metal–metal bond can occur without the metal atoms' being free to drift completely apart, at least two bonds having to be broken for that to be possible. The review begins, therefore, with a consideration of the kinetics of strongly bridged dimetal carbonyls and proceeds from there to trimetal clusters and on to clusters of higher nuclearity. Photochemical aspects are discussed separately.

4.2 STRONGLY BRIDGED DIMETAL CARBONYLS

4.2.1 Introduction

Strongly bridged dimetal carbonyls can be defined as those in which the bridging groups (1) cannot or do not exist alternatively as terminal ligands in the same molecule, and (2) are not easily displaced. The bridging groups can influence the chemistry of the dinuclear carbonyls in several ways. As a different type of substituent they may introduce specific electronic and steric effects at the metal atoms and they may greatly modify the extent to which such effects can be transmitted across the metal–metal bond from the coordination sphere of one metal to that of the other. In addition, they can provide a means of maintaining the integrity of the molecule even when the metal–metal bond is broken. This has the effect of stabilizing intermediates containing one or two reactive metal centers at which interesting reactions might occur. These reactions can be different from those at a single metal center either because the two centers could be radical in nature or because of the presence of a neighboring metal atom. Since it can be arranged that the metals are different, considerable potential exists for control of reaction type and reactivity through a suitable choice of metals. Thus there has recently been a pronounced development of interest in the reactions of strongly bridged complexes, although examples of kinetic studies on them date back to the beginning of mechanistic work on metal carbonyls (23,24).

4.2.2 Bridged Dicobalt Carbonyls

Early studies on CO exchange with $Co_2(CO)_8$ showed (24) that all the CO ligands exchanged rapidly and it was believed that this might be due to the presence, at least in one of the isomeric forms of the complex, of bridging CO ligands. This appeared to be confirmed by the observation that complex **4.1** also underwent rapid exchange, with rate parameters very similar to those for $Co_2(CO)_8$, whereas **4.2** did not (24). Although further kinetic studies have been carried out on **4.2** (25–27), nothing more has been reported on **4.1**. This is surprising since it must be one of the earliest "methylene-bridged" dinuclear complexes known.

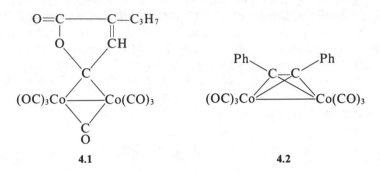

4.1   **4.2**

The complex **4.2** does undergo CO exchange under somewhat more forcing conditions (25), and extensive studies of its substitution reactions with P-donor nucleophiles have been reported (26,27). It reacts in decalin according to rate equation (4.5) which is characteristic of mixed associative and dissociative paths as shown in equations (4.6) to (4.8).

$$k_{obsd} = k_2[L] + k_1 k_3[L]/\{k_{-1}[CO] + k_3[L]\} \qquad (4.5)$$

$$(OC)_3\overline{Co(\mu\text{-}C_2Ph_2)Co}(CO)_3 + L \xrightarrow{k_2} L(OC)_2\overline{Co(\mu\text{-}C_2Ph_2)Co}(CO)_3 + CO \quad (4.6)$$

$$(OC)_3\overline{Co(\mu\text{-}C_2Ph_2)Co}(CO)_3 \underset{k_{-1}}{\overset{k_1}{\rightleftharpoons}} (OC)_2\overline{Co(\mu\text{-}C_2Ph_2)Co}(CO)_3 + CO \qquad (4.7)$$

$$(OC)_2\overline{Co(\mu\text{-}C_2Ph_2)Co}(CO)_3 + L \xrightarrow{k_3} L(OC)_2\overline{Co(\mu\text{-}C_2Ph_2)Co}(CO)_3 \qquad (4.8)$$

The dependence of k_2 on the nature of the nucleophile was moderately pronounced, placing the complex below $Mn(CO)_4(NO)$ and above $CpRh(CO)_2$ in discriminating power (27). The discriminating powers were quantified by the gradients of plots (7) of log k_2 against Δhnp, the relative half-neutralization potentials for titration of the nucleophilic ligands against perchloric acid in nitromethane. The constants k_3/k_{-1} were essentially constant at 0.33 ± 0.07, irrespective of the nucleophile, thus showing that the coordinatively unsaturated intermediate in the dissociative path is not at all discriminating and is, therefore, highly reactive.

A more complete study (26) of the reaction with P-n-Bu$_3$ provided activation parameters and kinetic data for the formation of the bis-substituted complex from the initially formed monosubstituted one. The presence of the P-n-Bu$_3$ substituent on one Co atom had virtually no effect on the ease of CO dissociation from the other Co atom, but it did have a pronounced effect on the ease of nucleophilic attack. $\Delta H_2^{\neq}$ increased from 12.5 to 25.7 kcal mol^{-1} on going from Co$_2$(CO)$_6$(μ-C$_2$Ph$_2$) to Co$_2$(CO)$_5$(P-n-Bu$_3$)(μ-C$_2$Ph$_2$), but this was offset to some extent by an increase from -29 to $+2$ cal K^{-1} mol^{-1} in $\Delta S_2^{\neq}$. These results are complemented in a very interesting way by studies of electrochemically induced reactions of such complexes (28). Reduction to the radical anions Co$_2$(CO)$_6$(μ-C$_2$R$_2$)$^-$ is followed by formation of a cluster formulated as **4.3** which contains a 17-electron Co atom at

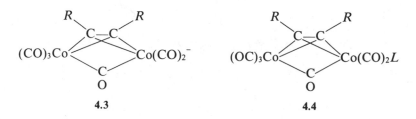

4.3   **4.4**

which a nucleophilic attack can occur. This can lead either to substitution, and formation of Co$_2$(CO)$_5$L(μ-C$_2$R$_2$) after reoxidation, or to fragmentation, the balance being strongly in favor of substitution when R = CF$_3$. Reduction potentials become much more favorable along the series R = ^{t}Bu(-1.03 V), Ph(-0.82 V), and CF$_3$(-0.51 V). Since nucleophilic attack at the LUMO of the complex is a form of reduction, it might be expected that susceptibility to associative substitution would increase along this series and such reactions might well proceed via intermediates such as **4.4**. Reduction potentials become much less favorable when CO is replaced by P donors, and this probably explains the much lower susceptibility of Co$_2$(CO)$_5$(P-n-Bu$_3$)(μ-C$_2$Ph$_2$) to nucleophilic attack compared with Co$_2$(CO)$_6$(μ-C$_2$Ph$_2$). The values of k_3/k_{-1} for the two complexes are quite similar so the effects do not apply in the formation and reaction of the coordinatively unsaturated complexes.

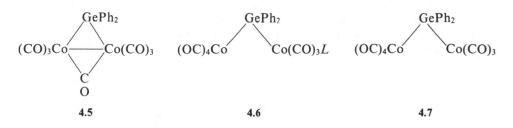

4.5   **4.6**   **4.7**

The complex **4.5** undergoes an interesting "cluster-opening" reaction with various ligands to form **4.6** (29). The reactions are strictly first order in [L] even when L = CO, but the second-order rate constants do not depend very much on the nature of L, P-n-Bu$_3$ reacting only twice as rapidly as PPh$_3$. It is therefore concluded that reaction involves initial highly reversible formation of complex **4.7** followed by the weakly competing addition of L to the vacant coordination site. The reverse reaction, **4.6** (L = CO) to **4.5**, must thus involve dissociative loss of CO to form **4.7** followed by very rapid clustering, so rapid indeed that the CO dissociation cannot be reversed even under 1 atm CO. The dissociative nature of the reaction is supported by the large value of $\Delta S^{\neq}$ (+21 cal K^{-1} mol^{-1}).

Complex **4.5** also undergoes a dissociative substitution reaction with PPh_3 to form **4.8** which then rapidly reacts with CO or PPh_3 to form **4.9**. The destabilizing effect of PPh_3 on the cluster is such that **4.9** is formed very rapidly from **4.8** at 20°C, whereas **4.6** (L = CO) is only formed from **4.5** quite slowly at 80°C. Complex **4.5** reacts also with alkynes to form **4.2** and polymeric $(GePh_2)_n$ (30). The kinetics

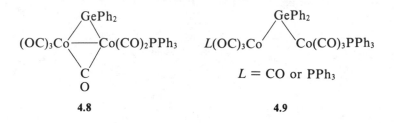

4.8   4.9

of the reaction with C_2Ph_2 suggest that reaction occurs via **4.6** (L = C_2Ph_2). It is proposed that this then undergoes intramolecular displacement of CO to form **4.10** followed by extrusion of $GePh_2$ and simultaneous formation of the Co—Co bond.

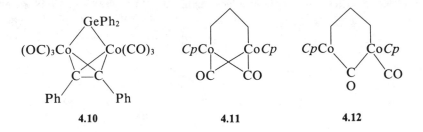

4.10   4.11   4.12

Bergman and his coworkers (31) have recently postulated similar processes to the reversible formation of **4.7** from **4.5**. Compound **4.11** reacts with L (L = CO, PPh_3, or PMe_3) at rates that increase with [L] to a limiting value; that is, **4.11** reacts

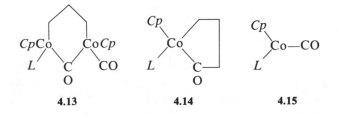

4.13   4.14   4.15

to form a reactive steady-state intermediate that can either revert to **4.11** or be scavenged by L. (The latter process is evidently more effective than the reaction of L with **4.7** since a limiting rate is reached at higher $[L]$ values.) Complex **4.12** is a sensible formulation for such an intermediate since L can add to the vacant coordination site, on the left-hand Co atom, and form **4.13**. This 18-electron intermediate has not been isolated or characterized but its formulation is reasonable in view of the product analysis. Thus, at high $[L]$ **4.14** and **4.15** are the main products, these being consistent with the nucleophilic attack of L on the Co atom on the right-hand side of **4.13**, whereas at low $[L]$ **4.13** spontaneously fragments to cyclopropane, **4.15**, and CpCo(CO). The latter adds L to form another molecule of **4.15**. This is a good example of how relative yields of products can be used to infer the nature of reactions occurring after the highest point on the reaction profile. It is interesting that spontaneous fragmentation of **4.13** does not lead simply to **4.14** and CpCo(CO). Formation of the cobaltacyclopentanone product must require an increase of electron density at a cobalt atom brought about by nucleophilic attack. While in these reactions the cobalt atoms are held together in the early stages by the $(CH_2)_3$ chain, another closely related set of reactions occurs without eventual fragmentation when the two cobalt atoms are linked as in **4.16** (32). This complex reacts with PPh$_3$ to form **4.17** (L = PPh$_3$) via a reversibly formed reactive isomer of **4.16** formulated as **4.18**, the ultimate product being **4.19**. An interesting feature

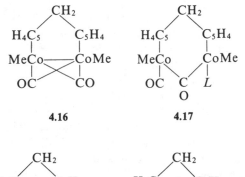

4.16   **4.17**

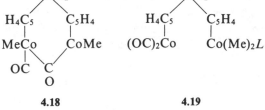

4.18   **4.19**

of its formation from **4.17** was revealed by labeling studies that included variation of [complex]. This showed that the transfer of Me groups from one metal to another can be either intramolecular or intermolecular. Although the exact nature of these processes is not clear, it is obvious that the need to consider them would not have arisen in the absence of such kinetic studies.

4.2.3 Bridged Di-iron Carbonyls

A wide variety of complexes $(OC)_3Fe(bridge)_2Fe(CO)_3$ is known in which the Fe—Fe
bond is bridged by two atoms, each of which donates 3 electrons to the Fe atoms
so that they conform to the 18-electron rule. The coordination around the Fe atoms
is quasi-octahedral if the Fe—Fe bond is considered to be bent. Substitution reactions
lead successively to $Fe_2(CO)_5L(bridge)_2$ and $Fe_2(CO)_4L_2(bridge)_2$ in which the substituents are trans to the bent Fe—Fe bond and on the same side of the molecule
as the bridging groups.

The kinetics of substitution of the complexes $Fe_2(CO)_6(\mu\text{-}SR)_2$ have been studied
in considerable detail. When R = Me, Et, or CH_2Ph the complexes **4.20** exist as

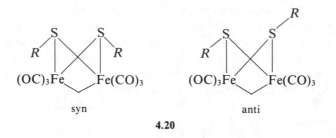

syn   anti

4.20

syn and anti isomers, the anti form being thermodynamically more stable and the
rates of interconversion being quite slow at room temperature. When R = Ph only
the anti form is identifiable (33). Rates of substitution were shown to be faster for
the less stable syn isomers and to be faster with P-n-Bu$_3$ than with PPh$_3$. The
substituted products showed a lesser relative stability of the anti isomer and the
rates of isomerism were greatly increased by the substituent.

Ellgen and Gerlach extended these observations by studying the kinetics of substitution reactions of the complexes containing the bridging groups $(\mu\text{-}SR)_2(R$ = Me,
Et, CH_2Ph, Ph, and p Tol), $(\mu\text{-}S_2C_2H_4)$, and $(\mu\text{-}S_2C_6H_3Me)$ (34). When the bridging
S atoms were joined by a rigid organic group the reactions were simply bimolecular
with k_{obsd} = $k_2[L]$, but reactions of the di-μ-mercapto complexes were often more
complicated because syn- $\rightleftharpoons$ anti-isomerization competed with substitution. The
reactions could all be represented as in Scheme 4.1.

$$\text{anti-Fe}_2(CO)_6(\mu\text{-}SR)_2 \xrightarrow[+L, -CO]{k_{2a}} \text{anti-Fe}_2(CO)_5L(\mu\text{-}SR)_2$$

$$k_{1a} \Big\Updownarrow k_{1s} \qquad\qquad\qquad\qquad\qquad \text{fast} \Big\Updownarrow \text{fast}$$

$$\text{syn-Fe}_2(CO)_6(\mu\text{-}SR)_2 \xrightarrow[+L, -CO]{k_{2s}} \text{syn-Fe}_2(CO)_5L(\mu\text{-}SR)_2$$

Scheme 4.1

Because of the slowness of the isomerization, when R = Me and Et the observed
reaction is simply the sum of the two independent reactions governed by k_{2a} and

k_{2s}, and for R = Me, k_{2s} = $20k_{2a}$. For R = Ph, p Tol, and CH_2Ph the rates of isomerization are greater and the concentration of the syn isomer is reduced to a steady-state value, its rate of formation from the anti form being equal to its rate of substitution plus its rate of isomerization back to the anti form. Under these conditions, the pseudo first-order rate constant for the loss of the anti isomer is given by equation (4.9). When $k_{2s}[L]/k_{1s}$ becomes $>> 1$ this reduces simply to equation (4.10).

$$k_{obsd} = k_{2a}[L] + k_{1a}(k_{2s}/k_{1s})[L]/\{1 + (k_{2s}/k_{1s})[L]\} \qquad (4.9)$$

$$k_{obsd} = k_{1a} + k_{2a}[L] \qquad (4.10)$$

A thorough and elegant study of all the reactions shown in equations (4.11) and (4.12) was carried out by Basato (35). The very complex rate behavior for the for-

$$\text{anti-Fe}_2(CO)_6(\mu\text{-SPh})_2 + PPh_3 \rightleftharpoons \text{anti-Fe}_2(CO)_5(PPh_3)(\mu\text{-SPh})_2 + CO \qquad (4.11)$$

$$\text{anti-Fe}_2(CO)_5(PPh_3)(\mu\text{-SPh})_2 + PPh_3 \rightleftharpoons Fe_2(CO)_4(PPh_3)_2(\mu\text{-SPh})_2 + CO \qquad (4.12)$$

ward and reverse reactions in (4.11) was interpreted as being due to a direct nucleophilic attack on the anti isomer together with rate determining anti → syn isomerization followed by rapid bimolecular substitution into the synform. Substitution into the syn isomers, whether by PPh_3 in the forward reaction or by CO in the reverse, is about 200 times faster than substitution into the anti form. The activation parameters for a nucleophilic attack on the anti forms by PPh_3 or CO in (4.11) are similar, and the activation parameters for the anti → syn isomerization of $Fe_2(CO)_6(\mu$-$SPh)_2$ suggest that significant Fe—S bond weakening is involved. The forward and reverse reactions in (4.12) are both simple ligand dissociation processes, the absence of associative paths being ascribed to the steric effect of the PPh_3 ligands.

Ellgen and coworkers extended the studies on the S-bridged complexes to include complexes with bridging N ligands such as **4.21** and **4.22** (36) and with some allene bridges **(4.23)** (37). The latter can be taken to be acting as two overlapping allylic groups. The N-bridged complexes undergo substitution with k_{obsd} = $k_2[L]$, and the dependence of k_2 on the nature of L shows a similar selectivity and similar steric

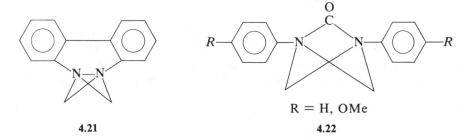

R = H, OMe

4.21   4.22

effects to those of the S-bridged complexes. The absolute reactivities seem to depend very little on the nature of the bridging atoms or on the solvent. The only clear-cut effect is a steric one. The complex with the allene bridge **4.23(a)** reacts with PPh_3

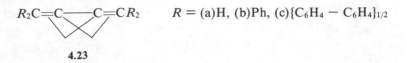

$R_2C{=}C{\underset{}{\diagdown}}C{=}CR_2$ $R = $ (a)H, (b)Ph, (c)$\{C_6H_4 - C_6H_4\}_{1/2}$

4.23

or $P(OPh)_3$ by a simple second-order path but for **4.23(b)** and **4.23(c)** a first-order term is also apparent. Reaction with $P\text{-}n\text{-}Bu_3$ is much faster, because of the much larger value of k_2, and an initial product is formed simply by addition of $P\text{-}n\text{-}Bu_3$ and insertion of a CO into an Fe—C bond. This is followed by a slower loss of a terminal CO and "de-insertion" to form the substituted product $Fe_2(CO)_5L(\mu\text{-}C_4R_4)$. This second-order substitution via an initial insertion-migration would not have been detected were it not for the high selectivity of the original complex, and it was proposed that all the second-order substitutions proceed in this way.

Successive substitution reactions of the P-bridged complex **4.24** proceed only by dissociative mechanisms, the absence of associative paths being ascribed to the greater steric inhibition consequent on the attachment of two groups to each of the bridging atoms (38). The activation parameters for the dissociative reactions of the complexes $Fe_2(CO)_6(\mu\text{-}C_4Ph_4)$ **4.23(b)**, $Fe_2(CO)_6(\mu\text{-}PPh_2)_2$, and $Fe_2(CO)_6(\mu\text{-}SPh)_2$ show an unusual "inverse isokinetic plot" of $\Delta H_1^{\neq}$ against $\Delta S_1^{\neq}$ in which the values of $\Delta S_1^{\neq}$ decrease linearly with increasing $\Delta H_1^{\neq}$ instead of increasing. The significance of this is not obvious but it does mean that the increase in $\Delta H_1^{\neq}$ is not accompanied by the increase in $\Delta S_1^{\neq}$ expected simple-mindedly if the degree of bond breaking increases along the series. Changes in the energy of the rest of the molecule evidently have to be considered as well as those involved simply in the Fe—CO bond-breaking process.

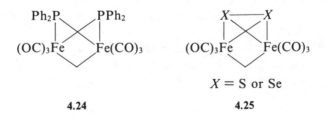

$X = $ S or Se

4.24   **4.25**

Some new features, not completely understood, were shown by substitution reactions of **4.25** (39). These reactions follow equation (4.13). For weak nucleophiles only the second-order term (i.e., first-order in $[L]$) is important but for stronger nucleophiles, and for $X_2 = Se_2$ rather than S_2, the third-order term becomes domi-

$$k_{obsd} = k_2[L] + k_3[L]^2 \qquad\qquad (4.13)$$

nant. The values of k_2 are very dependent on the nature of L, k_2 for PPh$_3$ being at least 10^3 times greater than that for AsPh$_3$. Both the k_2 and k_3 terms sometimes show negative temperature coefficients and an intermediate adduct Fe$_2$(CO)$_6$(PPh$_3$)(μ-S$_2$) has been detected spectroscopically. No deviations from simple second- or third-order behavior, of the sort expected for adduct formation, were observed. However, a scheme involving adduct formation in a preequilibrium step was proposed, the adduct either undergoing intramolecular substitution or further nucleophilic attack. The latter leads to the third-order term and also, it seems, to different but unspecified products. The S$_2$-bridged complex reacts much more rapidly than the Se$_2$-bridged one and this was ascribed to the greater positive charge on the Fe atoms resulting from the greater electronegativity of the S atoms.

4.2.4 Bridged Dimolybdenum Carbonyls

Only one kinetic study appears to have been carried out on a bridged dimolybdenum complex (40). Reaction of **4.26** with P-n-Bu$_3$ leads to successive substitution, first at one and then at the other Mo atom. Both reactions proceed only via CO dissociative paths which is not surprising in view of the behavior of the iron analog **4.24**. A complete set of rate and equilibrium data was obtained and the reactions found to be much more rapid than those of **4.24**.

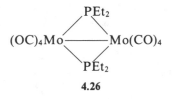

4.26

As with other bridged complexes, the effect of replacing the first CO by PPh$_3$ on the dissociative lability of a CO on the other metal atom is negligible. What is different about this system is that the values of the competition ratios $k_{+\text{CO}}/k_{+\text{PPh}_3}$ for a nucleophilic attack on the coordinatively unsaturated intermediates L(OC)$_3$ Mo(μ-PPh$_2$)$_2$Mo(CO)$_3$ are exceptionally large (60 and 2×10^4 at 70°C for $L = $ CO and PPh$_3$, respectively, as compared with normal values of ca. unity) and strongly dependent on the nature of L. This high selectivity in favor of CO suggests that the vacant coordination site is not nearly so electrophilic as most vacant sites and this can be ascribed to its high coordination number of 6. The origin of the transmitted effect of the substituent is not clear.

4.2.5 Bridged Dimanganese and Dirhenium Carbonyls

Dinuclear carbonyls with 3-electron donor bridging groups B as in M_2(μ-B)$_2$(CO)$_8$ are distinguished from those described earlier by not needing metal–metal bonds in order to maintain the 18-electron configuration of the metal atoms. Although a wide variety of such complexes has been known for many years, no kinetic studies of their reactions appear to have been reported.

When the decacarbonyls are substituted with good bridging ligands such as dppm ($Ph_2PCH_2PPh_2$) some unusual reactions are observed. Thus, the complex **4.27** is distinguished (41) by its losing a CO ligand to form **4.28** with its unusual 4-electron

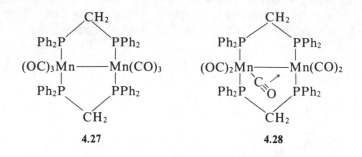

4.27   **4.28**

donor CO (16). The reverse reaction is unexpectedly slow and appears to be dependent on [CO] (42), but reaction with a wide variety of Lewis acids (CO, Br^-, CN^-, $4\text{-}MeC_6H_4NC$, MeCN) in the presence of protic acids leads rapidly to the Mn—H—Mn-bridged complexes $HMn_2(CO)_5L(dppm)_2^{+1.0}$ (43). Replacement of the two phenyl groups on one phosphorous atom by ethyl or cyclohexyl groups changes the relative stability of the hexa- and pentacarbonyls and both electronic and steric effects are important (44). Clearly, many mechanistic and kinetic studies await attention in this area.

Reactions of I_2 with somewhat analogous diarsine-bridged complexes $M_2(CO)_8(AsAs)$ ($M_2 = Mn_2$, MnRe, Re_2; AsAs represents various bidentate ligands

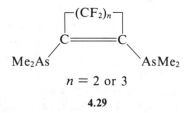

$n = 2$ or 3

4.29

such as **4.29**) lead to $I(OC)_4M(\mu\text{-}AsAs)M(CO)_4I$ by simple oxidative cleavage of the M—M bonds (45). The reactions are first order in [I_2], very fast, and probably occur by an electrophilic attack at the O atoms of the CO ligands (46).

4.2.6   Bridged Heterodimetallic Carbonyls

Vahrenkamp and his group have prepared a very wide range of complexes of the type **4.30** that react with many nucleophiles to form **4.31**, a process involving heterolysis of the Fe—M bonds and addition of L to the vacant coordination site that is created on M (47). Kinetic studies on the complexes with $M = Co(CO)_3$ (48) and $Fe(CO)_2(NO)$ (49) show that the reactions are first order in $[L]$ when $L = $ a variety of P donors as well as $AsPh_3$ and $SbPh_3$. No kinetic evidence is apparent for a two-step reaction in which initial highly reversible heterolysis is followed by

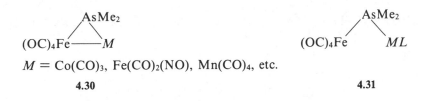

$M = Co(CO)_3, Fe(CO)_2(NO), Mn(CO)_4,$ etc.

4.30   **4.31**

addition of L. The dependence of the values of the second-order rate constants, k_2, on the nature of the nucleophiles suggests that there is a transition from an I_d mechanism for weak nucleophiles (k_2 fairly independent of the nature of L) to an I_a mechanism for stronger nucleophiles (k_2 increasing substantially with increasing basicity of L). Thus, for $M = Co(CO)_3$ the trend is $SbPh_3 \approx AsPh_3 \approx PPh_3 \approx P(OMe)_3 \approx P(OCH_2)_3CEt < PPh_2Et < PPhEt_2 < P\text{-}n\text{-}Bu_3 > P(C_6H_{11})_3$, the relative slowness of the reaction with $P(C_6H_{11})_3$ being due to the steric effect usually observed for this nucleophile in associative reactions. When $M = Fe(CO)_2(NO)$ the order is $SbPh_3 \approx AsPh_3 \approx P(OPh)_3 < PPh_3 < P(OMe)_3 < P(OCH_2)_3CEt \approx PPh_2Et < PPhEt_2 < P\text{-}n\text{-}Bu_3 > P(C_6H_{11})_3$, and it is believed that the presence of the NO ligand at the electrophilic center enhances the tendency to associative reaction so that the change from the I_d to the I_a mechanism occurs sooner along the series of ligands of increasing nucleophilicity.

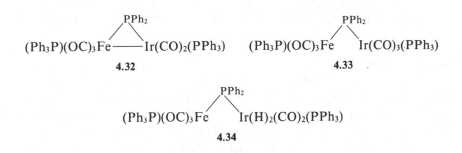

4.32   **4.33**

4.34

A similar heterolysis of a heteronuclear metal–metal bond occurs in the reaction of **4.32** with CO to form **4.33**, both metals having 18-electron configurations in both complexes (50). Although no kinetic data were reported, the reactions are reversible and quite rapid. Moreover, **4.32** reacts with H_2 very rapidly to form **4.34** probably via initial heterolysis to give **4.35** followed by oxidative addition at the 16-electron Ir(I) atom. A 16-electron Ir(I) can also be formed by facile loss of CO from **4.32** to give **4.36** which also undergoes oxidative addition with H_2 to form

4.37. These reactions show very clearly the capacity of such strongly bridged carbonyls to undergo interesting reactions by virtue of their ability to form reactive coordinatively unsaturated intermediates of various kinds.

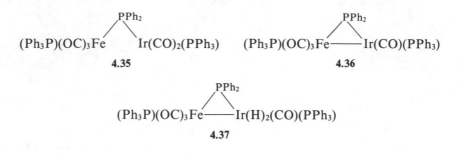

$$(Ph_3P)(OC)_3Fe \overset{PPh_2}{\diagup \diagdown} Ir(CO)_2(PPh_3)$$

4.35

$$(Ph_3P)(OC)_3Fe \overset{PPh_2}{\diagup \diagdown} Ir(CO)(PPh_3)$$

4.36

$$(Ph_3P)(OC)_3Fe \overset{PPh_2}{\diagup \diagdown} Ir(H)_2(CO)(PPh_3)$$

4.37

4.3 TRIMETAL CARBONYLS

4.3.1 Introduction

The triangular complexes $Fe_3(CO)_{12}$, $Ru_3(CO)_{12}$, and $Os_3(CO)_{12}$ are the smallest binary carbonyl clusters proper. The structures of $Ru_3(CO)_{12}$ and $Os_3(CO)_{12}$ in the solid state and in solution are straightforward, each metal being quasi-octahedrally coordinated by two equatorial and two axial carbonyls together, formally at least, with two metal–metal bonds (51). The structure of $Fe_3(CO)_{12}$ in the solid state includes two bridging carbonyls on one side of the triangle (51) and can be considered as an Fe_3 cluster in the middle of an icosahedron of CO ligands (52). The structure in solution is less clear and the nature of the bridging carbonyls is unusual (51). Substitution and the nature of the solvent change the tendency to form CO bridges and nonbridged derivatives are known (53).

4.3.2 $Fe_3(CO)_{12}$

The CO exchange reaction of this cluster was one of the first such reactions studied (23). It was found to be slow in room light ($t_{1/2} \approx 130h$ at room temperature) and even slower in the dark. Later workers (54) derived the parameters $\Delta H^{\neq} = 26.4$ kcal mol^{-1} and $\Delta S^{\neq} = +7$ cal K^{-1} mol^{-1}. Studies of reactions with other ligands are almost nonexistent. Reactions with phosphines tend to lead to fragmentation into mononuclear complexes but substitution is a major pathway for phosphites, usually leading to $Fe_3(CO)_9L_3$ complexes (53).

Only one clear-cut kinetic study seems to have been reported (55). Reaction with a pseudo first-order excess of PPh_3 in $CHCl_3$ led eventually to a mixture of $Fe(CO)_4(PPh_3)$ and $Fe(CO)_3(PPh_3)_2$. The reaction was followed by monitoring the electronic absorption band at 530 nm and showed an apparent induction period lasting several hours at 30 to 40°C. It seems likely that this was due to a substitution process, leading to $Fe_3(CO)_{11}(PPh_3)$, and that the subsequent first-order loss of absorbance was due to fragmentation. The rates seemed to be independent of [PPh$_3$] over a narrow range and the quite low value of $\Delta H^{\neq}$ (21.9 kcal mol^{-1}) suggests that the Fe$_3$ cluster is quite weakly bound. Earlier preparative experiments with much higher concentrations of PPh$_3$ showed that Fe$_3$(CO)$_{12}$ undergoes direct fragmentation under these conditions, presumably through an associative mechanism, to form Fe(CO)$_5$, Fe(CO)$_4$(PPh$_3$), and Fe(CO)$_3$(PPh$_3$)$_2$ (56). Reactions are likely to be very solvent dependent (57) and sometimes complicated by radical formation (58). Fe$_3$(CO)$_{12}$ spontaneously forms Fe$_3$(CO)$_{12}^{-}$ in THF, and this and other radical anion species can be generated by chemical or electrolytic reduction in THF and other solvents (58). The possibility of radical chain reactions such as that found in the reaction of P-n-Bu$_3$ with Co$_2$(CO)$_8$ (59) must obviously be considered.

The question of the relative amounts of Fe(CO)$_5$, Fe(CO)$_4$L, and Fe(CO)$_3$L$_2$ formed in the fragmentation reactions is an interesting one. The mononuclear products do not undergo substitution under the conditions obtaining and the product distribution must be determined by the nature of the reaction of the original trinuclear complex or of dinuclear intermediates. The ratio is dependent on the solvent since only Fe(CO)$_4$L is formed with P-t-Bu$_3$ in methanol, whereas only Fe(CO)$_3$L$_2$ is found in diethylether (57). The formation of dinuclear intermediates has been proposed to explain how Fe(CO)$_4$ [generated, e.g., by dissociation of olefin from Fe(CO)$_4$ (olefin) complexes] can react with PPh$_3$ to form Fe(CO)$_3$(PPh$_3$)$_2$ as well as Fe(CO)$_4$(PPh$_3$) (60). The sequence shown in equation (4.14) was proposed.

$$2Fe(CO)_4 \longrightarrow Fe_2(CO)_8 \xrightarrow{-CO} Fe_2(CO)_7 \xrightarrow{+L} Fe(CO)_4L + Fe(CO)_3L_2 \quad (4.14)$$

Fe$_2$(CO)$_8$ has the postulated structure shown in **4.38** for which there are known analogs. If this sequence does occur, then the amount of Fe(CO)$_3$L$_2$ formed should

$$(OC)_4Fe = Fe(CO)_4$$

4.38

be very dependent on the concentration of Fe(CO)$_4$ generated initially because of the competition between the second-order dimerization and the pseudo first-order direct addition of PPh$_3$. This does not seem to have been tested.

4.3.3 Ru$_3$(CO)$_{12}$ and Its Derivatives

Initial studies on substitution reactions of Ru$_3$(CO)$_{12}$ with group VA donors (61) showed that more basic ligands reacted according to $k_{obsd} = k_1 + k_2[L]$, whereas no second-order paths were seen with CO (54) and less nucleophilic ligands (61). A more extensive study showed that even the weak nucleophiles AsPh$_3$, P(OPh)$_3$, and PPh$_3$ showed a detectable $k_2[L]$ term in the rate equation if high enough concentrations were used (62). The values of k_2 increased with ligand basicity in a way consistent with a nucleophilic attack at a metal center or, perhaps, less specifically at the Ru$_3$ cluster itself. The selectivity of the complex was quantified by the gradient of the roughly linear plot of log k_2 against Δhnp. and found to be reasonably high. Deviations from the line of values of log k_2 for the larger ligands PPh_3 and $P(C_6H_{11})_3$ were pronounced and interpreted also to imply generally substantial bond making in the transition states (62). A good linear plot is also obtained when log k_2 is plotted against the enthalpy of formation, in THF, of the corresponding Pt—L bond in $MePt(PMe_2Ph)_2L$ (63). The values of ΔH_f° cover a range of 10 kcal mol^{-1}, whereas $\Delta G_2^{\neq}$ only covers ca. 3 kcal mol^{-1}. This must be due in part to the obviously incomplete bond formation in the transition states.

When the reactions with $P(OPh)_3$ and $P(OCH_2)_3CEt$ were monitored in the infrared the formation and decay of intermediates $Ru_3(CO)_{11}L$ and $Ru_3(CO)_{10}L_2$ were observed before final formation of $Ru_3(CO)_9L_3$. Reaction with PPh_3 shows no intermediates, whereas $P(OEt)_3$ and P-n-Bu_3 showed clear formation of $Ru(CO)_4L$ and $Ru(CO)_3L_2$ together with trisubstituted clusters (62). Examination of the reaction with P-n-Bu_3 in greater detail showed that the ratio $[Ru(CO)_3L_2]:[Ru(CO)_4L]$ in the product mixture was generally close to 1:2 (64). More $Ru_3(CO)_9L_3$ appeared to be formed at lower values of $[L]$ and at very much higher initial values of $[Ru_3(CO)_{12}]$. It was concluded that fragmentation occurred mainly after formation of $Ru_3(CO)_{11}L$ by an associative substitution step. Recent work suggests that fragmentation of $Ru_3(CO)_{11}L$ occurs when it undergoes a nucleophilic attack by P-n-Bu_3, CO dissociation leading to further substitution (65).

Since the study of effects of substituents on reactivity of clusters was then virtually unexplored, the system shown in Scheme 4.2 (L = PPh_3) was examined in some detail (66,67).

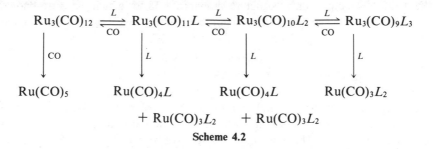

Scheme 4.2

Although the reaction of $Ru_3(CO)_{12}$ with PPh_3 proceeds via first- and second-order rate paths, all the other substitution reactions were found to be independent of the concentration of an entering ligand (66). The introduction of the first PPh_3 substituent increases the rate of CO dissociation by a factor of 60 at 50°C, but introduction of a second one leads to somewhat slower CO loss. Statistical effects could offset this but the labilizing effect of the second PPh_3 is obviously negligible. Activation enthalpies follow the same trend as the rate constants, the largest effect being the lowering of $\Delta H_1^{\neq}$ from 31.8 to 25.6 kcal mol^{-1} on introduction of the first PPh_3. The rate of PPh_3 dissociation increases steadily as the number of PPh_3 substituents increases. The formation of $Ru_3(CO)_{12}$ from $Ru_3(CO)_{11}(PPh_3)$ is complicated by coincident fragmentation and could occur via aggregation of $Ru(CO)_4$

intermediates. Equilibrium constants and competition ratios k_{+CO}/k_{+PPh_3} for a nucleophilic attack on the vacant coordination sites of the reactive intermediates were also obtained. Values of k_{+CO}/k_{+PPh_3} are normal and not very dependent on the nature of the complex, whereas the values of K, the equilibrium constant for displacement of CO, are >6, 8, and 0.4 at ca. 70°C, respectively, for the first, second, and third displacements.

Fragmentation of $Ru_3(CO)_{12}$ to form $Ru(CO)_5$ requires high pressures of CO for thermodynamic reasons (68) and no detailed study has been reported. The other fragmentation reactions are quite complex and, for $Ru_3(CO)_9(PPh_3)_3$, three paths lead to mononuclear products (67). All of them are first order in [complex] and none is accelerated by the presence of free CO or PPh_3. The fastest path ($k = 2 \times 10^{-1}$ s^{-1} at 100°C) is retarded by PPh_3 and must involve rate determining and reversible dissociation of PPh_3, whereas the second fastest ($k = 1 \times 10^{-5}$ s^{-1} at 100°C) involves CO dissociation. The slowest ($k = 6 \times 10^{-7}$ s^{-1} at 100°C) is not retarded by PPh_3 or CO, and since it is not accelerated by these ligands either, it must involve spontaneous fragmentation. The main product of each path is $Ru(CO)_3(PPh_3)_2$, although the yield of other more complex products increases as [PPh_3] decreases. Spontaneous fragmentation is believed to proceed through formation of $(Ph_3P)(OC)_3Ru=Ru(CO)_3(PPh_3)$ and $Ru(CO)_3(PPh_3)$, immediate formation of three mononuclear fragments being highly unlikely because the reverse reaction would then have to be trimolecular. Whether the initial fragmentation is concerted or whether it proceeds by the successive breaking of two Ru—Ru bonds is not known. The dissociative paths must involve Ru—PPh_3 or Ru—CO bond breaking at or before the transition state, but, as in any dissociative process, the changes in bonding that occur within the developing coordinatively unsaturated intermediate is not revealed by the form of the kinetics although they can have major effects on the rates. Thus, whether or not the Ru_3 cluster is still intact in the transition state is not known. This question might be resolvable by physical or chemical attempts to characterize intermediates formed before the transition states (e.g., **4.39**). What is at issue here is whether energy acquired by the cluster can be concentrated for the purpose of breaking just one bond or whether it is distributed between metal–metal and metal–ligand bonds in such a way that a metal–metal bond is broken first.

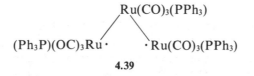

4.39

Spontaneous fragmentation processes were also observed for $Ru_3(CO)_{10}(PPh_3)_2$ and $Ru_3(CO)_{11}(PPh_3)$ (66). The activation parameters for $Ru_3(CO)_{12-n}(PPh_3)_n$ ($n = 1$, 2, 3) show a very pronounced dependence on n, $\Delta H^{\neq}$ increasing from 20.7 to 35.3 kcal mol^{-1} as n increases from 1 to 3. The rate constants at 100°C only decrease by a factor of 10^2, and there is an excellent isokinetic plot of $\Delta H^{\neq}$ against $\Delta S^{\neq}$ with an isokinetic temperature of ca. 200°C. The increase of $\Delta H^{\neq}$ with n is much greater than would be expected from electronic effects and is the opposite of what might be expected if steric effects weakened the Ru_3 cluster. However, if fragmentation involves initial formation of $Ru_2(CO)_{8-m}(PPh_3)_m (m = 0, 1, 2)$, the presence of increasing numbers of PPh_3 in this intermediate must decrease its stability. The two PPh_3 ligands in $Mn_2(CO)_8(PPh_3)_2$ weaken the Mn—Mn bond by *at least* 8 kcal mol^{-1} (9,69) and the corresponding steric effect in the Ru=Ru complexes could be larger than this because of the shorter multiple bond between the metal atoms. Some such effect as this must overcome any weakening in the Ru_3 cluster and the bond weakening in the $Ru_2(CO)_{8-m}(PPh_3)_m$ intermediate must presumably be accompanied by a pronounced increase in the entropy.

Yet another reaction path is available for $Ru_3(CO)_9(PPh_3)_3$ (70). Reaction with O_2 in the presence of sufficient PPh_3 to suppress the faster PPh_3 dissociative path leads to a product that is probably an analog of the known $Ru_3(CO)_6(O)(dppm)_2$ (71). However, the kinetics show an unusual dependence on $[Ru_3(CO)_9(PPh_3)_3]$ and $[O_2]$ that strongly suggests reversible fragmentation into two species. Thus, although the product is a Ru_3 cluster, it seems to be formed via fragmentation, reaction with O_2, and reaggregation. The possibility that the fragmentation is induced in some way by the presence of O_2 has to be considered, especially if scrambling reactions fail to demonstrate spontaneous reversible fragmentation. The fragmentation must produce different intermediates from those involved in the high-temperature process in the presence of CO and PPh_3.

Fragmentation of $Ru_3(CO)_9L_3$ (L = $P(OPh)_3$ and P-n-Bu$_3$) has also been studied (72) and found to involve L dissociative, CO dissociative, and spontaneous fragmentation processes, the rates decreasing along the sequence of mechanisms as for L = PPh_3 (67). Only the CO dissociative path received a thorough study. CO dissociation was necessary but not sufficient to cause fragmentation and the kinetics required that the initially formed $Ru_3(CO)_8L_3$ underwent a reversible isomerization to a form that fragmented when attacked by free L. The value of $\Delta H^{\neq}$ for the CO dissociative step decreases substantially along the series L = PPh_3 (33.0 kcal mol^{-1}), $P(OPh)_3$ (27.4 kcal mol^{-1}), and P-n-Bu$_3$ (19.5 kcal mol^{-1}) so that the nature of the substituents has a profound effect on this process. The effect is not what would be expected for simple steric or electronic effects on the reacting complex and must reflect major differences in the energies and, probably, the structures of the $Ru_3(CO)_8L_3$ intermediates.

Reactions of $Ru_3(CO)_{10}(dppm)$ (with the $Ph_2PCH_2PPh_2$ ligand bridging two Ru atoms in the plane of the Ru_3 cluster) have also been studied (73). With weak nucleophiles only a CO-dissociative path is observed, and the rate of this process is similar to that for $Ru_3(CO)_{10}(PPh_3)_2$. However, reactions with more nucleophilic ligands such as P-n-Bu$_3$ show a pronounced second-order term. It is found that only the k_1 term leads to substitution, the $k_2[L]$ term leading to fragmentation.

The complex $Ru_3(CO)_{10}(NO)_2$ contains a triangular array of Ru atoms but only two Ru—Ru bonds, the other side of the triangle being formed by two NO bridges without Ru—Ru bonding. Substitution into this complex by PPh_3 or $PMePh_2$ occurs sequentially at the two bridged Ru atoms but no evidence for any nucleophilic attack was obtained (74). Evidently, bridging NO groups do not enhance susceptibility to nucleophilic attack in the way terminal ones do. No evidence for fragmentation was reported.

Reactions of some halogens with $Ru_3(CO)_9L_3$ (L = PPh_3, $P(OPh)_3$, and $P\text{-}n\text{-}Bu_3$) have also been studied and found to proceed generally in two stages to form $Ru(CO)_3LX_2$ (75). Spectroscopic studies showed that the products of the first stage were $Ru(CO)_3LX_2$ and $Ru_2(CO)_6L_2X_2$ in equimolar amounts, and the second stage involved reaction of the dinuclear complex to more $Ru(CO)_3LX_2$. Reaction of $Ru_3(CO)_9L_3X_2$, containing an X—Ru—Ru—Ru—X chain and expected as an initial product, must therefore be very fast. Alternatively. concerted fragmentation and oxidative addition at one $Ru(CO)_3L$ unit might be occurring, the remaining $Ru_2(CO)_6L_2$ then reacting rapidly with X_2. The kinetics all follow the same pattern as those for reaction of substituted dimanganese and dirhenium carbonyls (46). They can be accounted for by rapid preequilibria, involving formation of a series of adducts, followed by relatively slow oxidative cleavage of a metal–metal bond as shown in equations

$$\text{Complex} + nX_2 \overset{\beta_n}{\rightleftharpoons} \text{Complex}(X_2)_n \tag{4.15}$$

$$\text{Complex}(X_2)_n \overset{k_n}{\longrightarrow} \text{products} \tag{4.16}$$

(4.15) and (4.16). This leads to the rate equation shown in (4.17). In several cases β_1 is so large that adduct formation is complete, even at the lowest value of $[X_2]$

$$k_{obsd} = \frac{k_1\beta_1[X_2] + k_2\beta_2[X_2]^2 + k_3\beta_3[X_2]^3 + \cdots}{1 + \beta_1[X_2] + \beta_2[X_2]^2 + \beta_3[X_2]^3 + \cdots} \tag{4.17}$$

used. Spectroscopic evidence is available for formation of adducts in some cases although the rapidity of the reactions makes this generally difficult. The most complicated rate equation observed is shown in (4.18) which applies to the reaction of $Ru_2(CO)_6\{P(OPh)_3\}_2I_2$ with I_2. The least-squares deviation of the 36 values of k_{obsd} from those calculated using derived values of the constants a, b, and c was only 7.2%.

$$k_{obsd} = a[X_2]^2/\{1 + b[X_2]^2 + c[X_2]^3\} \tag{4.18}$$

As with the dinuclear complexes (46), these reactions are envisioned to involve attachment of halogen molecules to the O atoms around the periphery of the complexes. This occurs to an extent that electron withdrawal from the metals in the cluster is sufficient to substantially weaken the metal–metal bonding.

4.3.4 $Os_3(CO)_{12}$ and Its Derivatives

Recent work (76) has shown that $Os_3(CO)_{12}$ reacts with phosphorus donors according to the same two-term rate equation found for $Ru_3(CO)_{12}$ (61,62). The value of k_2 for PPh_3 is negligible and $\Delta H_1^{\neq} = 32.9$ kcal mol^{-1} and $\Delta S_1^{\neq} = +8$ cal K^{-1} mol^{-1}

in decalin. CO exchange in benzene has led to similar values (54). These differ from those for $Ru_3(CO)_{12}$ mainly in the value of $\Delta S^{\neq}$ which is ca. 10 cal K^{-1} mol^{-1} less favorable. Reaction with $P-n-Bu_3$ shows a pronounced second-order path, and it appears that nucleophilic attack on the cluster leads to fragmentation, only the dissociative path leading to substitution. This has significance with respect to synthesis of $Os_3(CO)_9(P-n-Bu_3)_3$. Because of the lower value of $\Delta H_2^{\neq}$ compared with $\Delta H_1^{\neq}$, reaction at 70°C and high values of $[P-n-Bu_3]$ proceeds almost entirely by the second-order path and leads to complete fragmentation. At 160°C and with low values of $[P-n-Bu_3]$ the predominant path is CO-dissociation and yields of substituted cluster are high. The value of k_2 for $Os_3(CO)_{12-n}(P-n-Bu_3)_n$ ($n = 0, 1, 2$) increases from $n = 0$ to $n = 1$ but decreases to zero for $n = 2$. The rates of CO-dissociation at 124°C are barely affected by n, especially after allowance for statistical effects.

Reaction of $Os_3(CO)_{12}$ with $HC_2C_8H_{17}$ and C_2Ph_2 have also received some study (77). The eventual product is $Os_2(CO)_6(C_4H_2R_2)$ with the terminal alkyne at 125°C, and $Os_2(CO)_6(C_4Ph_4)$ with C_2Ph_2 at 170°C. The rate-limiting step with 1-decyne is CO dissociation from $Os_3(CO)_{12}$, whereas with C_2Ph_2 it appears to be CO dissociation from $Os_3(CO)_{11}(C_2Ph_2)$ formed in a labile preequilibrium. Direct attack by C_2Ph_2 on $Os_3(CO)_{12}$ is also indicated.

Reactions of $Os_3(CO)_{12}$ with halogens were studied some time ago and found to involve simple oxidative cleavage of one Os—Os bond to form $Os_3(CO)_{12}X_2$ (78). The rates in CH_2Cl_2 with Cl_2 and Br_2 are independent of $[X_2]$ but different ($k = 5.4$ and 1.8 s^{-1}, respectively, at 25°C). There is evidence for rapid adduct formation before the slow oxidative cleavage. Reaction in CCl_4 does show a dependence on $[X_2]$. Reaction with I_2 is more complex and much slower. These results are in accord with reactions of some substituted dinuclear complexes. Thus, $Mn_2(CO)_8(PEt_3)_2$ reacts with I_2 in CH_2Cl_2 in large part by a path independent of $[I_2]$, whereas in CCl_4 there is no path independent of $[I_2]$ (46). Reactions of $Ru_3(CO)_9L_3$ ($L = P(OPh)_3$ or PPh_3) with Br_2, I_2, and ICl in cyclohexane also show terms independent of $[X_2]$ that are assigned a similar significance (75). It seems that all these metal–metal-bonded carbonyls react with halogens according to the general mechanism shown in equations (4.15) and (4.16).

4.3.5   Nucleophilic Attack at CO Ligands

Nucleophilic attack at the C atom of coordinated CO is well known to occur rapidly when relatively hard nucleophiles are involved (3,4), and reactions of some clusters are also quite rapid and lead to interesting products (79). Very few kinetic studies, even of mononuclear carbonyls, have been reported. As far as clusters are concerned, only the fast reactions shown in equation (4.19) ($M = $ Fe, Ru, or Os) have been studied (80), the reactions being carried out in MeOH when $R = $ Me, and in

$$M_3(CO)_{12} + {}^{-}OR \underset{k_{-1}}{\overset{k_1}{\rightleftharpoons}} M_3(CO)_{11}(CO_2R)^{-} \qquad (4.19)$$

THF/OH$_2$ mixtures when R = H. Solvent effects appear to be small. The values of k_1 increase slightly along the series Os$_3$(CO)$_{12}$ < Ru$_3$(CO)$_{12}$ < Fe$_3$(CO)$_{12}$, a sequence opposite to that for the mononuclear pentacarbonyls, and reactions with OH$^-$ are a little faster than with MeO$^-$. The high reactivity of Fe$_3$(CO)$_{12}$ may be due to its bridging carbonyls but otherwise cluster effects are small.

4.4 STRONGLY BRIDGED TRIMETAL CLUSTERS

CO exchange kinetics of some methinyltricobaltenneacarbonyls, XCCo$_3$(CO)$_9$, were reported in 1965 and three CO ligands found to be more labile than the other six (81). The three distinctive ligands are axial and effectively trans to the C—Co bonds in the CCo$_3$ tetrahedral cluster. Rates decreased along the series X = F > Cl > Br > H and were significantly faster in heptane than in toluene. Reactions with some weakly basic P and As donors were later found to proceed in hexane at rates independent of the concentration and nature of L between 35 and 65°C (82). Activation enthalpies increased in the order F $\leqslant$ Ph < H < Me but entropies increased in that order to an almost compensating degree. All the values of $\Delta S^{\neq}$ were significantly negative and no retardation of the rates by CO was observed even when [L] < [CO]. The conclusion that a CO dissociative mechanism was operative implies that the XCCo$_3$(CO)$_8$ intermediates have a very low affinity for CO and a rather low intrinsic entropy. However, since the normal criteria for a CO dissociative path are missing, alternative mechanisms must be considered. Reversible formation of a reactive isomer of the complex followed by very rapid substitution could lead to the observed kinetics. The reactive isomer could be a more open coordinatively unsaturated form of the cluster as formed, for example, by insertion of a CO ligand into one of the C—Co bonds. This could lead to a negative value for $\Delta S_1^{\neq}$. Apart from X = Ph, which is anomalous, the labilizing effects of X run parallel with the Co—CO bond strengths implied by the C—O stretching frequencies (82). Although fairly small, the kinetic effects transmitted through the molecule are significant but the interpretation must be dependent on the as yet uncertain mechanism.

The complexes Fe$_3$(CO)$_9$$XY$ (X, Y = S, Se, and Te) contain a triangular array of iron atoms but only two Fe—Fe bonds. The two calcogens are above and below the Fe$_3$ isoceles triangle formed by iron atoms as shown in **4.40** where the three terminal CO groups on each Fe atom have been omitted (83). Substitution reactions with CO, AsPh$_3$, P(OPh)$_3$, and P-n-Bu$_3$ proceed in heptane (84) according to the

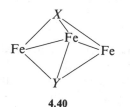

4.40

TABLE 4.1 Significant Terms in the Rate Equation: $k_{obsd} = k_1 + k_2[L]$ for Substitution Reactions of $Fe_3(CO)_9XY$ (84)

X	Y	L = CO	AsPh$_3$	P(OPh)$_3$	P-n-Bu$_3$
S	S	k_1	k_1	k_1	k_2
S	Se	k_1	k_1	k_1 and k_2	—
S	Te	k_1	k_1	k_1 and k_2	—
Se	Te	k_1 and k_2	k_1 and k_2	k_2	—
Te	Te	k_2	k_2	k_2	k_2

general rate equation $k_{obsd} = k_1 + k_2[L]$, but the significant terms are shown in Table 4.1. It was proposed that the k_2 terms involve attack at the calcogen atoms since it was expected that the more electronegative ones would encourage nucleophilic attack at a metal center, that is, the opposite of what was observed. However, it might be argued that the more polarizable calcogens would adjust better to a nucleophilic attack at a metal atom, in which case the k_2 terms would become more apparent with the heavier calcogens as observed. Another related possibility comprises highly reversible heterolysis of an Fe—X bond followed by a ligand interchange at the coordinatively unsaturated Fe atom. Weaker Fe—X bonds would facilitate this. It was found that reactions of $Fe_3(CO)_9(Te)_2$, even with CO, lead to quite stable intermediate adducts which were isolated but not fully characterized (85). Bonding of the extra ligand to one of the Te atoms was proposed but bonding to an Fe atom and the presence of one less Fe—Te bond seems quite possible.

Activation parameters for CO exchange by the k_1 path hardly depend on the natures of X and Y, $\Delta H_1^{\neq}$ being between 32.4 and 33.4 kcal mol^{-1} for $XY = S_2$, SSe, Se$_2$, STe, and SeTe, and $\Delta S_1^{\neq}$ was between 14 and 18 cal K^{-1} mol^{-1}. The values of $\Delta H_2^{\neq}$ and $\Delta S_2^{\neq}$ for CO exchange were 24.2 and 22.9 kcal mol^{-1}, and $+1$ and -1 cal K^{-1} mol^{-1}, for $XY = $ SeTe and Te$_2$, respectively. These studies were extended to compounds where X and Y were combinations of NH, NMe, NEt, and NPh (86). These all undergo CO exchange by [CO]-independent paths only and are considerably more labile than the calcogen complexes, probably because of the greater "hardness" of the nitrogen bridging atoms.

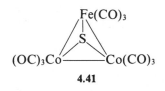

4.41

The complex $FeCo_2(CO)_9S$, **4.41**, has a tetrahedral FeCo$_2$S core and undergoes CO exchange with $k_{obsd} = k_1 + k_2[CO]$ (87). All CO ligands exchange at equal rates but there is rapid CO scrambling among all the positions in the complex. Substitution at a Co atom occurs with P(OPh)$_3$, PPh$_3$, P(OEt)$_3$, PEt$_2$Ph, and P-n-Bu$_3$ according to a simple bimolecular path alone. The complex is very susceptible to nucleophilic attack, the values of k_2 being as much as 10^3 times greater than corresponding ones for $Co(CO)_3(NO)$ (88). Although the value of k_2 for CO exchange is significant, it is 10^5 times smaller than that for P-n-Bu$_3$. A plot of log k_2 for $P(OPh)_3$, $P(OEt)_3$, PEt_2Ph, and P-n-Bu$_3$ against their Δhnp values is linear and log k_2 for PPh_3 falls ca. 2 log units below the line. The gradient is 3.8 V^{-1}. This all suggests a high degree of bond making in the transition states and attack was envisaged as occurring at the LUMO on the side of the FeCo$_2$ triangle opposite to the S atom. The values of $\Delta H_2^{\neq}$ decrease with increasing basicity of the nucleophile. For P-n-Bu$_3$, the value of $\Delta H_2^{\neq}$ may actually be negative, suggesting adduct formation in a labile preequilibrium. After introduction of a P-n-Bu$_3$ ligand on to one Co atom a second nucleophilic substitution occurs at the other Co atom, k_2 being 10^3 times smaller at 30°C. Substitution of a third CO occurs at the Fe atom, but by a dissociative path only. $\Delta H_1^{\neq}$ for CO exchange at the Co in **4.41** is 32 ± 4 kcal mol^{-1}, and $\Delta S_1^{\neq}$ = 19 ± 7 cal K^{-1} mol^{-1}. CO exchange at the (unsubstituted) Fe atom in $FeCo_2(S)(CO)_7(P-n-Bu_3)_2$ has the very different parameters $\Delta H_1^{\neq}$ = 24.2 kcal mol^{-1} and $\Delta S_1^{\neq}$ = −3.4 cal K^{-1} mol^{-1} although the values of k_1 at 70°C are about equal.

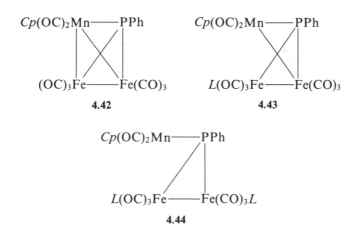

An interesting series of reactions involving transformation of **4.42** into **4.43** and then **4.44** has been given a partial kinetic study (89). The CO ligands on the Mn atom in **4.42** form "semibridges" to the Fe atoms. Reactions are generally reversible, and when L = PPh$_3$ reaction of **4.42** to **4.43** is first order in [PPh$_3$] with $\Delta H_2^{\neq}$ = 8 kcal mol^{-1}, the reverse reaction having a $\Delta H_1^{\neq}$ = 28 kcal mol^{-1}.

An interesting study of the kinetics of transformation of **4.45** into **4.46** has been reported (90). The bond indicated between the phenyl ring and the Os atom may involve C—H—Os trinuclear interaction or oxidative addition may have occurred so that there is a C—Os bond and an H bound to the Os$_3$ cluster somewhere. The rate is α [CO] with $\Delta H_2^{\neq}$ = 16 kcal mol^{-1} and $\Delta S_2^{\neq}$ = −6.5 cal K^{-1} mol^{-1}. The reaction is retarded by parasubstitution of the phenyl groups, and this was ascribed to repulsion between the parasubstituents and the entering CO molecule. Reaction with PF_3 proceeds at about the same rate as that with CO but Ph_3E (E = P, As, or Sb) react very much more rapidly. These parameters do not clearly favor either one of the two possible modes of bonding of the ortho carbon of the phenyl ring to the Os atom.

Complex **4.46** has also been shown to react with CO by a second-order reaction to form $Os_2(CO)_6(C_4Ph_4)$, and reaction with PPh_3 is very much faster (77).

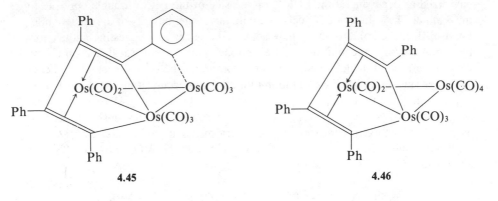

4.45   4.46

4.5 TETRAMETAL CARBONYL CLUSTERS

4.5.1 Tetracobalt Clusters

$Co_4(CO)_{12}$ represents the only case for which the rate of formation of a cluster has been measured. The reaction of $Co_2(CO)_8$ to form $Co_4(CO)_{12}$ is reported to be strongly inhibited by CO according to rates that are second order in $[Co_2(CO)_8]$ and inversely dependent on $[CO]^2$, $[CO]^3$, and $[CO]^4$ in heptane, heptene, and toluene, respectively (91). Although the reason for the particular solvent effects is not clear, the rate equations suggest formation of intermediates $Co_4(CO)_{14}$ and $Co_4(CO)_{13}$ in heptane and heptene respectively, and these can be formulated as **4.47** and **4.48** in which the Co atoms are tetrahedrally disposed. In toluene it seems that two CO ligands are lost from each of two $Co_2(CO)_8$ complexes before aggregation to form $Co_4(CO)_{12}$ directly. Once formed, **4.47** and **4.48** have to lose 2 and 1 CO ligands, respectively, before $Co_4(CO)_{12}$ is formed.

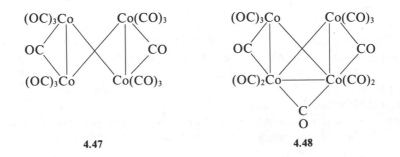

4.47   4.48

The kinetics in n-hexane of the reverse reaction, formation of $Co_2(CO)_8$ from $Co_4(CO)_{12}$ and CO, shows terms of the first and second order in [CO] (92). This suggests that **4.48** is formed first and that this either undergoes spontaneous fragmentation or is attacked by another CO to form **4.47** before fragmentation. These fragmentation reactions require 5 to 100 atm CO and temperatures between ca. 40 and 100°C. Values of $\Delta H°$ and $\Delta S°$ for this fragmentation are -29.5 kcal mol^{-1} and -64.71 cal K^{-1} mol^{-1}, respectively (93).

By contrast, $Co_4(CO)_{12}$ undergoes a quite facile CO exchange by a process independent of [CO] and probably CO dissociative in nature (94). Substitution reactions to form $Co_4(CO)_{11}L$ proceed extremely rapidly (95,96). These reactions must be [L]-dependent or they would proceed at the same rate as CO exchange. It is interesting that a value of 3.8×10^{-5} M^{-1} s^{-1} for second-order substitution of methyl methacrylate into $Co_4(CO)_{12}$ at 25°C was inferred from kinetic studies of the polymerization of methyl methacrylate in the presence of $Co_4(CO)_{12}$ and CCl_4 (97).

Substitution of $P(OMe)_3$ into $Co_4(CO)_{11}P(OMe)_3$ is dependent on [$P(OMe)_3$] but the rates are strongly retarded by the presence of CO (95). This must indicate a more complex mechanism than simple associative attack. Further substitution of $Co_4(CO)_{10}\{P(OMe)_3\}_2$ shows much less dependence on [$P(OMe)_3$] and is strongly retarded by O_2, an interesting but unexplained fact (98). Reaction of $Co_4(CO)_{11}(PPh_3)$ with PPh_3 leads directly to $Co_4(CO)_8(PPh_3)_4$ at rates that are dependent on [PPh_3], whereas reaction of $Co_4(CO)_{11}AsPh_3$ with $AsPh_3$ is very slow (99). Although CO exchange with $Co_4(CO)_{12-n}\{P(OMe)_3\}_n$ ($n = 0, 1, 2$) shows very little dependence on n (94), reaction of $Co_4(CO)_9\{P(OMe)_3\}_3$ with $P(OMe)_3$ is considerably slower (99), and clearly, the substitution reactions of these clusters are very dependent on the number and nature of the substituents.

A study of the fragmentation of $Co_4(CO)_8(PPh_3)_4$ to form $Co_2(CO)_6(PPh_3)_2$ shows features similar to that of $Ru_3(CO)_9(PPh_3)_3$ fragmentation (99). PPh_3 and CO dissociative paths are available as well as spontaneous fragmentation and the relative rates of these three processes are qualitatively the same as for the Ru_3 cluster. The values of $\Delta H^{\neq}$ for spontaneous fragmentation of the two complexes are about equal and the greater rate of fragmentation of $Co_4(CO)_8(PPh_3)_4$ is due to a more positive value of $\Delta S^{\neq}$. This indicates that the Co_4 cluster bonding is weaker than that of the Ru_3 cluster, the same enthalpy being sufficient to produce a much more disordered and more weakly bound transition state. This is reasonably ascribable to steric effects since $Co_4(CO)_8\{P(OMe)_3\}_4$ and, apparently, $Co_4(CO)_8(P-n-Bu_3)_4$ are much more stable toward fragmentation.

4.5.2 Tetrarhodium and Some Heterotetranuclear Clusters

No kinetic studies have yet been reported in full on $Rh_4(CO)_{12}$ but there are a number of semiquantitative observations (100–103) that suggest that kinetic studies would be rewarding. Fragmentation has been reported briefly (100) to occur at -30 to $+30$°C in hexane under pressures of 100 to 200 bar CO. The rates are first order in [CO] and about 500 times faster than those of $Co_4(CO)_{12}$ under similar conditions (92). The values of $\Delta H°$ and $\Delta S°$ for the fragmentation are -10.9 kcal mol^{-1} and

-63.7 cal K^{-1} mol^{-1}, respectively, showing that the greater thermodynamic stability of $Rh_4(CO)_{12}$ compared with $Co_4(CO)_{12}$ is entirely due to enthalpy factors (93,100). $Rh_4(CO)_{12}$ can be readily labeled with ^{13}CO and is therefore probably at least as labile to substitution as $Co_4(CO)_{12}$ (101). A very interesting observation is that a mixture of $Rh_4(CO)_{12}$ and $Rh_4(^{13}CO)_{12}$ undergoes very rapid redistribution of the ^{13}CO even at 25°C. A CO dissociative mechanism was considered to be most unlikely and either aggregation or fragmentation was preferred. Similar, but much slower, ligand redistribution was observed with Rh_4 clusters substituted with various P-donor ligands (101). Reactions with phosphites proceed readily at room temperature to form $Rh_4(CO)_8\{P(OR)_3\}_4$. Reaction with PPh_3 at room temperature leads to $Rh_4(CO)_{10}(PPh_3)_2$ in good yields in hexane, but $Rh_4(CO)_9(PPh_3)_3$ is formed quantitatively in CH_2Cl_2. Temperatures of 70 to 80°C are required to form $Rh_4(CO)_8(PPh_3)_4$ in benzene. Reactions in the presence of CO lead to dinuclear products as do reactions with more nucleophilic phosphines.

Reaction of $P(OMe)_3$ with $Co_2Ir_2(CO)_{12}$ leads to rapid substitution at the Co atoms. Substitution at the Ir atoms is slower but probably not so slow as in $Ir_4(CO)_{10}\{P(OMe)_3\}_2$ (104). Substitution reactions of $CoRh_3(CO)_{12}$ and $Co_2Rh_2(CO)_{12}$ are interesting in being sometimes accompanied by changes in the Co:Rh ratios in the clusters (105). Reversible fragmentation reactions are presumably occurring.

4.5.3   Tetrairidium Clusters

$Ir_4(CO)_{12}$ differs from $Co_4(CO)_{12}$ and $Rh_4(CO)_{12}$ in having T_d symmetry and no bridging carbonyls (102). However, substitution of one or more CO ligands by P donors leads to structures containing three bridging CO ligands in an Ir_3 face containing the substituted Ir atoms. No direct fragmentation of $Ir_4(CO)_{12}$ seems to occur (102) and even substituted derivatives fragment only under forcing conditions (106). No kinetics have been reported. The first kinetic studies of substitution reactions of $Ir_4(CO)_{12}$ showed that this structural change was accompanied by major change in rates and mechanism (107). Thus, $Ir_4(CO)_{12}$ forms $Ir_4(CO)_{11}(PPh_3)$ mainly through a nucleophilic attack on the cluster by PPh_3, whereas $Ir_4(CO)_{11}(PPh_3)$ and $Ir_4(CO)_{10}(PPh_3)_2$ react relatively rapidly by first-order and presumably CO dissociative processes. These studies have been extended quite systematically by Atwood and Darensbourg and their coworkers.

$Ir_4(CO)_{12}$ has been shown to be highly selective in its reactions with nucleophiles (108, 109). Reaction with $AsPh_3$ (108) shows a detectable $[AsPh_3]$-dependent term at 109°C in chlorobenzene. P-n-Bu$_3$ reacts quite rapidly at much lower temperatures although no kinetic data were reported. The second-order rate constant for attack by $P(OPh)_3$ at 109°C is ca. 3 times greater than that for attack by PPh_3 which shows that steric effects are important, PPh_3 being more basic but larger than $P(OPh)_3$. Reaction with t-BuNC (109) is also second order, being first order in $[t$-BuNC] at 29.8°C in chlorobenzene. It is very much faster than the reaction with PPh_3 due entirely to a significantly lower value of $\Delta H_2^{\neq}$. This must be ascribed to steric effects because $C_6H_{11}NC$ attacks $Co(CO)_3(NO)$ only a little more rapidly than does PPh_3 (88). The substituted complexes $Ir_4(CO)_{12-n}(CN$-t-Bu$)_n$ ($n = 1, 2, 3$), which do not contain any bridging CO ligands, also react with t-BuNC by second-order paths. Only quite small effects of different degrees of substitution are shown. The second-order reaction of $Ir_4(CO)_{11}(PPh_3)$ with t-BuNC has a value of k_2 5 times greater than that for $Ir_4(CO)_{11}(CN-t-Bu)$. If anything, the former cluster might be expected to react more slowly due to the greater size of its substituent, and this suggests the possibility that bridging carbonyls may encourage nucleophilic attack, presumably because of their greater capacity to withdraw electron density from the cluster.

$Ir_4(CO)_{11}(PPh_3)$ and $Ir_4(CO)_{11}(P-n-Bu_3)$ react with $P-n-Bu_3$ at room temperature solely by a second-order path, k_2 for $Ir_4(CO)_{11}(PPh_3)$ being ca. $5 \times 10^{-2} M^{-1} s^{-1}$ (110). This is ca. 10 times smaller than k_2 for reaction with t-BuNC, whereas $P-n-Bu_3$ reacts ca. 10 times faster with $Co(CO)_3(NO)$ than does $C_6H_{11}NC$ (88). Steric effects would seem to be implicated. Reaction with $P(OPh)_3$ at 89.5°C shows a value of $k_2 = $ ca. $1 \times 10^{-2} M^{-1} s^{-1}$ (110). Even disubstituted complexes show some susceptibility to nucleophilic attack. $Ir_4(CO)_{10}\{P(OPh)_3\}_2$ reacts with $P(OPh)_3$ at 89°C in chlorobenzene with $k_2 = $ ca. $5 \times 10^{-4} M^{-1} s^{-1}$, whereas $Ir_4(CO)_{10}(AsPh_3)_2$ reacts with PPh_3 at 59°C with $k_2 = $ ca. $3 \times 10^{-3} M^{-1} s^{-1}$ (111). These rather piecemeal observations deserve extension.

The effects of various substituents on the rates of CO dissociation from $Ir_4(CO)_{12-n}L_n$ ($n = 0, 1, 2, 3$) have been studied quite thoroughly, with results shown in Tables 4.2 and 4.3 (107,110–112). Data for dissociative loss of L from $Ir_4(CO)_8L_4$ (112) are also included. In general, the rates increase with increasing substitution. The results for $L = P(OPh)_3$ show that the effect of the structural change on introducing the substituents is not particularly large. The effect of the second $P(OPh)_3$ is about the same as that of the first and the activation parameters are virtually identical.

$AsPh_3$ and PPh_3 have larger effects on the rates, successive introduction of the first two ligands producing substantial acceleration. The introduction of the third PPh_3 produces only a small further enhancement and, in fact, the acceleration due to substitution decreases steadily and substantially with increasing n.

Introduction of one $P-n-Bu_3$ produces the largest observed acceleration due to a single substituent, but a second produces quite a small further acceleration and the third an even smaller acceleration. Although activation parameters are known only for $n = 1$ and 2 the values for $n = 3$ and $L = PEt_3$ are probably close to those for $L = P-n-Bu_3$. The acceleration due to the first $P-n-Bu_3$ is due entirely to the more positive value of $\Delta S^{\neq}$, the change in $\Delta H^{\neq}$ actually opposing the increase in rate. Further substitution decreases $\Delta H^{\neq}$ and $\Delta S^{\neq}$. Any complete explanation of substituent effects should take into account trends in activation parameters, although this will be difficult due to the low precision of some of them.

Restricting discussion to the relative rates, it can be seen that the acceleration caused by the first substituent is probably electronic in nature since the rates increase with the increasing electron donor character of the P donor substituents. (The σ donor character increases with decreasing values of Δhnp). This appears to be a transition-state effect since any ground-state effect would be expected to be in the opposite direction (110). No correlation with the size of the substituents is evident. Indeed, the fact that the accelerating effects of the substituents decrease with in-

TABLE 4.2 First-order Rate Constants ($10^6 k_1$, s^{-1}) for CO Dissociative Reactions of $Ir_4(CO)_{12-n}L_n$ at 80°C[a](110–112)

L	θ,[b] deg.	Δhnp mV	n							
			0[c]		1[d]		2[c]		3[e]	4[e,f]
P(OPh)$_3$	128	875	0.2	(5)[g]	1	(5)	5		—	Small
AsPh$_3$	ca. 127[h]	—	0.2	(55)	11	(30)	325		—	—
PPh$_3$	145	573	0.2	(75)	15	(22)	340	(4.6)	1562	—
PPh$_2$Me	136	400	0.2		—		—		—	250
PMe$_3$	118	111	0.2		—		—		35	0.04
PEt$_3$	132	111	0.2		—		—		266	120
P-n-Bu$_3$	132	131	0.2	(150)	29	(5)	143	(1.8)	254	53
P-i-Pr$_3$	160	131	0.2		—		—		1352	—

[a]For n = 0, 1, 2 the rate constants are for substitution reactions. Where rate constants at 80°C were not available they have been estimated by making use of the published temperature dependences of rate constants at other temperatures. The rate constants are observed values; that is, they have not been adjusted to allow for any statistical effects due to an assumed number of potentially reactive CO ligands. For n = 3 the data are for CO exchange reactions and the values quoted by Darensbourg et al. (Table III, Ref. 112) have been readjusted so as not to assume only two reactive CO ligands.

[b]Ligand cone angles (113).

[c]In chlorobenzene.

[d]In decane.

[e]In tetrachlorethylene.

[f]Rate constants for dissociative loss of L.

[g]Figures in parentheses represent the acceleration in rate caused by the introduction of an extra substituent.

[h]Ref. (114).

TABLE 4.3 Activation Parameters for CO Dissociative Reactions of $Ir_4(CO)_{12-n}L_n{}^a$

$Ir_4(CO)_{12}$
$\Delta H^{\neq} = 32.1 \pm 0.7; \Delta S^{\neq} = 6.4 \pm 1.8$

L	$Ir_4(CO)_{11}L$		$Ir_4(CO)_{10}L_2$		$Ir_4(CO)_9L_3$	
	$\Delta H^{\neq}$	$\Delta S^{\neq}$	$\Delta H^{\neq}$	$\Delta S^{\neq}$	$\Delta H^{\neq}$	$\Delta S^{\neq}$
$P(OPh)_3$	30.6 ± 0.8	7.0 ± 2.3	30.5 ± 0.9	7.6 ± 2.7		
$AsPh_3$	30.6 ± 1.3	9.7 ± 3.5	27.9 ± 0.9	8.6 ± 2.7		
PPh_3	32.0 ± 1.0	14.3 ± 2.7	32.0 ± 0.9	20.2 ± 2.8		
PEt_3					28.4 ± 2.7	3.5 ± 6.4
$P\text{-}n\text{-}Bu_3$	33.0 ± 0.4	18.4 ± 1.1	29.4 ± 1.1	11.3 ± 3.3		

$^a\Delta H^{\neq}$ in kcal mol^{-1}, $\Delta S^{\neq}$ in cal K^{-1} mol^{-1}, uncertainties are 95% confidence limits (110–112).

creasing numbers of substituents shows that steric effects certainly cannot be generally dominant and that a saturation of the electronic effect seems to be occurring. This may be due to a change in the relative contributions of ground-state and transition-state effects or to some detailed structural effect (see later discussion). It is reminiscent of the much larger effect in $Ru_3(CO)_{12-n}(PPh_3)_n$ ($n = 0,1,2,3$) where the first substituent produces a significant acceleration and the second and third produce increasing decelerations (66).

For $Ir_4(CO)_9L_3$ steric effects have become clearly detectable. The considerably faster rate for $L = P\text{-}i\text{-}Pr_3$ compared with the equally basic but smaller $L = P\text{-}n\text{-}Bu_3$ must be due to steric effects as must the much smaller rate for $L = PMe_3$. Electronic effects are still important since the more basic $L = P\text{-}i\text{-}Pr_3$ produces slower rates than the smaller but considerably less basic $L = PPh_3$. In these complexes high basicity seems to stabilize the cluster towards CO dissociation as expected for a ground-state effect. It is this that must be producing the saturation of the substituents' electronic effects mentioned earlier. The importance of steric effects is greater still in determining ease of loss of L from $Ir_4(CO)_8L_4$ as shown by the relative rates when $L = PMe_3$ and PEt_3. The steric effect must also account for the fact that $Ir_4(CO)_8(PPh_3)_4$ has not yet been isolated. Electronic effects are still evident, the less basic $P\text{-}n\text{-}Bu_3$ dissociating more slowly than the slightly more basic but equally sized PEt_3. This is likely to be due to effects transmitted from the remaining substituents rather than to the Ir-L bond strengths. The more weakly basic, and more strongly π-bonding, phosphites are very much slower to dissociate. The trends therefore demonstrate an electronic effect that begins by showing transition-state stabilization but ends, with the more highly substituted clusters, by showing ground-state stabilization. Superimposed on this is a steric effect that increases with increasing substitution and that has already become detectable when two substituents are present (111).

Any detailed understanding of these reactivity trends must take into account the different structures of the clusters. The substituents in $Ir_4(CO)_{11}L$ always seem to be axially attached to an Ir in the bridged Ir_3 basal plane, whereas the complexes $Ir_4(CO)_{10}L_2$ have one axial and one equatorial L. All the $Ir_4(CO)_9L_3$ clusters have one axial and two equatorial ligands, whereas the $Ir_4(CO)_8L_4$ clusters generally have one apical, one equatorial, and two axial substituents (112). Repulsion between the two neighboring axial PMe_3 groups in $Ir_4(CO)_8(PMe_3)_4$ leads to significant lengthening of the Ir—Ir bond and this must have kinetic consequences. Darensbourg (112) argues that the most readily dissociable CO ligands are the axial ones and has taken this into account in making a statistical adjustment to the observed rate constants quoted in Table 4.2. Atwood (110,111), on the other hand, suggests that it is always a CO on an already substituted Ir atom that dissociates in the rate determining step. Attack by L at the vacated coordination site cannot occur for steric reasons. The coordinative unsaturation is transferred rapidly to an unsubstituted Ir atom by CO migration after which attack by L can occur. This is shown schematically in equation (4.20) (no stereochemical details being implied) and has important kinetic implications. The limiting rates of substitution at high $[L]$ under CO would be governed by $k_1k_2/(k_{-1}[CO] + k_2)$ and would therefore be [CO] dependent. No attempt seems to have been made to investigate this consequence of the proposed mechanism which can, in any case, only be possible when ligands of the type being replaced are attached both to the substituted metal atom and to a nonsubstituted one. It cannot apply, for instance, to the substituent effects seen in the dissociation of PPh_3 from $Ru_3(CO)_{12-n}(PPh_3)_n (n = 1,2,3)$ where the rates increase with increasing n (66).

$$
\begin{array}{c}
L \quad\ \ CO \\
| \qquad | \\
M\!\!-\!\!-\!\!M \\
| \qquad | \\
CO \quad CO
\end{array}
\underset{\underset{+CO}{k_{-1}}}{\overset{\underset{k_1}{-CO}}{\rightleftharpoons}}
\begin{array}{c}
L \quad\ \ CO \\
| \qquad | \\
M\!\!-\!\!-\!\!M \\
| \qquad | \\
\quad\ \ CO
\end{array}
\underset{k_{-2}}{\overset{k_2}{\rightleftharpoons}}
\begin{array}{c}
L \quad\ \ CO \\
| \qquad | \\
M\!\!-\!\!-\!\!M \\
| \qquad | \\
CO
\end{array}
\overset{k_3}{\underset{L}{\longrightarrow}}
\begin{array}{c}
L \quad\ \ CO \\
| \qquad | \\
M\!\!-\!\!-\!\!M \\
| \qquad | \\
CO \quad L
\end{array}
\qquad (4.20)
$$

The statistical contributions to the relative rates of CO dissociation from $Ir_4(CO)_{12-n}L_n$ as n changes will obviously depend on the mechanistic details and no attempt to adjust the observed values in Table 4.2 seems justifiable as yet.

4.6 POLYNUCLEAR HYDRIDES

4.6.1 Protonation and Deprotonation

The ability of mononuclear metal carbonyls to undergo protonation (115), and for this to have kinetic effects (116), has been known for some time and is also a property of polynuclear carbonyls (117). Since the hydrogen in the clusters is hydridic in nature, protonations can be regarded as oxidative additions with the implication that quite large structural changes should occur. The slowness of hydrogen exchange between the cluster and protons in solution enabled Mays and coworkers to measure isotopic effects during protonation by competition studies. Thus, k_H/k_D was found to be ca. 16 for formation of $HMCo_3(CO)_{12}$ (M = Fe, Ru, and Os), ca. 11 for $HOs_3(CO)_{12}^+$, and ca. 7 for $Cp_2Fe_2(CO)_4H^+$ (118). Substitution of one CO in $HFeCo_3(CO)_{12}$ by $P(O\text{-}i\text{-}Pr)_3$ reduces the ratio from 16 to 8 (119). These very high values were ascribed to the importance of tunneling during the reaction (119).

Although the slow loss of hydrogen as protons from hydride clusters was evident quite early on, systematic studies of the rates and equilibria have only very recently been reported (120). The pK_a values in methanol for $H_4M_4(CO)_{12}$ (M_4 = Ru_4, Os_4, and $FeRu_3$) and $H_2M_4(CO)_{13}$ (M_4 = Ru_4 and $FeRu_3$) are all between 11 and 12. Substitution of $P(OMe)_3$ into $H_4Ru_4(CO)_{12}$ increases the pK_a by 3 units. Terminal hydrides in $H_2Os_3(CO)_{12}$ and $H_2Os(CO)_4$ are also considerably less acidic with pK_a values of 14.7 and 15.2, respectively, the greater acidity of the former being ascribed to a greater degree of charge delocalization in the polynuclear anionic product. Rates of deprotonation by MeO^- were studied and the results combined with the equilibrium data to give rates of proton transfer from MeOH to the anionic cluster. Deprotonation was first order in $[MeO^-]$, and the second-order rate constants for the complexes listed previously varied only from 2×10^3 to 16×10^3 M^{-1}s^{-1}, apart from H$_4$Ru$_4$(CO)$_{11}$(P-i-Pr$_3$) and H$_2$Os(CO)$_4$ for which $k_2 = 0.08 \times 10^3$ and 0.02×10^3 M^{-1} s^{-1}, respectively. Rates of proton transfer from the MeOH solvent varied more, from 0.02 s^{-1} for HRu$_4$(CO)$_{13}^-$ to 25 s^{-1} for HOs$_3$(CO)$_{12}^-$. The low acidity of H$_4$Ru$_4$(CO)$_{11}L$ when L = P-i-Pr$_3$ rather than CO is due to a 100-fold slower deprotonation and a ten-fold faster protonation, whereas H$_2$Os(CO)$_4$ is only slightly less acidic than H$_2$Os$_3$(CO)$_{12}$ because the 100-fold slower deprotonation is offset by a 30-fold slower protonation. For the clusters containing bridging hydrides the effect of different metals on deprotonation is small and the only significant effect is that accompanying replacement of CO by a P-donor ligand. Protonation of HOs$_3$(CO)$_{12}^-$ is 500 times faster than that of H$_3$Os$_4$(CO)$_{12}^-$, possibly because it involves formation of a terminal hydride rather than a bridging one. Only a small isotopic effect was observed for deprotonation of bridging hydrides, and this suggests that M—H bond breaking is not very important and that structural rearrangement is not, therefore, very advanced in the transition state. The slowness of deprotonation is probably determined by the difficulty the solvated MeO$^-$ ion has in approaching the hydridic hydrogen. This could be for both electronic and steric reasons. Apparently no attempt was made to see if there was a large isotope effect on the value of pK$_a$ required by the large effects on protonation rates described earlier.

4.6.2 Hydrogenation, Dehydrogenation, Fragmentation, and Substitution

An increasing number of clusters are being found to undergo hydrogenation at atmospheric pressure to form cluster hydrides in good yields. Kaesz and coworkers (121) showed that Os$_3$(CO)$_{12}$ formed H$_2$Os$_3$(CO)$_{10}$ at 120°C, whereas Ru$_3$(CO)$_{12}$ formed H$_4$Ru$_4$(CO)$_{12}$ at 90°C, H$_2$Ru$_3$(CO)$_{10}$ being unstable at this temperature. FeRu$_2$(CO)$_{12}$ also forms H$_4$Ru$_4$(CO)$_{12}$, showing the importance of fragmentation and aggregation in the formation of the tetranuclear cluster. H$_2$Ru$_4$(CO)$_{13}$ reacts very rapidly with H$_2$ at 70°C to form H$_4$Ru$_4$(CO)$_{12}$, whereas the closely related H$_2$FeRu$_3$(CO)$_{13}$ reacts much more slowly. HWOs$_3$(CO)$_{12}Cp$ forms H$_3$WOs$_3$(CO)$_{11}Cp$ in good yield by reaction in refluxing toluene (122), whereas HCoRu$_3$(CO)$_{13}$ reacts to form the very stable H$_3$CoRu$_3$(CO)$_{12}$ in refluxing hexane over $2h$ (123). Co$_2$Ru$_2$(CO)$_{13}$ gives H$_2$Co$_2$Ru$_2$(CO)$_{12}$ on reaction in hexane at 45 to 50°C (124) and the clusters HM_3(μ_2 – COMe)(CO)$_{10}$ react at 60°C (M = Fe, Ru) or 120°C (M = Os) to form H$_3M_3$(μ_3 – COMe)(CO)$_9$ (125,126). Although only one of these reactions has been studied kinetically, comparison with substitution reactions suggests that CO dissociation is an important rate-determining process.

Reaction (4.21) has been studied in detail over a wide range of pressures of CO and H$_2$ (127). The forward reaction was very strongly inhibited by CO but rates of

$$\text{HRu}_3(\mu_2 - \text{COMe}) \, (\text{CO})_{10} + \text{H}_2 \rightleftharpoons \text{H}_3\text{Ru}_3(\mu_3\text{-COMe}) \, (\text{CO})_9 + \text{CO} \quad (4.21)$$

approach to equilibrium coupled with the rate of the reverse reaction were in good agreement with rate equation (4.22), k_2/k_{-1} being small. This is characteristic of the simple dissociative mechanism shown in equations (4.23) and (4.24). Equilibrium constants, $K = k_1 k_2 / k_{-1} k_{-2}$, were found to be 2.6 ± 0.3 over the temperature

$$k_{obsd} = \frac{k_1 k_2 [H_2] / k_{-1} [CO] + k_{-2}}{1 + k_2 [H_2] / k_{-1} [CO]} \qquad (4.22)$$

range 60 to 80°C and, at 40°C, $k_2 / k_{-1} = 1.4 \times 10^{-3}$. The latter very small value

$$HRu_3(\mu_2 - COMe)\,(CO)_{10} \underset{k_{-1}}{\overset{k_1}{\rightleftharpoons}} HRu_3(\mu_2 - COMe)\,(CO)_9 + CO \qquad (4.23)$$

$$HRu_3(\mu_2 - COMe)\,(CO)_9 + H_2 \underset{k_{-2}}{\overset{k_2}{\rightleftharpoons}} H_3 Ru_3(\mu_3 - COMe)\,(CO)_9 \qquad (4.24)$$

was ascribed to the existence of $HRu_3(COMe)\,(CO)_9$ in the two isomeric forms **4.49** and **4.50**. Isomer **4.49** is formed by dissociation of a CO from a bridged Ru atom in $HRu_3(COMe)\,(CO)_{10}$; dissociation from the $Ru(CO)_4$ moiety is expected to be much slower. However, **4.49** must rearrange to **4.50** before oxidative addition of H_2 at the unbridged Ru atom can occur to form **4.51**. If **4.49** and **4.50** are in labile equilibrium, as seems likely, then k_2 / k_{-1} will be a function of their relative amounts. A very small value of k_2 / k_{-1} would result if **4.50** is much less predominant and, therefore, much less stable than **4.49**. The exact nature of the COMe bridging may also be different in **4.49** and **4.50** and so contribute to the value of k_2 / k_{-1} in some

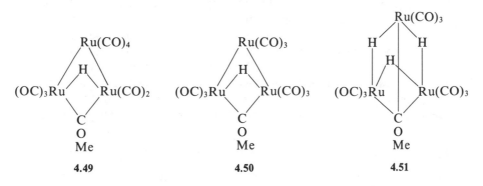

way. The activation parameters $\Delta H^{\neq}_{-2} = 31.0 \pm 0.8$ kcal mol^{-1} and $\Delta S^{\neq}_{-2} = 8 \pm 2$ cal K^{-1} mol^{-1} were obtained from the reverse reaction and k^D_{-2} / k^H_{-2} was found to be 1.4, the latter value being close to that found for reductive elimination of H_2 from some mononuclear complexes. A three-center RuH$_2$ transition state was proposed.

Geoffroy and coworkers (128) have studied the elimination of H_2 from a series of complexes $H_2 M_4 (CO)_{13}$ (M_4 = FeRu$_3$, FeRu$_2$Os, FeRuOs$_2$, Ru$_4$). Stoichiometrically the reactions are as shown in equation (4.25) so that fragmentation is involved as well as H_2 elimination. $M(CO)_5$ is always Fe(CO)$_5$ whenever Fe is present in the

$$H_2 M_4 (CO)_{13} + 4CO \longrightarrow M_3 (CO)_{12} + M(CO)_5 + H_2 \qquad (4.25)$$

TABLE 4.4   Rate Parameters for Fragmentation—H_2 Elimination Reaction of $H_2Ru_4(CO)_{13}$ and $H_2FeRu_3(CO)_{13}$ with CO in Hexane

M_4	$k_2(50°C)$ $M^{-1}\,s^{-1}$	$\Delta H^{\neq}$ kcal mol^{-1}	$\Delta S^{\neq}$ cal K^{-1} mol^{-1}
Ru_4	3×10^{-2}	12.5 ± 0.5	-36.6 ± 1.6
$FeRu_3$	5×10^{-5}	20.0 ± 2.0	-25.4 ± 5.8

cluster. Kinetic studies of reaction of CO with $H_2Ru_4(CO)_{13}$ and $H_2FeRu_3(CO)_{13}$ in hexane showed that the reactions are both first order in [CO]. The rate parameters in Table 4.4 show that the heteronuclear cluster is much more stable than the homonuclear one.

These results were rationalizable on the basis of a greater degree of M—CO bond making in the transition state when M is the larger Ru atom, attack on the apical $Fe(CO)_4$ moiety in $H_2FeRu_3(CO)_{13}$ being assumed. Overall second-order kinetics resulting from highly reversible M—M bond breaking (with formation of an unsaturated butterfly cluster) followed by a CO attack was deemed less likely because of the greater strength expected for the Ru—Ru bonds. It is also less likely because of the very negative values of $\Delta S^{\neq}$ (48). The initial product is proposed to be the butterfly cluster $H_2M_4(CO)_{14}$ which then fragments and forms $M(CO)_5$ and $H_2Ru_3(CO)_{11}$, the latter reacting very rapidly with CO to give $Ru_3(CO)_{12}$ and H_2. This is supported by the formation of $HRu_3(CO)_{11}^-$ and $Fe(CO)_5$ from $HFeRu_3(CO)_{13}^-$. The elimination of H_2 occurs, therefore, after fragmentation and has no effect on the kinetics. Some interesting effects of substituents were observed. $H_2FeRu_3(CO)_{12}(PMe_2Ph)$ exists in the two isomeric forms shown in **4.52** and **4.53** where the CO ligands are omitted and the dotted lines indicate semibridging carbonyls. In their reactions with CO, these clusters show a fragmentation pattern different from that of the unsubstituted cluster, $Ru(CO)_5$ and $FeRu_2(CO)_{11}(PMe_2Ph)$ being formed. $H_2FeRu_3(CO)_{12}PPh_3$ exists only as the C_s form and reacts under 1 atm CO to form $Fe(CO)_5$ and $Ru_3(CO)_{11}(PPh_3)$,

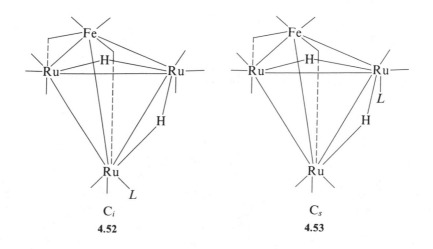

C_i

4.52

C_s

4.53

but at higher pressures it forms $Ru(CO)_5$ and $FeRu_2(CO)_{11}(PPh_3)$. These interesting observations remain unexplained but show that the site or sites at which a nucleophilic attack can occur are strongly affected by the nature of any substituents present. Reaction of PPh_3 with $H_2FeRu_3(CO)_{13}$ occurs by simple CO-dissociative reaction to form the C_s isomer at a rate much greater than the second-order fragmentation reaction with CO (129). The kinetic parameters are $k(50°C) = 7 \times 10^{-4}$ s^{-1}, $\Delta H^{\neq} = 25.3 \pm 0.9$ kcal mol^{-1}, and $\Delta S^{\neq} = 4.9 \pm 3.1$ cal K^{-1} mol^{-1} so that the greater rate is entirely due to the much more favorable value of $\Delta S^{\neq}$. Introduction of a second PPh_3 into the cluster occurs slightly more rapidly but the third substitution is much slower.

The relative thermodynamic stabilities of the C_s and C_i isomers depend on both the steric and electronic properties of the P-donor ligand (130). The isomerizations are rapid and proceed by rearrangement of the $FeRu_3$ framework coupled with migration of the H and CO ligands, a process that also occurs in the unsubstituted cluster.

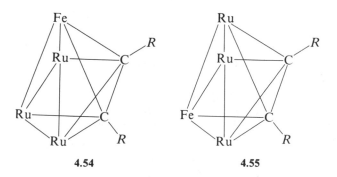

4.54   4.55

Reactions of $H_2FeRu_3(CO)_{13}$ with C_2R_2 (R = alkyl or aryl group) lead to $FeRe_3(CO)_{12}(C_2R_2)$ in the isomeric form shown by **4.54**, but this subsequently isomerizes to **4.55** (the three terminal CO ligands on each metal being omitted) (131). The initial reaction is believed to go via dissociative substitution to form $H_2FeRu_3(CO)_{12}(C_2R_2)$ followed by elimination of H_2. The isomerization is also believed to involve rate-determining CO dissociation, and this is followed by rearrangement of the $FeRu_3(CO)_{11}(C_2R_2)$ cluster and final reattachment of the CO ligand.

Rossetti and Stanghellini (132) have studied the elimination of H_2 from $H_4Ru_4(CO)_{12}$ according to the stoichiometric equation (4.26). An increase in rate to a limiting

$$H_4Ru_4(CO)_{12} + 5CO \longrightarrow Ru_3(CO)_{12} + Ru(CO)_5 + 2H_2 \qquad (4.26)$$

value with increasing [CO], and retardation by H_2, suggest the sequence of reactions (4.27) to (4.32). $H_4Ru_4(CO)_{12}^*$ is a reactive isomer of $H_4Ru_4(CO)_{12}$ present only in low, steady-state concentrations.

$$H_4Ru_4(CO)_{12} \underset{k_{-1}}{\overset{k_1}{\rightleftharpoons}} H_4Ru_4(CO)_{12}^* \qquad (4.27)$$

$$H_4Ru_4(CO)_{12}^* + CO \xrightleftharpoons[k_{-2}]{k_2} H_4Ru_4(CO)_{13} \qquad (4.28)$$

$$H_4Ru_4(CO)_{13} \xrightleftharpoons[k_{-3}]{k_3} H_2 + H_2Ru_4(CO)_{13} \qquad (4.29)$$

$$H_2Ru_4(CO)_{13} + CO \xrightarrow{k_4} H_2Ru_4(CO)_{14} \qquad (4.30)$$

$$H_2Ru_4(CO)_{14} + 2CO \longrightarrow H_2Ru_3(CO)_{11} + Ru(CO)_5 \qquad (4.31)$$

$$H_2Ru_3(CO)_{11} + CO \longrightarrow H_2 + Ru_3(CO)_{12} \qquad (4.32)$$

Reaction (4.30) was studied by Geoffroy et al. (128) (see preceding discussion) and reactions (4.31) and (4.32) are very fast (133). The mechanism leads to rate equation (4.33). This is exactly what is followed when $[H_2] = 0$, but the dependence

$$k_{obsd} = \frac{k_1k_2k_3k_4[CO]^2}{k_4[CO]\{k_{-1}(k_{-2} + k_3) + k_2k_3[CO]\} + k_{-1}k_{-2}k_{-3}[H_2]} \qquad (4.33)$$

on $[H_2]$ was only indicated qualitatively. This mechanism implies that the reaction of $H_2Ru_4(CO)_{13}$ with H_2 should be first order in $[H_2]$ just as its reaction with CO is first order in [CO]. The isotopic effect on k_1 of replacing the hydrides by deuterides was negligible at 60°C but significant at 80°C. Only a small change in the Ru—H bond energies in the first step is implied and the structure of $H_4Ru_4(CO)_{12}^*$ was suggested to be the unsaturated butterfly cluster **4.56**. This forms the saturated cluster **4.57** by addition of CO to the unsaturated Ru atom while **4.57** loses H_2 with concomitant closing of the cluster to form **4.58**. This then undergoes reaction (4.30). The activation parameters $\Delta H_1^{\neq}$ and $\Delta S_1^{\neq}$ are 32.5 ± 2.9 kcal mol^{-1} and 15 ± 5 cal K^{-1} mol^{-1}, respectively, and $k_1(50°C) = 1 \times 10^{-6}$ s^{-1}.

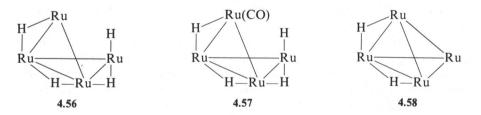

      4.56   **4.57**   **4.58**

Doi et al. (134) have studied the hydrogenation of C_2H_4 by $H_4Ru_4(CO)_{12}$. This is stoichiometric in the absence of H_2, 2 moles of C_2H_6 being formed per mole of cluster. In the presence of H_2 the reaction is catalytic and the rate of formation of C_2H_6 increases to limiting values with increasing $[C_2H_4]$ and increasing $[H_2]$, whereas the rate decreases with increasing [CO]. An unexpected feature was the observation that H_2 appears to react reversibly with $H_4Ru_4(CO)_{12}$ at 72°C, but this reaction was not studied quantitatively and the product was not identified. A mechanism for the catalytic reaction was proposed in which three rapid preequilibria lead to

$H_3Ru_4(CO)_{11}(C_2H_5)$ before a slow reaction with H_2 in a second-order process leads to C_2H_6 and reformation of $H_4Ru_4(CO)_{11}$. At high $[C_2H_4]$ and relatively low $[H_2]$ and $[CO]$, when the rate is limiting with respect to $[C_2H_4]$, this mechanism implies that all the complex should be present as $H_3Ru_4(CO)_{11}(C_2H_5)$. No evidence for this was offered and the studies of Rossetti and Stanghellini (132) discussed earlier suggest that the mechanism should probably be reformulated in terms of steady-state intermediates. Additional studies of the reaction of $H_4Ru_4(CO)_{12}$ with H_2 and its stoichiometric reaction with olefines would also be of relevance.

4.7 PHOTOCHEMICAL KINETICS

4.7.1 Introduction

The demonstration (135) that the low-energy electronic spectrum of $Mn_2(CO)_{10}$ can be assigned to $\sigma \rightarrow \sigma^*$ and $d\pi \rightarrow \sigma^*$ transitions has been associated with an extensive series of studies of the photochemical consequences of such excitations in dinuclear carbonyls (136). Although homolysis of the metal–metal bonds is clearly a major primary process, more recent studies (137) have shown that the electronic excitation of $Mn_2(CO)_{10}$ can also lead to CO dissociation. When $Mn_2(CO)_9$ is generated at 77K in an alkane matrix it has a bridging CO group and, on warming to room temperature in the presence of 10 mM CCl_4, it does not react to form $Mn(CO)_5Cl$. However, when it is generated by flash photolysis in CCl_4 as solvent at 20°C it does react to form $Mn(CO)_5Cl$ (138). In the presence of a variety of Lewis bases, $Mn_2(CO)_9$ forms $Mn_2(CO)_9L$ as expected (137). Quantum yields for disappearance of the dinuclear carbonyls are quite high but rarely approach unity (136). This can be accounted for if the primary fragments produced by photolysis are formed with unit quantum yield within a solvent cage but undergo reversion to the original complex in competition with diffusion out of the solvent cage. This is suggested by the decrease of ϕ, with increasing solvent viscosity observed for reaction of $Re_2(CO)_{10}$ with I_2 to form $Re(CO)_5I$ (139) and by the absence of formation of $Mn(CO)_5$ when $Mn_2(CO)_{10}$ is photolyzed in an alkane matrix at 77K (137).

Apart from formation of metal-centered radicals by homolysis, and of coordinatively unsaturated dinuclear products by CO dissociation, there are other types of intermediates that appear to be generated photochemically. Photolysis of $Cp_2Fe_2(CO)_4$ in THF in the presence of PPh_3 or $P(O-i-Pr)_3$ was concluded to proceed as in Scheme 4.3 via an intermediate with a bridging CO ligand but with no Fe—Fe bond (140).

On the other hand, it is proposed that flash photolysis at room temperature and matrix irradiation at 77K lead to $Cp_2Fe_2(CO)_3$ with three bridging carbonyls (141). Formation of a similar CO-bridged intermediate to that shown in Scheme 4.3 is believed to occur by photolytically induced metal migration in $Cp_2Mo_2(CO)_6$ as shown in equation (4.34) (142). The vacant coordination site on the right-hand

$$Cp(OC)_3Mo—Mo(CO)_3Cp \longrightarrow Cp(OC)_3Mo(\mu\text{-}CO)Mo(CO)_2Cp \quad (4.34)$$

Mo atom can be attacked by a nucleophile with eventual disproportionation and formation of ionic products in suitable solvents. Reaction (4.34) requires higher energies than those involved in the $d\pi \rightarrow \sigma^*$ and $\sigma \rightarrow \sigma^*$ transitions. These lower-

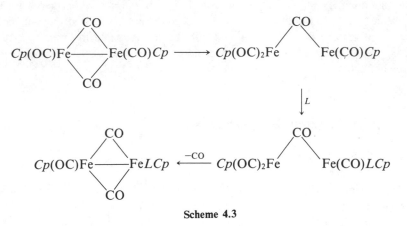

Scheme 4.3

energy excitations lead to a radical chain reaction pathway involving a very electron rich, 19-electron intermediate $CpMo(CO)_2L_2$. A dinuclear radical species that does react with CCl_4 has also been proposed to be formed from $Mn_2(CO)_{10}$ (143), and further work is obviously required to establish more clearly the composition and chemical properties of the various intermediates that can be generated in such reactions.

4.7.2 Photochemical Kinetics of $M_3(CO)_{12}$ (M = Fe, Ru, or Os)

The intense visible and near uv absorption of these complexes were initially assigned to transitions to metal–metal antibonding orbitals (144). It was concluded that the primary photoprocess is probably the formation of a diradical $M_3(CO)_{12}$ species formed by homolysis of one M—M bond (145). This could then undergo thermal reactions leading to eventual substitution or fragmentation as observed in preparative studies. Quantum yields were generally inferred to be low from the long reaction times required, and the few quantitative measurements led to values of $\lesssim 0.05$ for several reactions, little difference being shown between the various clusters (146).

Study of the reaction of $Ru_3(CO)_{12}$ with methylacrylate (*ol*) in hexane showed that the only product was mononuclear $Ru(CO)_4(ol)$ and that the quantum yield increased with [*ol*] to a maximum of 0.121 for irradiation at 313 nm and 0.038 for 395 nm light (147). A maximum value of 0.031 was found for reaction with ethyl acrylate in benzene at 366 nm (148). The 395 nm irradiation was concluded to lead to cleavage of at least one Ru—Ru bond, the intermediates either reverting to $Ru_3(CO)_{12}$ or reacting competitively with olefine. This reaction with olefine would have to involve direct attack on the intermediate(s) or the quantum yields would not depend on [*ol*] at lower concentrations. The much higher quantum yields at 313 nm were ascribed to CO dissociation as an additional primary photoprocess (147).

Reaction with CO to form $Ru(CO)_5$ proceeds with quantum yields that increase with [CO] toward a limiting value of 0.065 ± 0.006 that is independent of irradiating wavelength from 313 to 436 nm, the presumed CO dissociative process at 313 nm being reversed by the free CO present (148). The rate of increase to the limiting value appears to be very solvent dependent. Attack on the photo-produced intermediates is ca. 30 times faster in isooctane than in benzene so that the observed quantum yields for reaction under 1 atm CO in benzene are much lower than in isooctane.

An important feature of this reaction is that the less-than-limiting quantum yields are not dependent on the intensity of the absorbed light, a fact that shows that the CO is attacking a trinuclear intermediate and not mono- or binuclear ones formed by photofragmentation (148). Further, quantum yields for reaction with CCl_4 to yield chlorocarbonyl products are very small (ca. 0.001), which shows that either any radical species formed are unusually ineffective at chlorine atom abstraction or radicals are not primary reactive photoproducts (148,149). It was proposed that these could be species such as **4.59** formed by metal migration. Attack on the vacant coordination site by CO or an olefine could lead to fragmentation. The fact that the

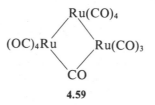

4.59

limiting quantum yield with CO is about twice that for reaction with alkyl acrylates is surprising since a common intermediate would be expected to give limiting quantum yields that are independent of the nature of the reactant (148). This may be connected with the fact that the addition of small amounts of THF or diglyme in octane reduces the less-than-limiting quantum yields for reaction under 1 atm of CO, $1/\phi$ being linearly dependent on the concentration of an added cosolvent (149). This is attributed to attack by THF on **4.59** causing reversion to $Ru_3(CO)_{12}$ rather than fragmentation. The dependence of the limiting quantum yield on the nature of the nucleophile might have a similar origin, the $Ru_3(CO)_{12}L$ formed by addition of L to **4.59** undergoing fragmentation or reversion to $Ru_3(CO)_{12}$ in proportions dependent on the nature of L.

Photoreaction of chlorocarbons with $Os_3(CO)_{12}$ to form $Os(CO)_4Cl_2$ also proceed with very low quantum yields (150). Reaction with PPh_3 leads to successive substitutions to form $Os_3(CO)_9(PPh_3)_3$ with quantitatively undefined but much higher quantum yields. Photofragmentation of $Os_3(CO)_9(PPh_3)_3$ to form $Os(CO)_3(PPh_3)_2$ in the presence of PPh_3 has $\phi = 0.005$ at 366 nm. The substitution of $Os_3(CO)_{12}$ was ascribed to photo-induced CO dissociation, the excited being state considered to be Os—CO antibonding to some extent. However, it does not seem unlikely that an osmium analog of **4.59** would undergo substitutional attack by PPh_3, rather than fragmentation, because of the greater strength of the Os—Os bonds. On the other hand, it is interesting that $Os_3(CO)_{12}$ undergoes photofragmentation with olefines (151) to form $Os(CO)_4(ol)$ and a dinuclear, C_2-bridged complex rather than substituted trinuclear products, and the limiting quantum yields are similar to those for $Ru_3(CO)_{12}$, namely, ca. 0.03 at 355 (146) and 436 nm (152). Photoreaction in CCl_4 of the open chain $Os_3(CO)_{12}Cl_2$ proceeds to form $Os(CO)_4Cl_2$ with quite high quantum yields (0.16 at 366 nm and 0.31 at 313 nm) that are characteristic of a primary photoprocess involving formation of quite reactive osmium-centered radicals (150). Osmium-centered radicals are not, therefore, intrinsically inefficient at chlorine atom abstraction.

Photochemically generated intermediates from $Fe_3(CO)_{12}$ (146), $Ru_3(CO)_{12}$ (146), and $Ru_3(CO)_9(PPh_3)_3$ (153) are capable of catalytically isomerizing 1-pentene to cis and trans 2-pentenes, but this does not occur with $Os_3(CO)_{12}$ (146). However, all three dodecacarbonyls catalyze reaction of 1-pentene with $HSiEt_3$ to form C_5H_{12}, (n-pentyl)$SiEt_3$, and various (pentenyl)$SiEt_3$ species (146). The distribution of products from $Fe_3(CO)_{12}$ is similar to that from $Fe(CO)_5$, suggesting a mononuclear catalyst. The pattern for $Ru_3(CO)_9(PPh_3)_3$ is quite different from that for $Ru(CO)_4(PPh_3)$, and the cluster is apparently maintained during catalysis by $Ru_3(CO)_9(PPh_3)_3$, although the nature of the catalytic species is not obvious from the results (153).

4.7.3 Photofragmentation of $H_3Re_3(CO)_{12}$

Irradiation of $H_3Re_3(CO)_{12}$ in degassed solutions leads quantitatively to $H_2Re_2(CO)_8$ with $\phi = 0.1$ irrespective of wavelength and light intensity (154). ϕ is reduced to 0.02 under an atmosphere of CO and $HRe(CO)_5$ is then formed in addition to $H_2Re_2(CO)_8$, the latter reacting thermally with CO to form $Re_2(CO)_{10}$. The results showed that the primary photochemical step could not involve fragmentation of the cluster. Possibilities consistent with the data were (1) CO-dissociation followed by thermal fragmentation to $HRe(CO)_4$ and $H_2Re_2(CO)_7$ or (2) formation of a chemically reactive isomer of $H_3Re_3(CO)_{12}$ followed by thermal dissociation of CO and fragmentation. The reactive isomer of $H_3Re_3(CO)_{12}$ was assumed to be formed by homolysis of one Re—Re bond. The effect of varying [CO] was not studied so it is possible that the quantum yield for reaction under CO applies to an additional, [CO]-independent path not involving CO dissociation at any stage.

4.7.4 Photofragmentation of $HCCo_3(CO)_9$ and $CH_3CCo_3(CO)_9$

The low-energy electronic absorptions of these tetrahedral alkylidyne clusters are assignable to transitions from nonbonding to antibonding orbitals located in the Co_3 triangle (155). Irradiation of $HCCo_3(CO)_9$ alone has no effect, but in the presence of H_2, CH_4 and $Co_4(CO)_{12}$ are formed quantitatively, all the CH_4 originating from the CH group (155). Quantum yields were not obtained because of "spectral overlap," but reaction in the presence of a 3:1 H_2-CO mixture leads quantitatively to $Co_2(CO)_8$ instead of to $Co_4(CO)_{12}$ and $\phi = 0.03$ at 366 nm. Photochemically induced substitution occurs at least as rapidly as photofragmentation in the presence of H_2 alone. Photodissociation of CO is possible, but the absence of any [CO] dependence of the quantum yield for reaction with H_2 and CO shows that this reaction, at least, probably occurs via photo-induced homolysis of a Co—Co bond followed by reaction of the diradical species with H_2. Formation of the diradical could also provide a path for substitution.

Photolysis of $CH_3CCo_3(CO)_9$ in the presence of H_2 leads to ethane and ethylene as well as to $Co_4(CO)_{12}$ but reaction is much slower than with $HCCo_3(CO)_9$. The presence of CO completely quenches any photolysis (155). It is possible that photolysis of $CH_3CCo_3(CO)_9$ proceeds entirely by a very readily reversed CO dissociative process, whereas photolysis of $HCCo_3(CO)_9$ proceeds partly by CO dissociation but with an additional path, not inhibited by CO and perhaps involving Co—Co homolysis. These studies do not provide information on the interesting steps that must follow the initial photolysis.

4.7.5 Photosubstitution of $H_2FeRu_3(CO)_{13}$, $H_2FeOs_3(CO)_{13}$, and $H_2Ru_4(CO)_{13}$

Photolysis (156) in the presence of PPh_3 leads to stepwise replacement of CO by PPh_3. For $H_2FeRu_3(CO)_{13}$ the quantum yield of 0.03 at 366 nm is independent of $[PPh_3]$ but is reduced by [CO] in a way quantitatively consistent with competition between CO and PPh_3 for a coordinatively unsaturated $H_2FeRu_3(CO)_{12}$ intermediate formed by photodissociation of CO. Photochemical cleavage of a metal–metal bond was rejected, not on purely kinetic grounds, but because the fragmentation of $H_2FeRu_3(CO)_{14}$ that would be formed subsequently would be expected, on the basis of its known thermal reactivity, to fragment readily. However, photo-induced fragmentation is very slow indeed. Photosubstitution of $H_2FeOs_3(CO)_{13}$ and $H_2Ru_4(CO)_{14}$ have quantum yields of 0.06 and 0.016, respectively.

Photolysis in the presence of H_2 leads initially to $H_4M_4(CO)_{12}$, and the quantum yield for $H_2FeRu_3(CO)_{13}$ is 0.015. It is not clear whether or not the difference between this value and that for photosubstitution (0.03) can be considered significant. If it is not, then photodissociation of CO can be considered a common primary process.

4.7.6 Photolysis of $H_4Ru_4(CO)_{12}$ (157)

$H_4Ru_4(CO)_{12}$ undergoes photosubstitution with PPh_3 and $P(OMe)_3$ with $\phi = 0.005$ irrespective of L, $[L]$, or λ. Photodissociation of CO is considered the most likely primary step. $H_4Ru_4(CO)_{12}$ is a very effective catalyst for photoisomerization of 1-pentene and reduces it stoichiometrically in the absence of H_2 and catalytically in its presence. In the presence of H_2 1- and 2-pentyne are reduced catalytically to 1-pentene and cis 2-pentene, respectively, showing that displacement of the coordinated pentene products from an intermediate by pentyne is more rapid than their further reaction.

$H_2Ru_4(CO)_{13}$ also photoisomerizes 1-pentene (a process that occurs thermally) and photoreduces it (a process that does not occur thermally). These catalytic reactions are all considered to occur via $H_4Ru_4(CO)_{11}L$ and $H_2Ru_4(CO)_{12}L$ (L = alkene or alkyne) and insertion of L into a Ru—H bond.

4.7.7 Photooxidation of $Cp_4Fe_4(CO)_4$ (158)

This complex undergoes smooth photooxidation in halocarbon solvents. The process is wavelength dependent in a way that is consistent with a charge-transfer-to-solvent transition and the quantum yields increase with the increasing ease of the reduction of the halocarbon, namely, from 0.0007 in C_2Cl_4 to 0.21 in CCl_4 at 313 nm. The importance of such processes in the photochemistry of clusters in general has yet to be investigated.

4.7.8 Excited State Decay of $H_4Re_4(CO)_{12}$ and $H_6Re_4(CO)_{12}^{2-}$ (159)

$H_4Re_4(CO)_{12}$ is found to emit at ca. 1400 cm^{-1} from a state with a lifetime varying from <0.02 μs at 298K in 3-methyl pentane to 16 μs at 77K in a 3-methyl pentane glass. The quantum yields are 0.003 and 0.25, respectively. Emission occurs from the lowest excited state of the complex. This is quenched by anthracene but not by stilbene, an observation consistent with the low-lying excited state being a triplet associated with the H_4Re_4 cluster. $H_6Re_4(CO)_{12}^{2-}$ emits at ca. 18,000 cm^{-1}, but only at 77K, and with a lifetime of ca. 2 μs and a quantum yield of ca. 0.01. Quenching of the $H_4Re_4(CO)_{12}$ emission in toluence at 298K by anthracene leads to a rate constant for triplet energy transfer of 1.5×10^9 $M^{-1}s^{-1}$.

   $Ru_3(CO)_{12}$, $Os_3(CO)_{12}$, and $Ru_3(CO)_9(PPh_3)_3$ are also reported to emit (160), and these results are interesting in view of the effective quenching of triplet biacetyl by $Ru_3(CO)_{12}$, $Os_3(CO)_{12}$, and $Os_3(CO)_9(PPh_3)_3$ and a number of other metal carbonyl clusters (161). The triplet state of $H_4Re_4(CO)_{12}$ does not appear to be chemically reactive in solution but the possible reactivity of trinuclear clusters such as $Ru_3(CO)_{12}$ has yet to be investigated.

4.8 LARGE CARBONYL CLUSTERS

No systematic studies of the kinetics of carbonyl clusters with more than four metal atoms have been reported. $Rh_6(CO)_{16}$ is the one most readily available and the half-life for CO exchange has been found to be ca. 1 month in chloroform at 20°C (162). This contrasts with the relatively rapid substitution reactions found with P-donor ligands (103a,163). Phosphites and bidentate phosphines lead to replacement of up to six of the CO ligands, and this can also occur with monodentate phosphines. However, reaction with phosphines in the presence of CO leads to dinuclear products in high yields, or to mononuclear $Rh(CO)L_2Cl$ when the reactions are carried out in chlorocarbon solvents (103a). $Rh_6(CO)_{16}$ itself is not at all easily fragmented by CO (164), but some substituted intermediates evidently are very susceptible to fragmentation. The rapidity of the initial substitutions compared with the rate of CO exchange suggests very strongly that substitution occurs via nucleophilic attack.

$Rh_6(CO)_{12}(dppb)_2$ (dppb = $Ph_2P(CH_2)_4(PPh_2)$ undergoes CO exchange with a half-life of $18h$ at 20°C (162) so that substitution appears to enhance the rate of CO exchange as it does with most other clusters.

$Os_6(CO)_{18}$ is an 84-electron cluster with a bicapped tetrahedral structure. Reduction with I^- produces the 86-electron $Os_6(CO)_{18}^{2-}$ which has an octahedral arrangement of Os atoms. Reaction in dichloromethane (165) is first order in $[Os_6(CO)_{18}]$ and second order in $[I^-]$, and this is explicable by the sequence of reactions shown in equations (4.35) and (4.36). Reaction in (4.35) is a labile preequilibrium that

$$Os_6(CO)_{18} + I^- \overset{K}{\rightleftharpoons} Os_6(CO)_{18}I^- \qquad (4.35)$$

$$Os_6(CO)_{18}I^- + I^- \overset{k}{\longrightarrow} Os_6(CO)_{18}^{2-} + I_2 \qquad (4.36)$$

lies well to the left. The nature of the intermediate $Os_6(CO)_{18}I^-$ is not known but it is believed to undergo a slow bimolecular reductive elimination reaction with I^- as shown in (4.36). The observed third-order rate constant is given by the product Kk, and there is some kinetic evidence for formation of significant amounts of $Os_6(CO)_{18}I^-$, a high $[I^-]$.

4.9 FINAL COMMENTS

The foregoing account is a reasonably comprehensive review of the current knowledge of kinetic behavior of metal carbonyl clusters. Although it may no longer be true to say, as Johnson (1) did in 1980, that the study of cluster reactivities is "still in its infancy," it cannot be said to be much past the "toddler" stage. Even though the crude distinction between dissociative and associative activation in substitution or fragmentation reactions is empirically clear, knowledge of the timing of ligand dissociation and of the nature of the intermediates formed is still very limited. How important is formation of a reactive isomer before dissociative or associative activation? Where in the cluster does such activation occur and what are the detailed consequences? What factors control the rates of fragmentation of clusters and what are the properties and functions of radical or other intermediates? Answers to these questions are by no means adequate. Data for reactions with small molecules (such as H_2, halogens, NO, etc.) and with unsaturated organic molecules are still very limited as are data for clusters formed by such reactions.

What kinetic data there are that relate to these questions are generally available only for thermal or photochemical reactions of quite small neutral clusters in nonpolar, inert solvents. Kinetic studies of larger clusters must be undertaken. Studies of reactions in a much wider range of solvents have to be developed. In this way, ionic clusters, and reactions with ionic reagents, can be studied and the interplay among the strength of bonding in the cluster, the nature of reagents, the redox properties of the system, the nature of photo-generated intermediates, participation of homolysis and heterolysis of metal–metal bonds, and the nature of the solvent can be investigated. Much impetus for these studies comes particularly from the growing number of synthetic studies involving promotion or catalysis by various reagents (166).

If we are to obtain anything approaching a thorough knowledge and understanding of the mechanisms and energetics of reactions of this very diverse and important class of compounds, then many more basic, classical, kinetic studies will have to be undertaken. There must also be concurrent and substantial development of studies of reactive intermediates produced by electrochemical and flash-photolytic techniques, for example. A considerably greater degree of application, expertise, and ingenuity will have to be exerted than heretofore for this to be accomplished and for our knowledge of reactivities of carbonyl clusters to begin to compare with that of their syntheses and structures.

REFERENCES

1. B. F. G. Johnson, Ed., *Transition Metal Clusters,* Wiley, Chichester, 1980.

2. B. F. G. Johnson and J. Lewis, *Adv. Inorg. Radiochem.,* **24**, 225 (1981).

3. For example, *Organometallic Chemistry, Specialist Periodical Reports, R. Soc. Chem., London,* **1–11**, (1972–1983); Sir Geoffrey Wilkinson, Ed., *Comprehensive Organometallic Chemistry,* Pergamon, Oxford, 1982.

4. For example, *Inorganic Reaction Mechanisms, Specialist Periodical Reports, R. Soc. Chem., London,* **1–7**, (1971–1981).

5. M. V. Twigg, Ed., *Mechanisms of Inorganic and Organometallic Reactions,* Vol. 1, 2, and 3, Plenum, New York, 1983, 1984, and 1985.

6. G. W. Parshall, *Homogeneous Catalysis,* Wiley, New York, 1980, p. v.

7. F. Basolo and R. G. Pearson, *Mechanisms of Inorganic Reactions,* 2nd ed., Wiley, New York, 1967, Ch. 7.

8(a). D. J. Darensbourg, *Adv. Organometal. Chem.,* **21**, 113 (1982); (b) J. A. S. Howell and P. M. Burkinshaw, *Chem. Rev.,* **83**, 557 (1983).

9(a). A. J. Poë, *Chem. Br.,* **19**, 997 (1983); (b) A. J. Poë, in M. H. Chisholm, Ed., *Reactivity of Metal–Metal Bonds,* American Chemical Society, Washington, DC, ACS Symp. Ser. No. 155, 1981, Ch. 7; (c) A. J. Poë and C. V. Sekhar, *J. Am. Chem. Soc.,* **107**, 4874 (1985).

10. E. L. Muetterties, R. R. Burch, and A. M. Stolzenberg, *Ann. Rev. Phys. Chem.,* **33**, 89 (1982).

11. G. L. Geoffroy and M. S. Wrighton, *Organometallic Photochemistry,* Academic Press, New York, 1979, Ch. 2.

12. S. P. Schmidt, W. C. Trogler, and F. Basolo, *Inorg. Chem.,* **21**, 1698 (1982); A. M. Stolzenberg and E. L. Muetterties, *J. Am. Chem. Soc.,* **105**, 822 (1983); N. J. Coville, A. M. Stolzenberg, and E. L. Muetterties, *J. Am. Chem. Soc.,* **105**, 2499 (1983).

13. A. Marcomini and A. J. Poë, *J. Am. Chem. Soc.,* **105**, 6952 (1983).

14. A. J. Poë, *Inorg. Chem.,* **20**, 4029, 4032 (1981).

15. J. D. Atwood, *Inorg. Chem.,* **20**, 4031 (1981).

16. R. Colton and C. J. Commons, *Aust. J. Chem.,* **28**, 1673 (1975).

17. R. H. Hooker, K. A. Mahmoud, and A. J. Rest, *J. Chem. Soc., Chem. Commun.*, 1022 (1983); A. F. Hepp, J. P. Blaha, C. Lewis, and M. S. Wrighton, *Organometallics*, **3**, 174 (1984).

18. M. Basato and A. J. Poë, *J. Chem. Soc., Dalton Trans.*, 607 (1974); M. Basato, J. P. Fawcett, and A. J. Poë, *J. Chem. Soc., Dalton Trans.*, 1856 (1974).

19. S. Amer and A. J. Poë, *J. Organometal. Chem.*, **209**, C31 (1981).

20. S. Breitschaft and F. Basolo, *J. Am. Chem. Soc.*, **88**, 2702 (1966).

21. P. F. Barrett and A. J. Poë, *J. Chem. Soc. A*, 429 (1968).

22. P. F. Barrett, *Can. J. Chem.*, **52**, 3773 (1974) and references therein.

23. D. F. Keeley and R. E. Johnson, *J. Inorg. Nucl. Chem.*, **11**, 33 (1959).

24. A. Wojcicki and F. Basolo, *J. Am. Chem. Soc.*, **83**, 520 (1961).

25. G. Cetini, O. Gambino, P. L. Stanghellini, and G. A. Vaglio, *Inorg. Chem.*, **6**, 1225 (1967).

26. M. Basato and A. J. Poë, *J. Chem. Soc., Dalton Trans.*, 456 (1974).

27. M. A. Cobb, B. Hungate, and A. J. Poë, *J. Chem. Soc., Dalton Trans.*, 2226 (1976).

28. M. Arewgoda, P. H. Rieger, B. H. Robinson, J. Simpson, and S. J. Visco, *J. Am. Chem. Soc.*, **104**, 5633 (1982).

29. M. Basato, J. P. Fawcett, and A. J. Poë, *J. Chem. Soc., Dalton Trans.*, 1350 (1974).

30. M. Basato, J. P. Fawcett, S. A. Fieldhouse, and A. J. Poë, *J. Chem. Soc., Dalton Trans.*, 1856 (1974).

31. K. H. Theopold and R. G. Bergman, *Organometallics*, **1**, 1571 (1982).

32. H. E. Bryndza and R. G. Bergman, *J. Am. Chem. Soc.*, **101**, 4766 (1979); K. H. Theopold and R. G. Bergman, *J. Am. Chem. Soc.*, **102**, 5694 (1980).

33. L. Maresca, F. Greggio, G. Sbrignadello, and G. Bor, *Inorg. Chim. Acta*, **5**, 667 (1971).

34. P. C. Ellgen and J. N. Gerlach, *Inorg. Chem.*, **12**, 2526 (1973).

35. M. Basato, *J. Chem. Soc., Dalton Trans.*, 911 (1975).

36. P. C. Ellgen and J. N. Gerlach, *Inorg. Chem.*, **13**, 1944 (1974).

37. J. N. Gerlach, R. M. Wing, and P. C. Ellgen, *Inorg. Chem.*, **15**, 2959 (1976).

38. J. N. Gerlach, S. L. McMullin, and P. C. Ellgen, *Inorg. Chem.*, **15**, 1232 (1976).

39. S. Aime, G. Gervasio, R. Rossetti, and P. L. Stanghellini, *Inorg. Chim. Acta*, **40**, 131 (1980).

40. M. Basato, *J. Chem. Soc., Dalton Trans.*, 1678 (1976).

41. R. Colton and C. J. Commons, *Aust. J. Chem.*, **28**, 1663 (1975).

42. K. G. Caulton and P. Adair, *J. Organometal. Chem.*, **114**, C11 (1976).

43. H. C. Aspinall and A. J. Deeming, *J. Chem. Soc., Chem. Commun.*, 724 (1981).

44. T. E. Wolff and L. P. Klemann, *Organometallics*, **1**, 1667 (1982).

45. W. R. Cullen and G. L. Hou, *Inorg. Chem.*, **14**, 3121 (1975).

46. G. Kramer, J. Patterson, A. J. Poë, and L. Ng, *Inorg. Chem.*, **19**, 1161 (1980).

47. H. J. Langenbach and H. Vahrenkamp, *Chem. Ber.*, **110**, 1195 (1977); *ibid.*, **112**, 3390, 3773 (1979).

48. R. A. Jackson, R. Kanluen, and A. J. Poë, *Inorg. Chem.*, **20**, 1130 (1981).

49. R. A. Jackson, R. Kanluen, and A. J. Poë, *Inorg. Chem.*, **23**, 523 (1984).

50. M. J. Breen, M. R. Duttera, G. L. Geoffroy, G. C. Novotnak, D. A. Roberts, P. M. Shulman, and G. R. Steinmetz, *Organometallics*, **1**, 1008 (1982).

51. F. A. Cotton and G. Wilkinson, *Advanced Inorganic Chemistry*, 4th ed., Wiley-Interscience, New York, 1980, Ch. 25.

52. C. M. Wei and L. F. Dahl, *J. Am. Chem. Soc.*, **91**, 1351 (1969).

53. S. M. Grant and A. R. Manning, *Inorg. Chim. Acta*, **31**, 41 (1978).

54. G. Cetini, O. Gambino, E. Sappa, and G. A. Vaglio, *Atti della Accad. Sci. Torino*, **101**, 855 (1966–1967).

55. R. Kumar, *J. Organometal. Chem.*, **136**, 235 (1977). But see also: A. Shojei and J. D. Atwood, *Organometallics*, **4**, 187 (1985).

56. R. J. Angelici and E. D. Siefert, *Inorg. Chem.*, **5**, 1457 (1966).

57. H. Schuman and J. Optiz, *J. Organometal. Chem.*, **166**, 233 (1979).

58. P. A. Dawson, B. M. Peake, B. H. Robinson, and J. Simpson, *Inorg. Chem.*, **19**, 465 (1980).

59. M. Absi-Halabi, J. D. Atwood, N. P. Forbus, and T. L. Brown, *J. Am. Chem. Soc.*, **102**, 6248 (1980).

60. I. Fischler, K. Hildenbrand, and E. K. von Gustorf, *Angew. Chem. Int. Ed. Engl.*, **14**, 54 (1975).

61. J. P. Candlin and A. C. Shortland, *J. Organometal. Chem.*, **16**, 289 (1969).

62. A. J. Poë and M. V. Twigg, *J. Chem. Soc., Dalton Trans.*, 1860 (1974).

63. M. V. Twigg, *Inorg. Chim. Acta*, **21**, L7 (1977).

64. A. J. Poë and M. V. Twigg, *Inorg. Chem.*, **13**, 2982 (1974).

65. N. Brodie, V. Desikan, and A. J. Poë, unpublished observations.

66. S. K. Malik and A. J. Poë, *Inorg. Chem.*, **17**, 1484 (1978).

67. D. P. Keeton, S. K. Malik, and A. J. Poë, *J. Chem. Soc., Dalton Trans.*, 233 (1977).

68. R. Huq, A. J. Poë, and S. Chawla, *Inorg. Chim. Acta*, **38**, 121 (1980).

69. J. P. Fawcett and A. J. Poë, *J. Chem. Soc., Dalton Trans.*, 1302 (1977); A. Marcomini and A. J. Poë, *J. Chem. Soc., Dalton Trans.*, 95 (1984).

70. D. P. Keeton, S. K. Malik, and A. J. Poë, *J. Chem. Soc., Dalton Trans.*, 1392 (1977).

71. G. Lavine, N. Lugan, and J. J. Bonnet, *Nouv. J. Chim.*, **5**, 423 (1981).

72. S. K. Malik and A. J. Poë, *Inorg. Chem.*, **18**, 1241 (1979).

73. B. Ambwani, S. K. Chawla, and A. J. Poë, *Inorg. Chem.*, **24**, 2635 (1985) and unpublished results.

74. J. R. Norton and J. P. Collman, *Inorg. Chem.*, **12**, 476 (1973).

75. G. Kramer, A. J. Poë, and S. Amer, *Inorg. Chem.*, **20**, 1362 (1981).

76. A. J. Poë and V. C. Sekhar, *Inorg. Chem.*, **24**, 4376 (1985).

77. A. J. Poë and R. Smith, unpublished results.

78. J. P. Candlin and J. Cooper, *J. Organometal. Chem.*, **15**, 230 (1968).

79. A. Mayr, Y. C. Lin, N. M. Boag, and H. D. Kaesz, *Inorg. Chem.*, **21**, 1704 (1982).

80. D. C. Gross and P. C. Ford, *Inorg. Chem.*, **21**, 1702 (1982).

81. G. Cetini, R. Ercoli, O. Gambino, and G. Vaglio, *Atti Accad. Sci. Torino*, **99**, 1123 (1965); obtained from *Chem. Abstr.*, **68**, 72774g; Ref. (7), p. 551.

82. A. Cartner, R. G. Cunninghame, and B. H. Robinson, *J. Organometal. Chem.*, **92**, 49 (1975).

83. L. F. Dahl and P. W. Sutton, *Inorg. Chem.*, **2**, 1067 (1963); C. H. Wei and L. F. Dahl, *Inorg. Chem.*, **4**, 493 (1965).

84. R. Rossetti, P. L. Stanghellini, O. Gambino, and G. Cetini, *Inorg. Chim. Acta*, **6**, 205 (1972); G. Cetini, P. L. Stanghellini, R. Rossetti, and O. Gambino, *Inorg. Chim. Acta*, **2**, 433 (1968).

85. G. Cetini, P. L. Stanghellini, R. Rossetti, and O. Gambino, *J. Organometal. Chem.*, **15**, 373 (1968).

86. R. Rossetti and P. L. Stanghellini, *J. Coord. Chem.*, **3**, 217 (1974).

87. R. Rossetti, G. Gervasio, and P. L. Stanghellini, *J. Chem. Soc., Dalton Trans.*, 222 (1978).

88. E. M. Thorsteinson and F. Basolo, *J. Am. Chem. Soc.*, **88**, 3929 (1966).

89. G. Huttner, J. Schneider, H. D. Müller, G. Mohr, J. von Seyerl, and L. Wohlfahrt, *Angew. Chem. Int. Ed. Engl.*, **18**, 76 (1979).

90. O. Gambino, G. A. Vaglio, R. P. Ferrari, and G. Cetini, *J. Organometal. Chem.*, **30**, 381 (1971); R. P. Ferrari, G. A. Vaglio, O. Gambino, M. Valle, and G. Cetini, *J. Chem. Soc., Dalton Trans.*, 1998 (1972).

91. F. Ungvary and L. Marko, *J. Organometal. Chem.*, **71**, 283 (1974).

92. G. Bor, U. K. Dietler, P. Pino, and A. J. Poë, *J. Organometal. Chem.*, **154**, 301 (1978).

93. G. Bor and U. K. Dietler, *J. Organometal. Chem.*, **191**, 295 (1980).

94. D. J. Darensbourg, B. S. Peterson, and R. E. Schmidt, *Organometallics*, **1**, 306 (1982).

95. D. J. Darensbourg and M. J. Incorvia, *Inorg. Chem.*, **19**, 2585 (1980).

96. Ref. 9(b), p. 158.

97. C. H. Bamford, G. C. Eastmond, and W. R. Maltman, *Trans. Faraday Soc.*, **60**, 1432 (1964).

98. D. J. Darensbourg and M. J. Incorvia, *J. Organometal. Chem.*, **171**, 89 (1979).

99. R. Huq and A. J. Poë, *J. Organometal. Chem.*, **226**, 277 (1982).

100. F. Oldani and G. Bor, *J. Organometal. Chem.*, **246**, 309 (1983).

101. B. T. Heaton, L. Longhetti, D. M. P. Mingos, C. E. Briant, P. C. Minshall, B. R. C. Theobald, L. Garleschelli, and U. Sartorelli, *J. Organometal. Chem.*, **213**, 333 (1981).

102. P. Chini and B. T. Heaton, *Topics Curr. Chem.*, **71**, 1 (1977).

103. (a) B. L. Booth, M. J. Else, R. Fields, and R. N. Haszeldine, *J. Organometal. Chem.*, **27**, 119 (1971); (b) R. H. Whyman, *J. Chem. Soc., Dalton Trans.*, 1375 (1972).

104. D. Labroue, R. Queau, and R. Poilblanc, *J. Organometal. Chem.*, **233**, 359 (1982).

105. D. Labroue, R. Queau, and R. Poilblanc, *J. Organometal. Chem.*, **186**, 101 (1980).

106. A. J. Drakesmith and R. Whyman, *J. Chem. Soc., Dalton Trans.*, 363 (1973).

107. K. J. Karel and J. R. Norton, *J. Am. Chem. Soc.*, **96**, 6812 (1974).

108. D. C. Sonnenberger and J. D. Atwood, *Inorg. Chem.*, **20**, 3243 (1981).

109. G. F. Stuntz and J. R. Shapley, *J. Organometal. Chem.*, **213**, 389 (1981).

110. D. C. Sonnenberger and J. D. Atwood, *J. Am. Chem. Soc.*, **104**, 2113 (1982).

111. D. C. Sonnenberger and J. D. Atwood, *Organometallics*, **1**, 694 (1982).

112. D. J. Darensbourg and B. J. Baldwin-Zuschke, *J. Am. Chem. Soc.*, **104**, 3906 (1982).

113. C. A. Tolman, *J. Am. Chem. Soc.*, **92**, 2956 (1970); C. A. Tolman, W. C. Seidel, and L. Gosser, *ibid.*, **96**, 53 (1974); C. A. Tolman, *Chem. Rev.*, **77**, 313 (1977); Ref. 114, ftn. 8.

114. R. A. Jackson and A. J. Poë, *Inorg. Chem.*, **18**, 3331 (1979).

115. A. Davison, W. McFarlane, L. Pratt, and G. W. Wilkinson, *J. Chem. Soc.*, 3653 (1962).

116. F. Basolo, A. T. Brault, and A. J. Poë, *J. Chem. Soc.*, 676 (1964); K. Noak and M. Ruch, *J. Organometal. Chem.*, **17**, 309 (1969).

117. J. Knight and M. J. Mays, *J. Chem. Soc., Chem. Commun.*, 384 (1969); A. J. Deeming, B. F. G. Johnson, and J. Lewis, *J. Organometal. Chem.*, **17**, P40 (1969).

118. J. Knight and M. J. Mays, *J. Chem. Soc. A*, 711 (1970); M. J. Mays and R. N. F. Simpson, *J. Chem. Soc. A.*, 1444 (1968).

119. C. G. Cooke and M. J. Mays, *J. Organometal. Chem.*, **74**, 449 (1974).

120. H. W. Walker, R. G. Pearson, and P. C. Ford, *J. Am. Chem. Soc.*, **104**, 1255 (1982).

121. S. A. R. Knox, J. W. Koepke, M. A. Andrews, and H. D. Kaesz, *J. Am. Chem. Soc.*, **97**, 3942 (1975).

122. M. R. Churchill, F. J. Hollander, J. R. Shapley, and D. S. Foose, *J. Chem. Soc., Chem. Commun.*, 534 (1978).

123. W. L. Gladfelter, G. L. Geoffroy, and J. Calabrese, *Inorg. Chem.*, **19**, 2569 (1980).

124. E. Roland and H. Vahrenkamp, *Organometallics*, **2**, 183 (1983).

125. J. B. Keister, *J. Chem. Soc., Chem. Commun.*, 214 (1979).

126. J. B. Keister, M. W. Payne, and M. J. Muscatella, *Organometallics*, **2**, 219 (1983).

127. L. M. Bavaro, P. Montangero, and J. B. Keister, *J. Am. Chem. Soc.*, **105**, 4977 (1983).

128. J. R. Fox, W. L. Gladfelter, and G. L. Geoffroy, *Inorg. Chem.*, **19**, 2574 (1980).

129. J. R. Fox, W. L. Gladfelter, T. G. Wood, J. A. Smegal, T. K. Foreman, G. L. Geoffroy, I. Tavanaiepour, and V. W. Day, *Inorg. Chem.*, **20**, 3214 (1981).

130. W. L. Gladfelter, J. R. Fox, J. A. Smegal, T. G. Wood, and G. L. Geoffroy, *Inorg. Chem.*, **20**, 3223 (1981).

131. J. R. Fox, W. L. Gladfelter, G. L. Geoffroy, I. Tavanaiepour, S. Abdel-Mequid, and V. W. Day, *Inorg. Chem.*, **20**, 3230 (1981).

132. R. Rossetti and P. R. Stanghellini, *Inorg. Chim. Acta*, **70**, 121 (1983).

133. J. B. Keister, *J. Organometal. Chem.*, **190**, C36 (1980).

134. Y. Doi, K. Koshizuka, and T. Keii, *Inorg. Chem.*, **21**, 2732 (1982).

135. R. A. Levenson, H. B. Gray, and G. P. Ceasar, *J. Am. Chem. Soc.*, **92**, 3653 (1970).

136. Ref. 11, pp. 82–87, 136–141.

137. A. F. Hepp and M. S. Wrighton, *J. Am. Chem. Soc.*, **105**, 5934 (1983).

138. H. Yesaka, T. Kobayashi, K. Yasufuku, and S. Nagakura, *J. Am. Chem. Soc.*, **105**, 6249 (1983).

139. M. S. Wrighton and D. S. Ginley, *J. Am. Chem. Soc.*, **97**, 2065 (1975).

140. D. R. Tyler, M. A. Schmidt, and H. B. Gray, *J. Am. Chem. Soc.*, **105**, 6018 (1983); *ibid.*, **101**, 2753 (1979).

141. R. H. Hooker, K. A. Mahmoud, and A. J. Rest, *J. Chem. Soc., Chem. Commun.*, 1022 (1983).

142. A. E. Stiegman, M. Stieglitz, and D. R. Tyler, *J. Am. Chem. Soc.*, **105**, 6032 (1983).

143. A. Fox and A. J. Poë, *J. Am. Chem. Soc.*, **102**, 2497 (1980).

144. D. R. Tyler, R. A. Levenson, and H. B. Gray, *J. Am. Chem. Soc.*, **100**, 7888 (1978).

145. Ref. 11, p. 152.

146. R. G. Austin, R. S. Paonessa, P. J. Giordano, and M. S. Wrighton, *Adv. Chem. Ser.*, **168**, 189 (1978).

147. F.-W. Grevels, J. G. A. Reuvers, and J. Takats, *J. Am. Chem. Soc.*, **103**, 4069 (1981).

148. J. Malito, S. Markiewicz, and A. J. Poë, *Inorg. Chem.*, **21**, 4335 (1982).

149. M. F. Desrosiers and P. C. Ford, *Organometallics*, **1**, 1715 (1982).

150. D. R. Tyler, M. Altobelli, and H. B. Gray, *J. Am. Chem. Soc.*, **102**, 3022 (1980).

151. M. R. Burke, J. Takats, F.-W. Grevels, and J. G. A. Reuvers, *J. Am. Chem. Soc.*, **105**, 4092 (1983).

152. V. C. Sekhar and A. J. Poë, unpublished observations.

153. J. L. Graff, R. D. Sanner, and M. S. Wrighton, *J. Am. Chem. Soc.*, **101**, 273 (1979).

154. R. A. Epstein, T. R. Gaffney, G. L. Geoffroy, W. L. Gladfelter, and R. S. Henderson, *J. Am. Chem. Soc.*, **101**, 3847 (1979).

155. G. L. Geoffroy and R. A. Epstein, *Inorg. Chem.*, **16**, 2795 (1977); *Adv. Chem. Ser.*, **168**, 132 (1978).

156. H. C. Foley and G. L. Geoffroy, *J. Am. Chem. Soc.*, **103**, 7176 (1981).

157. J. L. Graff and M. S. Wrighton, *Inorg. Chim. Acta*, **63**, 63 (1982).

158. C. R. Bock and M. S. Wrighton, *Inorg. Chem.*, **16**, 1309 (1977).

159. M. S. Wrighton and J. L. Graff, *J. Am. Chem. Soc.*, **103**, 2225 (1981).

160. Ref. 159, ftn. 13.

161. M. Kucharska-Zon and A. J. Poë, unpublished observations.

162. K. Nomiya and H. Suzuki, *J. Organometal. Chem.*, **168**, 115 (1979).

163. P. Chini, G. Longoni, and V. G. Albano, *Adv. Organometal. Chem.*, **14**, 285 (1976); R. Mutin, W. Abboud, J. M. Basset, and D. Sinou, *Polyhedron*, **2**, 539 (1983); A. Ceriotti, G. Ciani, L. Garlaschelli, U. Sartorelli, and A. Sironi, *J. Organometal. Chem.*, **229**, C9 (1982).

164. J. L. Vidal and W. E. Walker, *Inorg. Chem.*, **19**, 896 (1980).

165. G. R. John, B. F. G. Johnson, J. Lewis, and A. L. Mann, *J. Organometal. Chem.*, **171**, C9 (1979).

166. M. O. Albers and N. J. Coville, *Coord. Chem. Rev.*, **53**, 227 (1984).

5

ORGANOMETALLIC CLUSTER CHEMISTRY

JOHN S. BRADLEY
Corporate Research Laboratories
Exxon Research and Engineering Co.
Annandale, New Jersey

5.1 INTRODUCTION

The organometallic chemisry of transition-metal cluster compounds occupies a tan-talizingly promising position in the area of overlap between organic chemistry and surface chemistry. The attraction of this hybrid science lies in the ostensible simi-larities between the accepted (or postulated) notions of the chemisorbed state of organic molecules on metal sufaces and the multiply coordinated state of organic fragments covalently bound to the faces of the metal atom polyhedra that make up the core of cluster molecules. This analogy has been widely cited by researchers in both specialities(1,2). The corollary of organocluster chemistry mimicking surface chemistry is as yet undemonstrated for any but the simplest of reactions (e.g., ligand association and chemisorption), but the impetus provided by these provocative com-parisons has had an undoubtedly stimulating effect on the study of organic chemistry of cluster molecules.

The purpose of this chapter is to present examples of the reactions that interconnect various cluster-bound organic molecules (or molecular fragments). Limitations of space prevent the inclusion of the organometallic chemistry of dinuclear metal complexes, which themselves exhibit reactivity markedly different from that found in mononuclear complexes. Most of the examples chosen exemplify the reactions of organic species bound to two or more atoms in trinuclear and tetranuclear clusters. I have excluded such processes as ligand exchange or reactions occurring at single metal atoms in cluster molecules.

A survey of cluster structures appears elsewhere in this volume, and no attempt will be made here to give more than occasional examples of the types of coordination geometry found for small organic fragments coordinated to the edges and faces of cluster cores. These examples will arise in the discussion of organometallic reactivity which is the principal goal of this chapter.

The molecules to be considered are those in which one or more atoms of an or-ganic ligand are bound to two or more metal atoms of an array of three or more such atoms in the cluster core. The accepted structural designation for the molecules with which we are concerned is μ_m-η^n which connotates a bonding geometry where the organic species is bound via n of its atoms to m of the metal atoms in the cluster core.

It will be clear to anyone who has perused even quite narrow areas of the literature on organometallic clusters that the field is a huge one. I have restricted my coverage to examples of reactions in which the basic units of organic chemistry—C—H, C—C, and C—O bonds—in organic fragments or molecules are formed and broken under the influence of a transition-metal cluster.

5.2 CARBON—HYDROGEN BOND CHEMISTRY

Examples of reactions in which C—H bonds are made and broken are numerous in cluster chemistry, and many of them predate the analogous mononuclear reactions, o-metallation, α-elimination, and C—H activation, which have become such an important branch of organometallic chemistry. Quite clearly, any reaction in which

C—H bonds are labilized is of fundamental importance in organometallic chemistry and by implication in catalysis. It is then quite appropriate that it was just this aspect of the chemistry of clusters that projected triosmium clusters into the preeminent position that they hold in this field.

It was discovered several years ago that C_1 hydrocarbyl fragments could be introduced into the coordination sphere of triosmium clusters by reaction of diazomethane with $H_2Os_3(CO)_{10}$, **5.1** (Eq. 5.1) (3).

$$H_2Os_3(CO)_{10} + CH_2N_2 \xrightarrow{25°} H_2Os_3(CO)_{10}(\mu\text{-}CH_2) \qquad (5.1)$$
$$\textbf{5.1} \qquad\qquad\qquad\qquad\qquad\qquad \textbf{5.2}$$

The methylene group in **5.2** is located in a bridging position on an Os—Os bond, by both X-ray and neutron diffraction studies (4,5). The latter also established that the hydrogen atoms are situated as shown schematically in Scheme 5.1. In solution

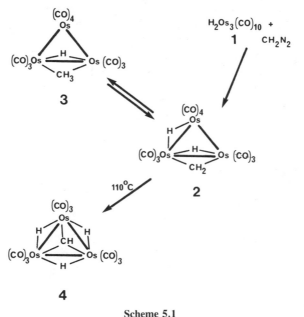

Scheme 5.1

an equilibrium is established between **5.2** and the isomeric hydridomethyl complex **5.3** by the migration of a hydrogen atom between the Os—Os edge of the cluster

$$H_2Os_3(CO)_{10}(\mu_2\text{-}CH_2) \rightleftharpoons HOs_3(CO)_{10}(\mu_2\text{-}CH_3) \qquad (5.2)$$
$$\textbf{5.2} \qquad\qquad\qquad\qquad \textbf{5.3}$$

and the bridging methylene group. The structure of **5.3** was inferred from spectroscopic data, with the methyl group being assigned a $(\mu_2\text{-}\eta^2)$-bridging position, a relatively unusual coordination mode for this ligand (6). The interaction between the C—H σ bond and an osmium atom (an example of an agostic interaction) has analogs in many other transition metal-hydrocarbyl compounds, and the lability of this bond in **5.3** is confirmation of the importance of such interactions in C—H activation. Thus, the triosmium framework provides a suitable site for the making and breaking of C—H bonds. Indeed, **5.2** is itself susceptible to C—H bond cleavage if CO loss is induced in the cluster to provide the necessary unsaturation at the reaction site. Heating **5.2** at 110°C in toluene under a nitrogen purge for 24 hours results in the migration of a second hydrogen atom from the methylene group to the hitherto unbridged Os—Os bond, yielding $H_3Os_3(CO)_9(\mu_3\text{-}CH)$, **5.4**. The μ_3-CH group now bridges all three osmium atoms.

These observations are of obvious significance when we consider the possible mechanisms of C—H bond making and breaking on catalytically active metal surfaces, as found in CO methanation, hydrocarbon hydrogenolysis and other heterogeneously catalyzed processes.

Perhaps the most fundamental example of C—H formation in cluster chemistry comes from the reactions of the μ_4-carbido tetrairon clusters (7). These molecules contain a single carbon atom bound only to four iron atoms arranged in an open butterfly configuration, as in $Fe_4(CO)_{12}(\mu\text{-}CO)C$, **5.5**, (8) and $[Fe_4(CO)_{12}C]^{2-}$, **5.6** (9). The unique carbon atom is susceptible to C—H bond formation from either **5.5** by heating under hydrogen or from **5.6** by protonation or oxidation under hydrogen to yield $HFe_4(CO)_{12}(CH)$, **5.7** (Scheme 5.2). The methylidyne group in **5.7** takes part in a three center [FeHC] interaction (10).

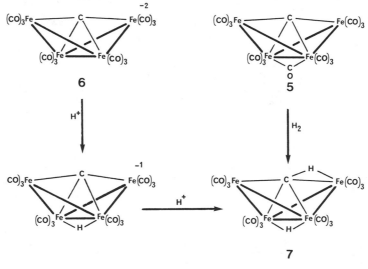

Scheme 5.2

A large number of reactions is known involving hydrogen transfer between organic molecules and trinuclear clusters. Briefly summarized, several coordinatively unsaturated clusters, produced by ligand dissociation from saturated molecules such as $Os_3(CO)_{12}$ and $Ru_3(CO)_{12}$, are reactive toward olefinic, acetylenic, and aromatic C—H bonds. The products from these reactions generally contain the dehydrogenated organic molecule bound to the cluster in a μ_2- or μ_3-bridging mode, the several metal atoms providing adjacent reaction sites (the distinguishing feature of cluster chemistry). Some of these reactions are summarized in Scheme 5.3 which encompasses the reactions of monoolefins with $M_3(CO)_{12}$ (M = Os and/or Ru) and

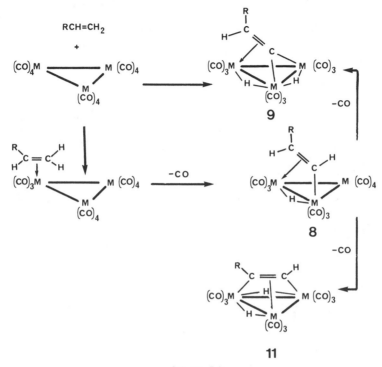

Scheme 5.3

its derivatives. The reacting olefin coordinates to a vacant site on the triatomic core of the cluster, and, held in proximity to the metal atoms, the C—H bonds are susceptible to cleavage to produce the μ_2-alkenyl cluster, **5.8**. The formation of **5.9** (M = Os) directly from $Os_3(CO)_{12}$ and olefin (11) occurs only at temperatures that are sufficiently high so that the intermediates are not detected, but **5.8** may be isolated by alternative low-temperature reactions, which involve C—H bond *formation* in one of two ways (Scheme 5.4).

The trinuclear dihydride $H_2Os_3(CO)_{10}$, **5.1**, provides the hydrogen atoms needed to form **5.8** from an acetylene at room temperature—in this case, hydrogen migration occurs from an Os—Os bond to the organic ligand (12). Similarly, the hydrogen atoms in **5.1** can be transferred to an olefin molecule, eliminating alkane and generating a vacant site on the cluster under relatively mild conditions. This allows the isolation of **5.8** by reaction with a second molecule of olefin (12,13).

On heating **5.8** (*M* = Os) to 125°C hydrogen transfer again occurs via initial loss of CO from the cluster. The vacant site thus generated activates the α-C—H bond, and the vinyl group in **5.8** is transformed into the vinylidene group in **5.9**.

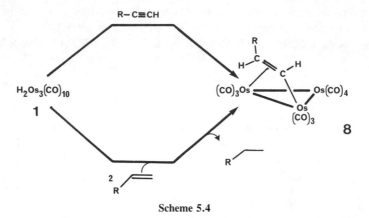

Scheme 5.4

Intermediates have been isolated in the reaction between **5.1** and diethyl fumarate to yield the hydridoalkyl **5.10** which slowly decomposes to the corresponding alkane (14). This reaction forms the basis for a catalytic cycle for the hydrogenation of alkenes by **5.10**, one of very few proven examples of cluster catalysis (Scheme 5.5).

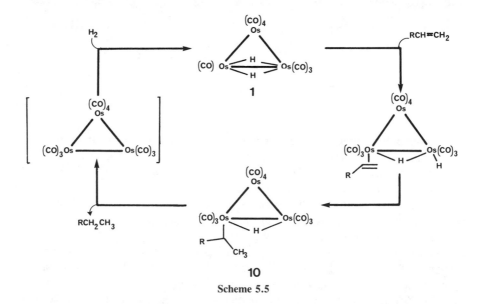

Scheme 5.5

An alternative isomer to **5.9**, **5.11**, is also formed in the reaction between $Os_3(CO)_{12}$ and propene and internal olefins, but not with ethylene, which does give both isomers with $Ru_3(CO)_{12}$, (15) (Scheme 5.3). The latter isomer is, of course, the only one available for cyclic olefins, and also for benzene which reacts with $Os_3(CO)_{12}$ via migration of two *ortho* hydrogens to the Os_3 core, to give a $(\mu_3-\eta^2)$-benzyne cluster **5.12** (11). The synthesis of **5.12** in high yield is also achieved by reaction of $Os_3(CO)_{10}(MeCN)_2$ with benzene (16).

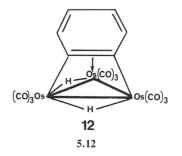

12

5.12

Analogous reactions have been observed for the tetranuclear analogs $H_4Os_4(CO)_{12}$, **5.13**, with the added variation provided by the possibility of alternative geometries for the Os_4 core—a closed tetrahedron or an open butterfly. The number of skeletal bonding electron pairs required for each of these alternatives is six and seven, and so the tetrahedron can open to a butterfly by the addition of a two-electron donor to the cluster. The site geometry for a cluster-bound organic species can thus be radically altered. In the domain of C—H chemistry this is nicely illustrated by the reactions of the tetrahedral cluster $H_4Os_4(CO)_{12}$, **5.13**, with alkenes (17,18) (Scheme 5.6).

The reaction of **5.13** with a variety of alkenes is similar to that of $Os_3(CO)_{12}$, in that a μ_2-alkenyl complex $H_3Os_4(CO)_{11}(C_2H_2R)$, **5.14**, is formed, via migration of one vinylic hydrogen to the cluster core. Treating **5.14** with CO induces the transfer of a hydrogen atom from the Os_4 core to the alkenyl ligand to give **5.15**. A second C—H bond is ruptured on heating **5.14** to give the alken-1,2-diyl complex, $H_2Os_4(CO)_{11}(C_2HR)$, **5.16**, in which the organic group is bound to a triangular face of the tetrahedral Os_4 core, in a manner similar to that found in **5.9**. Under moderate pressures of carbon monoxide, one Os—Os bond in **5.16** is broken, yielding the butterfly cluster $Os_4(CO)_{12}(C_2HR)$, **5.17**, in which the organic fragment is now bonded to all four of the osmium atoms. Similar M_4C_2 cores are found for ruthenium and cobalt clusters.

C—H cleavage reactions have also been observed extensively in the reactions of alkynes with trinuclear clusters. Typically, a terminal alkyne reacts via acetylenic C—H cleavage to give products of the structure **5.18** in which the acetylenic C_2 fragment is σ-bonded to one metal atom and π-bonded to the other two (19).

Internal alkynes can undergo C—H cleavage at a methylene group adjacent to the C≡C bond. For example, 2-pentyne reacts with $Ru_3(CO)_{12}$, yielding **5.19** and **5.20**. In the former a dimetallo-allyl group π bonds to the third ruthenium atom (20); in the latter the organic ligand is an allenyl group σ-bonded to one ruthenium atom and π-bonded to the other two (an arrangement that requires considerable

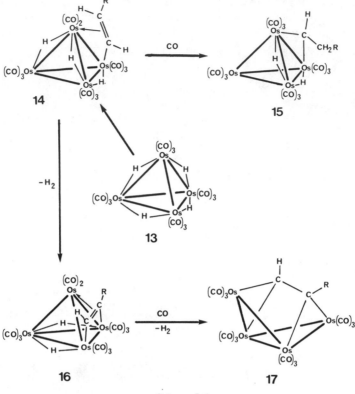

Scheme 5.6

bending of the allenyl group) (21). In each case C—H cleavage and hydrogen migration have occurred. Similar structures result from reactions of olefins with allylic C—H bonds.

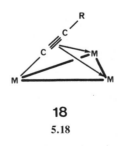

18

5.18

C—H bond chemistry is also found in the cluster chemistry of other unsaturated organic molecules. In a particularly apposite example of catalysis-related organo-cluster chemistry it has been shown that the stepwise transfer of hydrogen atoms to a cluster-coordinated nitrile occurs (22). [HFe$_3$(CO)$_{11}$]$^-$ reacts with acetonitrile to give the μ^3-C-iminyl cluster [Fe$_3$(CO)$_9$(CH$_3$=NH)]$^-$, **5.21**, which on protonation yields the neutral analog **5.22**. Thermal dehydrogenation results in isomerization of the cluster by hydrogen transfer from nitrogen to carbon, forming the N-iminyl

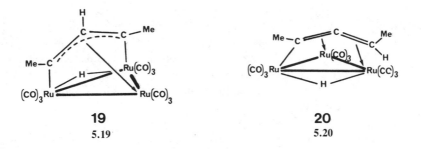

19

5.19

20

5.20

isomer **5.23**, which smoothly adds hydrogen to give $H_2Fe_3(CO)_9(\mu_3\text{-}NCH_2CH_3)$, **5.24**. In this sequence of reactions the C≡N triple bond has been completely reduced (Scheme 5.7). The similarity to the chemistry of hydrocarbyl analogs such as **5.9** and **5.10** is apparent.

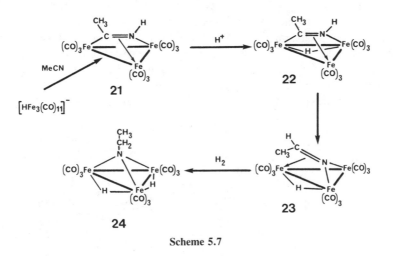

Scheme 5.7

The making and breaking of C—H bonds in organic molecules coordinated to clusters is a common phenomenon, presumably for reasons similar to the ubiquity of *o*-metallation in mononuclear triarylphosphine complexes. The primary coordination of the organic ligand to the metal atom or cluster results in neighboring C—H groups being oriented toward reactive sites, either an adjacent coordination site on the same metal, as in *o*-metallation, or on an adjacent metal atom on the edge or face of a cluster. The recent exciting advances in the intermolecular activation of aliphatic C—H bonds by mononuclear organometallics has as yet no counterpart in cluster chemistry, but it takes little more than the usual faith in the perspicacity of organometallic chemists to feel confident that this gap in the repertoire of cluster chemistry will be filled before long.

5.3 CARBON—CARBON BOND CHEMISTRY

5.3.1 Hydrocarbyl Chemistry

The chemistry of C—H bonds is obviously of great significance in the organic chemistry of clusters, but possibly of even greater importance in this field are those reactions in which C—C bonds are formed or disrupted. The delineation of such transformations is crucial in our attempts to relate cluster chemistry to the chemistry of chemisorbed organic molecules on metal surfaces, with all the ramifications for catalysis such a relationship implies.

Clusters have proved to be adept at the coupling of alkynes and the array of alkyne complexes of polynuclear carbonyls is huge. The organic ligands in these molecules run the gamut from π-bound alkynes to extensively coupled oligomers, coordinated to several metal atoms. One example will serve to illustrate this aspect of cluster chemistry, and once again it is drawn from the rich organometallic chemistry of the triosmium family.

As described previously, terminal alkynes characteristically react with triosmium dodecacarbonyl (and similar clusters) by undergoing cleavage of the terminal C—H bond. However, with internal acetylenes this option is not available, and so diphenylacetylene, for example, reacts with $Os_3(CO)_{12}$ by a different route, two alkyne molecules coupling across one edge of the Os_3 triangle to form an osmiacyclopentadiene ring π-bonded to the adjacent osmium atom in **5.25** (Scheme 5.8). (23,24) In **5.25** the μ_2-η^4 cyclic organic ligand is bound on one Os—Os edge in such a way so as to facilitate o-C—H cleavage in one of the phenyl rings by the third

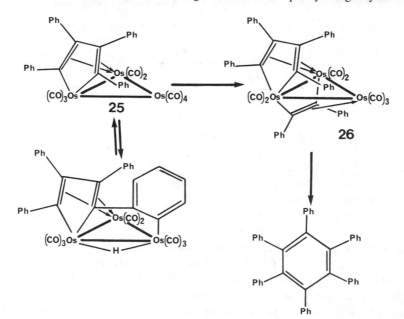

Scheme 5.8

osmium atom. This side reaction, which involves loss of CO by the metallating $Os(CO)_3$ group, is reversible. Addition of a third diphenylacetylene to **5.25**, with loss of two CO molecules, occurs in the familiar fashion, with the acetylene donating four electrons to the cluster, **5.26**. Pyrolysis of **5.26** leads to the elimination of hexaphenylbenzene.

Alkynes can also undergo C—C forming reactions with other carbon containing ligands. Carbon monoxide is incorporated into an osmiacyclohexadienone ring in the reaction of $H_2Os_3(CO)_{10}$ with but-2-yne to form $Os_3(CO)_9(C_4(CH_3)_4CO)$, **5.27** (25).

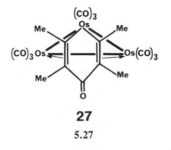

27

5.27

An interesting group of reactions is found in the coupling of C_1 hydrocarbyl fragments with alkynes. Both alkylidene and alkylidyne ligands have been observed to undergo C—C bond-forming reactions with alkynes on M_3 clusters. The reaction of diazomethane with the coordinatively unsaturated μ_3,η^2-diphenyl acetylene complex **5.28** results in the formation of the μ_2-methylene-μ_3,η^2-acetylene cluster, **5.29**, via an isolable diazomethane adduct of **5.28**. Heating **5.29** in xylene to 135°C induces coupling between the methylene and diphenylacetylene ligands. The coordinatively unsaturated intermediate, **5.30**, thus produced, undergoes C—H oxidative addition at the methylene group to form the 1,3-dimetalloallyl cluster **5.31** (Scheme 5.9). In this reaction the C—C bond-forming step (**5.29** → **5.30**) was identified as the rate-determining step (26).

A coupling reaction between μ_3-alkylidynes and alkynes has been observed on triruthenium clusters. The clusters $H_3Ru_3(\mu_3\text{-}CX)(CO)_9$, **5.32**, ($X$ = OMe, Me, Ph) react with alkynes HC_2R (R = Ph, n-Bu) to form 1,3-dimetalloallyl clusters, **5.33**, in two isomeric forms (a) and (b) (Scheme 5.10) (27). When R = ^{t}Bu only isomer (b) is formed. Hydrogenation of **5.33** under mild conditions (1–4 atm, 80–90°C) yields the μ_3-alkylidynes **5.34** in which the original alkylidyne group has been extended by the alkyne C_2 group. As we saw earlier in an analogous reaction of $H_2Os_3(CO)_{10}$ with alkenes, the cluster **5.32** reacts first with one molecule of alkyne, which is hydrogenated to the corresponding alkene. The cluster, which is now coordinatively unsaturated, reacts with a second mole of alkyne to form $HRu_3(\mu_3\text{-}CX)$ (alkyne)$(CO)_9$, which then undergoes C—C bond formation between the μ_3-C-X and the alkyne, to form **5.33**.

A facile coupling reaction between alkylidene groups on a trinuclear cluster has been achieved (Scheme 5.11). Two alkylidene groups can be introduced sequentially

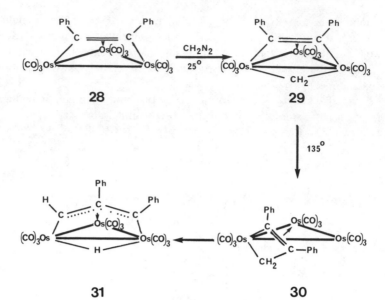

28 → (CH₂N₂, 25°) **29**

31 ← **30**

Scheme 5.9

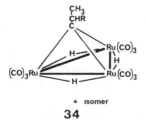

32

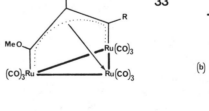

33 (a)

(b)

34

+ isomer

Scheme 5.10

onto an Os_3 triangle by treatment of $Os_3(CO)_{12}$ with trimethylamine oxide in acetonitrile to give $Os_3(CO)_{11}(MeCN)$, which reacts with diazomethane to give $Os_3(CO)_{10}$
$(\mu_2CO)(\mu_2CH_2)$, **5.35**. Repeating this procedure with more CH_2N_2 yields $HOs_3(CO)_{10}$
$(CH = CH_2)$, **5.36** (cf. **5.8**) by C=C bond formation followed by C—H cleavage.
Thus, the Os_3 cluster mimics in this respect the ability of metal surfaces to couple
CH_2 fragments to a C_2 species (28).

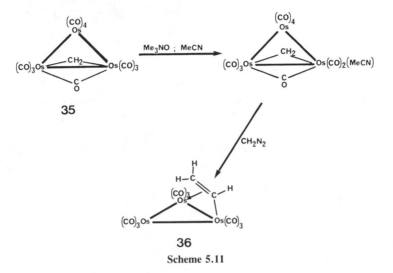

35

36

Scheme 5.11

Two of the previous examples involve the coupling of alkylidene and alkylidyne
fragments with alkynes. These C_1 units can themselves be cleaved from organic
molecules by clusters, reactions that involve the breaking of C≡C triple bonds and
C=C double bonds, respectively. The tetranuclear cluster $CpWOs_3(CO)_{12}H$ reacts
with alkynes, *RCCR'*, on treatment with trimethylamine oxide (to remove CO from
the cluster), forming $CpWOs_3(CO)_{10}(RCCR')$ **5.37**. Further decarbonylation with
Me_3NO, to produce an open coordination site on the cluster, followed by mild
pyrolysis in refluxing toluene, results in C≡C bond cleavage and the formation of
the bis-alkylidyne cluster $CpWOs_3(CO)_9(RC)(R'C)H$, **5.38** (29). This is a striking
example of organic reactivity on a cluster face.

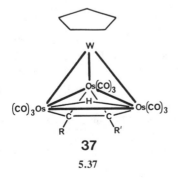

37

5.37

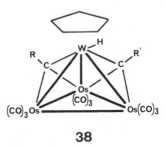

38

5.38

The cleavage of a C=C double bond on a cluster is exemplified by the reaction of ketene with $Os_3(CO)_{10}(MeCN)_2$ (30). Although on the basis of the alkene C—H bond chemistry described earlier we might expect the formation of a μ_3,η^2-ketenylidene cluster, the reaction takes a different course. The C=C bond in ketene is broken with the formation of a μ_2-CH$_2$ ligand in $Os_3(CO)_{11}(CH_2)$ **5.35** (Eq. 3).

$$Os_3(CO)_{12} \xrightarrow[\text{MeCN}]{\text{Me}_3\text{NO}} Os_3(CO)_{10}(MeCN)_2 \xrightarrow{\text{CH}_2\text{CO}} Os_3(CO)_{11}(CH_2) \qquad (5.3)$$

This cluster has been prepared by other routes and will recur in the discussion of C—C bond-forming reactions.

Single C—C bonds are also susceptible to cleavage on clusters. The propensity for cleavage of C—H bonds adjacent to alkyne C≡C bonds was mentioned earlier in the formation of **5.19** and **5.20** from 2-pentyne. If the internal alkyne has no C—H bond α to the C≡C triple bond, C—C cleavage is a possibility, and this has been realized for the α-hydroxy alkynes $HOCR_2C≡CCR_2OH$ (31). For example, $HOCMePhC≡CCMePhOH$ reacts with $Os_3(CO)_{12}$ or with $Ru_3(CO)_{12}$ to give clusters of the type $M_3(CO)_9(\mu_3,\eta^2$-C$_2$CMePhOH)H, **5.39** plus acetophenone. These molecules have structures similar to those observed for the products of the reaction of terminal alkynes with $Ru_3(CO)_{12}$ as shown for **5.18**.

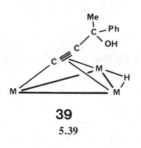

39

5.39

These three examples demonstrate the ability of cluster sites to facilitate the scission of C≡C, C=C, and C—C bonds.

5.3.2 Carbon Monoxide and Carbides in Carbon—Carbon Bond Chemistry

The chemistry of CO in metal cluster complexes has been the subject of much interest especially during the increased awareness of the importance of CO as a potential starting material for fuels and chemicals. Although the initial high hopes for cluster-based homogeneous catalysts for CO hydrogenation have not been realized to any practically useful extent, the intensity of research in this area has yielded much fascinating chemistry. In this section we explore some of the more fundamental aspects of CO-related cluster chemistry.

A sequence of reactions involving the ligands CH, CH$_2$, and CO, their interconversion and coupling reactions, has been identified on a triosmium cluster site.

The central molecule in this reaction manifold is the bridging methylene complex $Os_3(CO)_{11}(CH_2)$, **5.35** (32) prepared by protonation of $[Os_3(CO)_{11}(CHO)]^-$ or in the reaction of diazomethane with $Os_3(CO)_{11}(MeCN)$ (see preceding discussion). The cluster contains a μ_2-CH_2 group bridging the same edge of the Os_3 triangle as a μ_2-CO ligand.

Treatment of **5.35** with CO for several hours at room temperature results in the formal insertion of CO into one of the Os-CH_2 bonds, and the isolation of $Os_3(CO)_{12}(\mu_2,\eta^2$-$CH_2CO))$, **5.40** (Scheme 5.12) (33). The new cluster contains a ketene ligand bridging an open Os—Os edge of the Os_3 core. The ketene carbonyl group derives from coordinated CO in **5.35** and not from added CO. As expected for such an organic ligand, the coordinated ketene reacts with hydroxylic reagents such as water and methanol to give acetic acid and methyl acetate. There is evidence for a reversal of the CO insertion, that is, **5.40** $\rightarrow$ **5.35** on gentle heating.

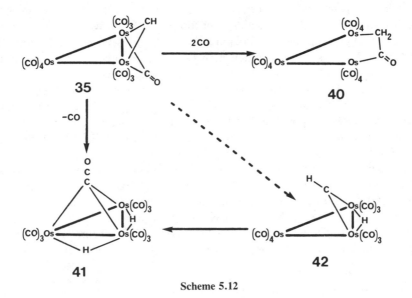

Scheme 5.12

Heating **5.35** in the absence of added CO induces a second C—C bond-forming reaction in quite an unexpected fashion. In a nitrogen-purged refluxing toluene solution **5.35** undergoes loss of CO, producing $H_2Os_3(CO)_9(\mu_3$-$CCO)$, **5.41**, in which the μ_3-ketenylidene group is bonded axially to the face of the Os_3 core (34). The formation of **5.41** from **5.35** may well proceed by sequential hydrogen atom migration from the μ_2-CH_2 group to the Os_3 core, in a manner analogous to the reaction **5.2** $\rightarrow$ **5.4**. Indeed, a possible intermediate in this process, $HOs_3(CO)_{10}(\mu CH)$, **5.42**, has been found to undergo a facile rearrangement to **5.41** under conditions much less stringent than those required for the reaction **5.35** $\rightarrow$ **5.41** (Scheme 5.12) (35). In going from **5.42** to **5.40** it is not clear whether or not C—H bond cleavage occurs prior to C—CO bond formation, to give a μ_3-carbide complex such as **5.43**. Alternatively, CO might insert into one Os—CH bond to give a μ_2-ketenyl cluster, **5.44**, analogous to the allenyl cluster **5.20**.

The C—C bond in the μ_3,η^1-ketenylidene ligand in **5.41** is readily cleaved. Treatment of **5.41** with hydrogen generates $H_3Os_3(CO)_9CH$, **5.4**, which implies the intermediate formation of **5.42**, but once again, it is not clear whether this reaction, like **5.35** → **5.40** and **5.42** → **5.43**, proceeds via a ketenyl or carbide intermediate.

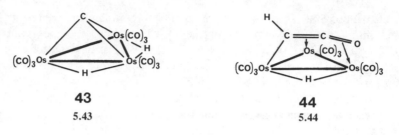

43

5.43

44

5.44

Analogous reactivity has been observed in a triiron system. Treatment of $[Fe_3(CO)_{10}(\mu_2\text{-COCH}_3)]^-$ with sodium benzophenone ketyl results in the removal of the methoxy group and the isolation of $[Fe_3(CO)_9(CCO)]^{2-}$, **5.45** (36). The structure of **5.45**, as yet undetermined, may well be similar to the isoelectronic osmium cluster **5.41** with an axial μ_3,η^1-ketenylidene ligand. The lability of the C—CO bond in **5.45** is clearly demonstrated by the incorporation of ^{13}CO into both the terminal carbonyl sites on the $Fe_3(CO)_9$ cluster core *and* the CCO carbonyl group. This raises the possibility of the transient existence of a μ_3-carbide intermediate, which may also be implicated in the protonation and alkylation reactions of **5.45** to $[Fe_3(CO)_{10}(CH)]^-$ and $[Fe_3(CO)_{10}(CCH_3)]^-$.

The migration of CO to a μ_3-carbide may also be involved in analogous syntheses of $[Co_3(CO)_9(CCO)]^+$ from $Co_3(CO)_9(\mu_3\text{-CBr})$ and of $[H_3Ru_3(CO)_9(CCO)]^+$ from $(H_3Ru_3(CO)_9(\mu_3\text{-COMe}))$, both by treatment with aluminum halides (37,38).

A ketenylidene intermediate **5.47** has also been proposed in the formation of $[Fe_4(CO)_{12}(\mu_4\text{-C(O)OCH}_3)]^-$, **5.46**, from $Fe_4(CO)_{13}C$, **5.5** (Scheme 5.13) (39). This reaction, which was the first example of reactivity at the carbon atom in a tetranuclear

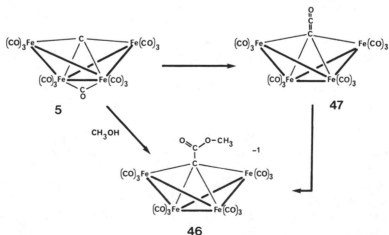

46

Scheme 5.13

carbide, and others in this family of clusters have been analyzed in some detail by molecular orbital calculations, the results of which allowed the recognition of both metal-centered and organic reaction pathways and the relationship between the two (40). This study revealed that the frontier orbitals in butterfly Fe_4C clusters such as **5.5** are metal centered and contain no contribution from the carbide carbon atomic orbitals. This explains well the metal-centered chemistry of this family of clusters such as protonation of $[Fe_4C(CO)_{12}]^{2-}$, **5.6**, to $[HFe_4C(CO)_{12}]^-$ (see Scheme 5.2), where protonation is controlled by the geometry of the HOMO (localized on the backbone of the butterfly), and such as the carbonylation of $[Fe_4C(CO)_{12}]^{2-}$ on oxidation, in which the added CO bridges the two backbone iron atoms. On the basis of these calculations, no carbon-centered chemistry would be expected since neither the energetically accessible empty orbitals (LUMO's) nor filled orbitals (HOMO's) contained any carbon character. However, it is well established that C—C bond-forming reactions will occur readily at the carbide carbon (e.g., Scheme 5.13). This apparent difference between theoretical prediction and experimental observation was reconciled by recognizing that in the reaction of **5.5** to **5.46** the Fe_4C cluster undergoes a change in core geometry, the wings of the butterfly opening from a dihedral of ca. 101 to ca. 130°. Removing the backbone CO from **5.5** and allowing the hypothetical intermediate $Fe_4C(CO)_{12}$ to distort to the more open geometry disrupts some of the iron-carbon orbital overlaps and imparts to the LUMO a significant carbon $2p_z$ character. This provides an energetically accessible route for C—C bond formation by addition of the CO to the carbide carbon, giving the ketenylidene **5.47**. Although this postulated intermediate has not been detected directly, its involvement in the formation of **5.46** from **5.5** and other analogous C—C bond-forming reactions (7) is supported by the existence and reactivity of $H_2Os_3(CO)_9CCO$, **5.41**, which gives the carbomethoxymethylidyne cluster $H_2Os_3(CO)_9$ $(\mu_3\text{-}CCO_2CH_3)$ on treatment with methanol (34).

It is interesting to note that attempts to alkylate directly the exposed carbide carbon in $[Fe_4C(CO)_{12}]^{2-}$, using methyl fluorosulfonate result in the closure of the butterfly core to a tetrahedron, the μ^4-carbide being methylated to a μ^3-ethylidyne in $[Fe_4(CO)_{12}(\mu^3\text{-}CCH_3)]^-$ (41). Indeed, there are no examples of C-derivitized Fe_4C clusters that have sp^3 carbons bound to the methylidyne carbon. In all the reported examples an sp^2 carbon is adjacent to the cluster-bound carbon atom, and this allows for a significant stabilization of the cluster by a π-overlap between a methylidyne carbon $2p$ orbital and the sp^2 carbon.

5.4   CARBON—OXYGEN BOND CHEMISTRY

5.4.1   Carbon—Oxygen Bond Breaking in Clusters

Carbon Monoxide Scission Reactions

Some of the significance of the chemistry described in the preceding section derives from the potential similarities between the C—C bond-forming reactions between small organic fragments on clusters and their counterparts on metal surfaces during heterogeneously catalyzed Fischer-Tropsch synthesis. This significance is enhanced when it is realized that the C_1 fragments that undergo these coupling reactions are often derived from CO (in the form of carbonyl ligands on the cluster) as they are during CO hydrogenation catalysis.

The disproportionation of CO to C and CO_2 has been observed in the synthesis of carbidocarbonyl clusters from metal carbonyl precursors (42). In this reaction the oxygen is lost as CO_2 and the carbon atom is encapsulated in a metal atom polyhedron, as in the syntheses of $[Fe_6C(CO)_{16}]^{2-}$ from $Fe(CO)_5$, and of $Ru_6C(CO)_{17}$ from $Ru_3(CO)_{12}$. Both of these hexanuclear clusters contain distorted octahedral M_6C cores, the carbon atom residing near the center of the octahedron. The origin of the carbide carbon as CO has been proved by the use of ^{13}CO in the starting materials.

The stepwise cleavage of coordinated CO under acid conditions has been elegantly demonstrated by Shriver and coworkers in the proton-induced reduction of the triply bridging carbonyl in $[Fe_4(CO)_{13}]^{2-}$, **5.48** (Scheme 5.14) (43,44). The first protonation of the tetrahedral dianion proceeds to give the butterfly cluster $[HFe_4(CO)_{13}]^-$, **5.49**. This molecule contains a unique bridging carbonyl, which may be protonated at the oxygen to form $HFe_4(CO)_{12}(COH)$, **5.50**, identified at low temperature by ^1H and ^{13}C nmr. A second protonation occurs at the hydroxyl oxygen, and after loss of water (concomitant with a sacrificial redox reaction generating Fe^{2+} and electrons) the carbide cluster $[HFe_4C(CO)_{12}]^-$, **5.51**, is formed, which further protonates to the methylidyne cluster **5.7**.

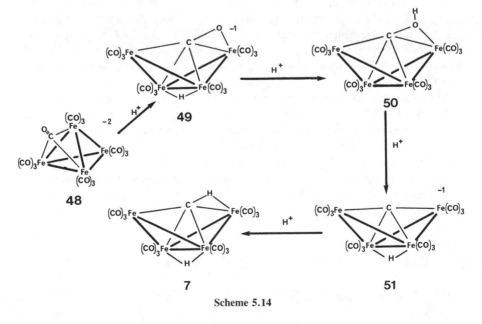

Scheme 5.14

Carbon—Oxygen Bond Cleavage in Organic Ligands

Another class of prototypical cluster-mediated reactions involves the breaking of C—O bonds in cluster-bound organic fragments. In an example of this reaction the isolation of intermediates has established that C—O cleavage occurs after C—C coupling between coordinated CO and a μ_3-methylidyne ligand (Scheme 5.15). Electrochemical reduction of $[Fe_3(CO)_9(\mu_3\text{-CO})(\mu_3\text{-CCH}_3)]^-$ **5.52** results in the formation of $[Fe_3(CO)_9(\mu_3,\eta^2\text{-CH}_3C\equiv CO)]^{2-}$, **5.53** (45), which has been characterized by X-ray diffraction (46). Coupling of the μ_3-CO and the μ_3-CCH$_3$ groups on the face of the cluster has produced μ_3,η^2-propynolate (CH$_3$C≡CO$^-$) anion bonded to the Fe$_3$ cluster in a manner reminiscent of alkyne cluster complexes. Reaction of **5.53** with proton sources results in C—O cleavage by loss of OH$^-$ to form the (μ_3-C≡CCH$_3$) cluster **5.54**. Thus coupling between CO and an alkylidyne renders the CO bond susceptible to cleavage by electrophiles.

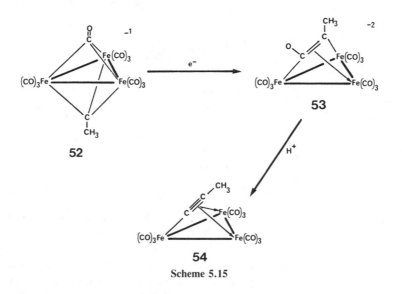

Scheme 5.15

Electrophillic attack on a cluster-bound acetyl can also result in C—O bond cleavage, and in this case both the C- and O-containing fragments remain bound to the cluster. The reaction of $[Fe_3(CO)_9(\mu_3\text{-COMe})]^-$, **5.56**, with methylfluorosulfonate at room temperature results in methylation of the acetyl oxygen and formation of $Fe_3(CO)_9(\mu_3\text{-CMe})(\mu_3\text{-OMe})$, **5.58** (47). Analogy with a similar reaction with HBF$_4$ suggests that **5.58** forms from **5.56** via a μ_3-MeCOMe complex **5.57** (Scheme 5.16) in which the C—O bond is considerably weakened and lengthened. In the analogous protonation of **5.56** a μ_3-MeCOH cluster has been observed spectroscopically, but subsequent protonation leads to loss of water presumably from the doubly protonated analog of **5.58**.

Cleavage of C—O bonds in cluster-bound organic species can also be induced by nucleophiles. As described earlier, treatment of $[Fe_3(CO)_{10}(\mu_2\text{COCH}_3)]^-$ with sodium benzophenone ketyl results in the formation of a reactive μ_3-carbide cluster by loss of methoxide. A similar reaction has also been observed for a tetranuclear case. Sodium benzophenone ketyl reacts readily with $[Fe_4(CO)_{12}(\mu\text{-COR}]^-$ to give the μ_4-carbide cluster $[Fe_4(CO)_{12}C]^{2-}$, **5.6** (36).

An example has also been reported of the cleavage of a cluster-bound acyl C—O bond unassisted by added electrophiles, and in which the oxygen atom is retained

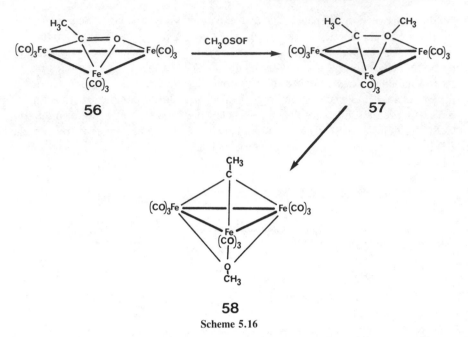

56   **57**

58

Scheme 5.16

by the cluster as an oxo ligand. The reaction of $CpW(CO)_2(CTol)$ with $H_2Os_3(CO)_{10}$
produces the tetranuclear acetyl cluster $CpWOs_3(CO)_{11}(\mu_3,\eta^2\text{-}C(O)CH_2Tol)$, **5.59**,
in which the triply bridging acyl ligand shows a marked elongation of the C—O
bond (1.37 Å). On refluxing in toluene, **5.59** undergoes a C—O scission to give
the oxo alkylidyne cluster $CpWOs_3(CO)_9(O)(\mu_3\text{-}CCH_2Tol)$, **5.60** (Scheme 5.17), in
which the oxo ligand bridges a W—Os bond (48).

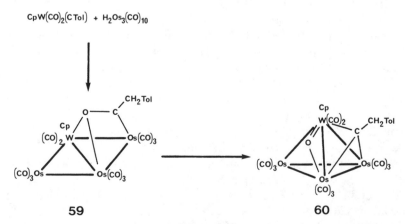

59   **60**

Scheme 5.17

5.5 CLUSTERS AND HOMOGENEOUS CATALYSIS

It is probably fair to say that the discovery of soluble molecular catalysts of practical utility has been the principal driving force in organometallic chemistry for the past 20 years. With mononuclear organometallics this goal has been achieved in many instances, and the hope of a similar role for cluster complexes has been the stated incentive for much of the cluster chemistry research performed in the last decade or so. This coincided with an intensification of research in CO hydrogenation catalysis, and the proven utility of transition-metal catalysts for this process, coupled with the notion that clusters in some way mimicked metal surfaces, gave to the search for cluster CO hydrogenation catalysts something of the intensity of the quest for the Holy Grail. Indeed, quite early on, the apparent activity of rhodium carbonyl clusters for ethylene glycol synthesis from CO/H_2 was reported by Union Carbide chemists, and this provided the sanction of potential industrial utility to the entire field of cluster chemistry (49).

Considerable momentum was given to research in this area by some dramatic reports of hydrocarbon synthesis from CO/H_2, catalyzed by organic solutions of simple cluster carbonyls, both from industrial laboratories (50) and academic research groups (51), and much was made of the requirement for multiple sites in CO hydrogenation catalysts. However, it was soon demonstrated that the observed activity was the result of thermal decomposition of the dissolved cluster to catalytically active metal under reaction conditions (52,53), and the focus shifted to oxygenate synthesis (alcohols, glycols, etc.) using cluster catalysts. The most prominent work in this area was that of Union Carbide mentioned earlier. Under quite severe conditions of temperature and pressure (e.g., 230°C, 1000 atm), solutions containing rhodium carbonyl clusters catalyzed the synthesis of ethylene glycol and methanol (plus higher polyhydric alcohols in smaller quantities) from synthesis gas. The clusters were detected during catalysis by in situ high-pressure infrared spectroscopy. However, it is not clear whether or not the catalytically active species was polynuclear or a mononuclear complex formed by fragmentation of the spectroscopically predominant clusters. Just such a fragmentation was shown to be operating in the catalytic synthesis of methanol and methyl formate from CO/H_2 in which ruthenium clusters of various sizes served as precursors for the actual mononuclear catalyst based on $Ru(CO)_5$ (53). In another case, solutions containing the ruthenium carbonyl cluster $[HRu_3(CO)_{11}]^-$ and iodide have been found to catalyze the conversion of CO/H_2 to organic oxygenates (54), but again, it is unclear whether the clusters actually catalyze one or more steps in the reaction in concert with mononuclear complexes, or simply serve as a reservoir of the mononuclear complexes that are the catalytic species. A similar cluster fragmentation process was involved in the erroneous assignment of cluster catalytic activity to polymer-supported $RCCo_3(CO)_9$ (55), which had been reported to catalyze the hydroformylation of olefins. In fact, under reaction conditions the cluster fragments to the extent of ~40% giving spectroscopically identifiable $Co_2(CO)_8$, the known precursor to $HCo(CO)_4$ in cobalt-catalyzed hydroformylation (56).

The preceding examples make clear two universal problems associated with any effort to identify soluble clusters as homogeneous catalysts:

1. Is the reaction catalyzed by molecular species in solution or by metallic pyrolysis products?
2. If homogeneously catalyzed, is the catalyst a multinuclear cluster or a mononuclear fragment?

The first question, which is common to all homogeneous catalysis with metal complexes, is relatively easy to answer (although if it remains unasked, problems will arise); the second is very difficult to answer unequivocally. A series of criteria for assigning the nuclearity of soluble catalysts has been proposed (57), but this will remain a contentious issue.

The number of reactions that seem to be well established as examples of homogeneous cluster catalysis is small (58), and it is unfortunately true to say that even if clusters are involved in some of the catalytic processes in which their presence has been detected, in no case has the catalyst performance been sufficiently superior, qualitatively or quantitatively, to mononuclear homogeneous catalysts to warrant high expectations for cluster catalysis.

There may be a good reason for this rather pessimistic outlook, and one inherent in that very aspect of cluster chemistry that gave rise to the initial burst of optimism. As any researcher can attest who has attempted to construct or synthesize a catalytic process from individual organometallic reactions, a transition-metal complex has a variety of options in its interactions with an organic fragment, and the vast majority of these alternatives lead to stable (i.e., catalytically useless) products. Ideally, the most energetically accessible pathway from any point on the reaction coordinate will involve only those transformations required for the catalytic cycle, including product dissociation from the catalyst molecule. Excursions from this path, leading to stable species, are anathema to the catalytic sequence. (Of course, just such diversions as these are often contrived in attempts to unravel the mechanism of a catalytic cycle, e.g., by trapping reactive intermediates.) Even with mononuclear catalysts there can be many nonproductive pathways (and even one is sufficient to cause problems), but for a cluster catalyst, with its array of reaction sites, each providing potential secondary reaction paths, the problem is enhanced enormously. As the examples in this chapter show, clusters can interact with organic fragments in a variety of ways, with one or more of the atoms of the organic group binding to vertices, edges, or faces of the cluster. For the synthesist, this is a rich and rewarding diversity; for the catalysist, it is an unwelcome complication. If a catalytic cycle is to operate on a cluster, each of those possible bonding interactions must be of little significance in stabilizing the organometallic intermediates. One means of destabilizing all the noncatalytic intermediates is to work at temperatures that are high enough so that kinetic stability is no longer a determining factor. This alternative, which is widely available to the heterogeneous catalysist, is unfortunately of limited utility in cluster chemistry, since clusters have a limited thermal stability and must be held within temperature limits such that fragmentation and metal deposition are avoided.

5.6 CLUSTERS AND SURFACES

In the section dealing with the making and breaking of bonds to hydrogen, it was pointed out that several of the reactions described provided models for possible steps in catalytic hydrogenation reactions, although, as a class, clusters are dramatically inferior to mononuclear catalysts in homogeneous hydrogenations and to elemental metal as heterogeneous catalysts. Similarly, the chemistry of C—O and C—C bonds described earlier lends itself to speculation on the possible pathways followed in CO hydrogenation in the catalytic synthesis of organic molecules from syngas. Despite some initially high hopes, the impact of cluster chemistry on syngas conversion has been limited to a small number of examples of specialized interest. In fact, the organometallic chemistry of clusters, outlined in this chapter, has seemed more relevant to heterogeneous catalysis in providing models for CO hydrogenation. The initiation of the process by carbide formation from adsorbed CO has been mimicked by carbide formation in clusters from coordinated CO. Hydrogenation of cluster-bound carbides to alkylidyne clusters has been demonstrated, allowing comparisons with methanation catalysis. The variety of chain growth intermediates postulated for surface-catalyzed Fischer-Tropsch synthesis is matched by the wealth of organometallic cluster chemistry involving C—C bond formation or C—O bond cleavage, and organocluster chemists have not been slow to point out these possible similarities. Of course, the existence of a molecular cluster analog to a postulated surface species or reaction is by no means sufficient confirmation of the accuracy of the postulate, and these examples simply provide food for thought. Such comparisons of structures and reactivities as these have been the mainstay of the "surface-cluster analogy," but it is important to attempt to define just how far this analogy can be taken.

At first sight it seems quite reasonable to propose that an array of n adjacent metal atoms on a surface might provide an environment similar in chemically significant ways to an n-vertex face of a polyhedral molecular cluster, and that the latter might provide a model for the former. Certainly in terms of site symmetry and simple coordination geometry similar options seem to be available both for molecular clusters and surface metal atom arrays. It is, however, much more difficult to point to specific examples in which similar geometries have been *demonstrated* (and not just postulated) for a simple ligand in a cluster molecule and adsorbed on a surface. This is because in contrast to the wealth of structural information provided by single crystal X-ray diffraction and collateral spectroscopic techniques on crystalline cluster compounds, there are few examples of the unequivocal assignment of the structure of an adsorbed molecule on a surface. The last decade has seen major advances in the sophistication and precision of surface spectroscopic techniques such as low-energy electron diffraction (LEED) and electron energy loss spectroscopy (EELS) which have greatly advanced the state of knowledge about the structures of surface species. However, the last step in arriving at a hypothesis for the geometry of an adsorbate-substrate complex is often the citing of an organometallic analog with a similar structure whose very existence is then supposed to add credibility to the postulate. A survey of the structural organometallic cluster literature quickly reveals that a wide variety of bonding modes and geometries can be adopted by small organic molecules in clusters, and so some diagnostic data other than those derived from single crystal X-ray studies are necessary, and to be useful they must be transferrable between molecular and surface chemistries. High-resolution nmr for example, although uniquely revealing for molecular systems, is of limited utility for surface analysis.

In those areas where transferable data could be obtained, there is a paucity of results for *molecular* systems, despite the relative ease of data acquisition and analysis. An example is vibrational spectroscopy. For molecular organometallics this is probably the most commonly applied technique, especially for metal carbonyl complexes, and a number of surface spectroscopies can be applied to provide vibrational information on adsorbed molecules or molecular fragments. However, the organometallic chemist seldom reports frequencies assignable to just those modes that are most important for the surface spectroscopist, for example, metal-carbon and carbon-hydrogen modes for small hydrocarbons coordinated to clusters. This lack of relevant data is being remedied by the appearance of compilations of such information from the laboratories of spectroscopists (59,60), and the situation will be alleviated further if a recent request is heeded for the reporting of organometallic vibrational frequencies over wider frequency ranges (2).

REFERENCES

1. E. L. Muetterties, *Pure Appl. Chem.*, **54**, 83 (1982).

2. N. D. S. Canning and R. J. Madix. *J. Phys. Chem.*, **88**, 2437 (1984).

3. R. B. Calvert and J. R. Shapley, *J. Am. Chem. Soc.*, **99**, 5225 (1977).

4. R. B. Calvert, J. R. Shapley, A. J. Schultz, J. M. Williams, S. L. Suib, and C. D. Stuckey, *J. Am. Chem. Soc.*, **100**, 6240 (1978).

5. A. J. Schultz, J. M. Williams, R. B. Calvert, J. R. Shapley, and G. D. Stucky, *Inorg. Chem.*, **18**, 319 (1979).

6. R. B. Calvert and J. R. Shapley, *J. Am. Chem. Soc.*, **100**, 7726 (1978).

7. J. S. Bradley, *Phil. Trans. R. Soc. London*, **A308**, 103 (1982).

8. J. S. Bradley, G. B. Ansell, M. E. Leonowicz, and E. W. Hill, *J. Am. Chem. Soc.*, **103**, 4968 (1981).

9. J. H. Davis, M. A. Beno, J. M. Williams, J. Zimmie, M. Tachikawa, and E. L. Muetterties, *Proc. Natl. Acad. Sci. USA*, **78**, 668 (1981).

10. M. A. Beno, J. M. Williams, M. Tachikawa, and E. L. Muetterties, *J. Am. Chem. Soc.*, **103**, 1485 (1981).

11. A. J. Deeming and M. Underhill, *J. Chem. Soc., Dalton Trans.*, 1415 (1974).

12. A. J. Deeming, S. Hasso, and M. Underhill, *J. Chem. Soc., Dalton Trans.*, 1614 (1975).

13. J. B. Keister and J. R. Shapley, *J. Organometal. Chem.*, **85**, C29 (1975).

14. J. B. Keister and J. R. Shapley, *J. Am. Chem. Soc.*, **98**, 1056 (1978).

15. A. J. Deeming, S. Hasso, M. Underhill, A. J. Canty, B. F. G. Johnson, W. G. Jackson, J. Lewis, and T. M. Matheson, *J. Chem. Soc. Chem. Commun.*, 807 (1974).

16. R. J. Goudsmit, B. F. G. Johnson, J. Lewis, P. R. Raithby, and M. J. Rosales, *J. Chem. Soc., Dalton Trans.*, 2257 (1983).

17. S. Bhaduri, B. F. G. Johnson, J. W. Kelland, J. Lewis, P. R. Raithby, S. Rehani, G. M. Sheldrick, K. Wong, and M. McPartlin, *J. Chem. Soc., Dalton Trans.*, 562 (1979).

18. B. F. G. Johnson, J. W. Kelland, J. Lewis, A. L. Mann, and P. R. Raithby, *J. Chem. Soc. Chem. Commun.*, 547 (1980).

19. See, for example, M. Catti, G. Gervasio, and S. A. Mason, *J. Chem. Soc., Dalton Trans.*, 2260 (1967).

20. M. Evans, M. Hursthouse, E. W. Randall, E. Rosenburg, L. Milone, and M. Valle, *J. Chem. Soc. Chem. Commun.*, 545 (1972).

21. G. Gervasio, D. Osella, and M. Valle, *Inorg. Chem.*, **15**, 1221 (1976).

22. M. A. Andrews and H. D. Kaesz, *J. Am. Chem. Soc.*, **99**, 6763 (1967).

23. G. A. Vaglio, O. Gambino, R. P. Ferrari, and G. Cetini, *Inorg. Chim. Acta*, **7**, 193 (1973).

24. R. P. Ferrari, G. A. Vaglio, O. Gambino, and G. Cetini, *J. Chem. Soc. A*, 1998 (1972).

25. W. G. Jackson, B. F. G. Johnson, J. W. Kelland, J. Lewis, and K. T. Schorpp, *J. Organometal. Chem.*, **88**, C17 (1975).

26. A. D. Clauss, J. R. Shapley, and S. R. Wilson, *J. Am. Chem. Soc.*, **103**, 7387 (1981).

27. L. R. Beaman, Z. A. Rahman, and J. B. Keister, *Organometallics*, **2**, 1062 (1983).

28. J. R. Shapley, A. C. Sievert, M. R. Churchill, and H. J. Wasserman, *J. Am. Chem. Soc.*, **103**, 6975 (1981).

29. J. T. Park, J. R. Shapley, M. R. Churchill, and C. Bueno, *J. Am. Chem. Soc.*, **105**, 6182 (1983).

30. A. J. Arce and A. J. Deeming, *J. Chem. Soc. Chem. Commun.*, 364 (1982).

31. S. Aime, L. Milone and A. J. Deeming, *J. Chem. Soc. Chem. Commun.*, 1168 (1980).

32. G. R. Steinmetz, E. D. Morrison, and G. L. Geoffroy, *J. Am. Chem. Soc.*, **106**, 2559 (1984).

33. E. D. Morrison, G. R. Steinmetz, G. L. Geoffroy, W. C. Fultz, and A. L. Rheingold, *J. Am. Chem. Soc.*, **106**, 4783 (1984).

34. A. C. Sievert, D. S. Strickland, J. R. Shapley, G. R. Steinmetz, and G. L. Geoffroy, *Organometallics*, **1**, 214 (1982).

35. J. R. Shapley, D. S. Strickland, G. M. St. George, M. Churchill, and C. Bueno, *Organometallics*, **2**, 185 (1983).

36. J. W. Kolis, E. M. Holt, M. Drezdon, K. H. Whitmire, and D. F. Shriver, *J. Am. Chem. Soc.*, **104**, 6134 (1982).

37. D. Seyferth, J. E. Hallgren, and C. S. Eshbach, *J. Am. Chem. Soc.*, **96**, 1730 (1974).

38. J. B. Keister and T. L. Horling, *Inorg. Chem.*, **19**, 2304 (1980).

39. J. S. Bradley, G. B. Ansell, M. E. Leonowicz, and E. W. Hill, *J. Am. Chem. Soc.*, **103**, 4968 (1981).

40. S. Harris and J. S. Bradley, *Organometallics*, **3**, 1086 (1984).

41. E. M. Holt, K. H. Whitmire, and D. F. Shriver, *J. Am. Chem. Soc.*, **104**, 5621 (1982).

42. J. S. Bradley, *Adv. Organomet. Chem.*, **22**, 1 (1984).

43. E. M. Holt, K. H. Whitmire, and D. F. Shriver, *J. Organometal. Chem.*, **213**, 125 (1981).

44. K. H. Whitmire and D. F. Shriver, *J. Am. Chem. Soc.*, **102**, 1456 (1980).

45. D. deMontauzon and R. Mathieu, *J. Organometal. Chem.*, **252**, C83 (1983).

46. F. Dehan and R. Mathieu, *J. Chem. Soc. Chem. Commun.*, 432 (1984).

47. W.-K. Wong, K. W. Chiu, G. Wilkinson, A. M. R. Galas, M. Thornton-Pett and M. B. Hursthouse, *J. Chem. Soc., Dalton Trans.*, 1557, (1983).

48. J. R. Shapley, J. T. Park, M. R. Churchill, J. W. Ziller, and L. R. Beanan, *J. Am. Chem. Soc.*, **106**, 1144 (1984).

49. See, for example, R. L. Pruett, *Ann. N.Y. Acad. Sci.*, **295**, 239 (1977).

50. C. Masters and J. A. Van Doorn (to Shell), *German Offen.*, **185**, 2644 (1975).

51. M. G. Thomas, B. F. Beier, and E. L. Muetterties, *J. Am. Chem. Soc.*, **98**, 1296 (1976).

52. M. J. Doyle, A. P. Kouwenhoven, C. A. Schaap, and B. van Oort, *J. Organometal. Chem.*, **174**, C55 (1979).

53. J. S. Bradley, *J. Am. Chem. Soc.*, **101**, 7419 (1979).

54. B. D. Dombek, *J. Organometal. Chem.*, **250**, 467 (1983).

55. R. C. Ryan, C. U. Pittman and J. P. O'Connor, *J. Am. Chem. Soc.*, **99**, 1986 (1977).

56. H. P. Withers and D. Seyferth, *Inorg. Chem.*, **22**, 2931 (1982).

57. R. M. Laine, *J. Mol. Catal.*, **14**, 137 (1982).

58. See, for example, R. Whyman in B. F. G. Johnson, Ed., *Transition Metal Clusters*, Wiley, New York, 1980, p. 391.

59. I. A. Oxton, *Rev. Inorg. Chem.*, **4**, 1 (1981).

60. N. Sheppard and T. T. Nguyen, *Adv. Infrared Raman Spectrosc.*, **5**, 67 (1978).

6

THE GAS-PHASE CHARACTERIZATION OF THE MOLECULAR ELECTRONIC STRUCTURE OF SMALL METAL CLUSTERS AND CLUSTER OXIDATION

JAMES L. GOLE
High Temperature Laboratory
Center for Atomic and Molecular Science
School of Physics
Georgia Institute of Technology
Atlanta, Georgia

6.1 INTRODUCTION

The companion chapters of this text outline the potential use of "metal clusters" and the behavior of these unique species in environments in which they (1) come into intimate contact with an appropriate support surface (chemisorption, physisorption), (2) are present as an integral part of a metal cluster compound, or (3) are characterized after being trapped in a low-temperature rare gas or organic matrix. In the following discussion we focus on progress and future possibilities for the characterization of naked, freely moving, gas-phase clusters and the nature of their oxidation.

The solid-state physicist, the surface chemist, and the metallurgist have expended considerable effort toward our understanding of both the micro-and macromolecular features associated with quasi-infinite conducting metal crystals (1). The atomic and molecular physicist and the physical chemist have provided a wealth of information on the microstructure of atoms and diatomic molecules. These efforts have been impressive yet they leave open the middle ground whose exploration is still in its infancy. Despite considerable recent effort, the region encompassing naked, freely moving clusters of those elements that will eventually form bulk metals and that range in size from 2 to 100 atoms remains largely unexplored. This, of course, attests to the intrinsic difficulties associated with both experimental and theoretical work on this topic.

Because their characterization is believed to represent an important link in our understanding of the fundamental mechanisms of catalysis and numerous chemical conversions, the basic properties (geometry, bond strength, reactivity) of small "metal aggregates" have become the subject of intense theoretical (2) and experimental study (3). In addition, their characterization contributes to our understanding of the nucleation and growth of small metal particles (3) and the development of features inherent in the bulk metallic phase (1). We are afforded the opportunity to study the transition from vapor to liquid or solid as a function of chemical or electronic properties.

In studying gas-phase cluster oxidation, we have the opportunity to characterize the intermediate region bordered on the one side by the gas-phase oxidation of metallic atoms and dimers and on the other by the surface oxidation of the bulk metallic phase. In fact, it is thought that these studies will provide information that may be useful for the assessment of short- and long-range factors affecting surface oxidation (4).

Although it is an intuitive assumption that the electronic properties of small metal clusters lie intermediate between atomic properties and bulk metal characteristics, this picture is as yet only semiquantitative. That is, although several theoretical studies have emerged in the last 5 years (2), precise experimental data with which to compare them have only been obtained recently (5), and the fit to these data by current theoretical models, even though very promising (5,6), will require considerable fine tuning. Here, it will be especially important to establish the structural properties for small polyatomic clusters as an aid to the parameterization of calculations on much larger groupings.

The definition of when a cluster becomes a metal is by no means precise. From the standpoint of chemistry, the transition from molecule-like to bulk-like properties appears to occur in the region around 13 atoms (7); however, the many properties whose behavior can be used to signal the onset of metallic character converge to those of the bulk phase at very different rates. The electron affinity and the ionization potential that are respectively lower and higher than that of the bulk metal both converge to the work function of the bulk metal yet the ionization potential converges much more rapidly. The binding energy per atom as cluster size increases appears to converge very slowly to the larger cohesive energy of the bulk metal (8). In the transition series, the binding energy per atom is approximately four times as large in the bulk metal as in the dimer (5). This suggests that structural, electronic, and reactive properties vary widely with increasing cluster size. For the alkali metals, where the cohesive energy of the bulk is small and the reactivity high, the binding energy per atom still converges slowly with increased cluster size being in the ratio 3 to 3.5 bulk to dimer (5,9). In seeming contrast, the rotational constants for the ground states of alkali diatomics are characterized by a constraction of the nuclear distance (12% in Li to 20% in Cs) relative to the separation of nearest neighbors in the stable body-centered cubic (bcc) structure of the elements (5,9). The latter trend continues for larger clusters where surface tension and curvature initiate sizable compression of the aggregate (5,9). Starting from the bulk, one might ask what size of crystallite will lead to the discrete electron states associated with a molecular system. Here, it is intriguing that some evidence suggests that the density of states for relatively small clusters (Ni_8, Cu_{13}, . . .) are believed to be very close to that of the bulk (7).

We have introduced but a few of the properties with which one must be concerned in dealing with metal clusters. Although one expects a gradation of molecule-bulk differences as aggregate size increases, the exact point at which the aggregate becomes bulk-like is the subject of dispute. One point is apparent. In proceeding through the buildup of a metal cluster, one anticipates a range of chemical properties strongly dependent on cluster size.

It is significant that the electronic properties of small metal clusters are substantially different from the bulk. The nature of these differing electronic properties is believed to play an important role in the adsorption of organic molecules and the cleavage of bonds. It is thought that metal atom clusters of suitable size distribution may be able to produce most of the geometries and unusual electronic features of the convex crystal defects responsible for catalytic effects. For this reason, researchers have assumed that small metal clusters may have unusual catalytic properties based on both electronic and geometric structure. In compiling physical and electronic data on these metal clusters, one hopes in the future to model the types of reactions occurring on the surface of larger metal crystallites. This effort may also lead to the development of more specific and efficient catalysts. For a given element, is there a particular number of clustered atoms and a particular arrangement that maximizes not only catalytic efficiency but also the selective formation of one product over another (10)?

The subject of free-metal atom-cluster spectroscopy is a very young and developing field. The difficulties inherent in doing such work on free gaseous cluster systems have caused the majority of activities to be confined to the matrix isolation technique where studies of electronic transitions (11), Raman and resonance Raman effects (12), and electron spin resonance (13) account for nearly all of our detailed knowledge of triatomic and larger bare metal clusters. Although matrix isolation has provided a great deal of valuable information, such studies lack the detail possible in gas-phase work. Questions do arise concerning (1) the carrier of an observed spectroscopic transition and (2), more often, the magnitude of matrix-solute interactions. Moreover, the finer details of rotational and vibrational structure in electronic transitions are obscured in many cases (14). The study of naked metal clusters in the gas phase not only provides greater detail but also serves as a benchmark from which the metal-atom/rare-gas interaction may be evaluated. These trends once understood and correlated with the behavior of small metal clusters on support surfaces can provide valuable insights into the nature of chemi-and physisorption.

Throughout this chapter, we will refer to several molecular species as small metal clusters; however, we must caution that this reference is nebulous for diatomics and small polyatomics. We emphasize that it is by no means certain what combination of those elements referred to as metals will constitute a metal. It might be better to refer to these species as small "metal molecules."

6.2 TECHNIQUES FOR PROBING NAKED METAL CLUSTERS

Several approaches have now emerged whose aim is the characterization of metal clusters in the size range 1 to 4000 atoms. For the smaller clusters corresponding primarily to diatomic and triatomic metal molecules but including some larger groupings, mass spectroscopy has yielded a wealth of information on bond energies. Gas-phase optical spectroscopy on smaller metal clusters (diatomics and polyatomics) is now emerging, resulting in large part from a sophisticated coupling of near state-of-the-art laser technology and methods that allow for the simultaneous sampling and cooling of high-temperature, high-pressure systems and mass spectroscopy. Using this combination, one is now able to produce and interrogate metal molecules at temperatures less than 100K, facilitating the interrogation of their structure and molecular electronic spectra. The cooling is necessary since most metal cluster molecules are characterized by a significant density of states. Through attendant cooling, one not only simplifies molecular spectra but also ensures a substantial increase in the population of probed levels.

Again, in part using those techniques whose goal it is to produce cold molecules, researchers have obtained some information in the small to intermediate cluster range using electron spin resonance spectroscopy. Concentrations of clusters that apparently in certain instances extend to 500 atoms have been produced and studied employing a combination of effusive sources, inert gas quench-flow techniques, and time-of-flight mass spectroscopy. Means for generating reasonably controlled metal cluster size distributions have also emerged. Extremely large clusters have now been characterized using electron diffraction techniques and electron microscopy, and, in the near future, it appears that information on the polarizability of large clusters will be forthcoming. Finally, we should note that there has been some information obtained on the oxidation of small metal clusters. In the following discussion we will consider all of these topics in greater detail.

6.3 MASS SPECTROSCOPY—METAL CLUSTER ENERGETICS

As of this writing, the most prevalent data on small metal clusters emanate from mass spectrometry where Knudsen cell effusion techniques (applied at temperatures that may approach 3000K) have been used to study the energetics of the equilibrium vapors above a variety of metallic elements and metalloids (15,16). These studies have provided measurements of the enthalpies of formation for several dimers of the elements of Group IA, IIIA, IVA, IB, and IIB and a significant, although less accurately determined, sampling of the transition elements Sc_2 through Ni_2, Y_2, La_2, Nb_2, Mo_2, Rh_2, Pd_2, and Pt_2. A very diverse group of intermetallic dimers has also been investigated. Although a considerably smaller catalog has been compiled for polyatomic homonuclear metal molecules and intermetallic compounds, the list grows at a notable rate. A summary of experimentally determined diatomic dissociation energies and the atomization energies of several polyatomic metal molecules is given in Table 6.1 (15,16). Several of the homonuclear diatomic transition metal bond energies that are not yet measured and hence not given in Table 6.1 have been estimated by Brewer and Winn (17).

Experimentally determined reaction enthalpies are obtained from measurements of the partial pressures above a liquid or solid for the monitored diatomic and/or polyatomic molecule under study. One may employ the "third-law" method using the relation (15,16)

$$\Delta H° = -RT \ln K_P - T\Delta(G_T° - H_0°/T) \qquad (6.1)$$

Alternatively, when a large enough temperature range can be covered, the "second-

TABLE 6.1 Dissociation Energies and Atomization Energies for Diatomic and Polyatomic Metal Molecules

Molecule	Dissoc. or At. Energy (KJ/mole)	Ref.	Molecule	Dissoc. or At. Energy (KJ/mole)	Ref.	Molecule	Dissoc. or At. Energy (KJ/mole)	Ref.
Ag_2	159.0 ± 6.3	43	AuV	238 ± 12	64	CuHo	139 ± 19	79
Ag_3	253.0 ± 13	44	AuY	304 ± 8.2	65	CuLi	189.3 ± 8.8	51
AgAl	172 ± 17	45*	Au_2Ba	552 ± 21.0	66	CuNa	172.4 ± 16.7	80
AgAu	200.8 ± 10.5	45	Au_2Eu	549.4 ± 16.7	48	CuNi	201 ± 21	45
AgBi	192 ± 42	45	$AuGe_2$	531.9 ± 10	23	CuSn	168 ± 10	81
AgCu	169.5 ± 10.5	45	$AuGe_3$	897 ± 20	23	CuTb	187 ± 19	79
AgDy	124 ± 19	47	$AuGe_4$	1295 ± 30	23	$CuGe_2$	506.0 ± 25	81
AgEu	123.0 ± 12.5	48	Au_2Ge	534.9 ± 12	23	$CuSn_2$	391 ± 25	81
AgGa	177.8 ± 6.3	43	Au_2Ge_2	927 ± 14	23	Cu_2Sn	452 ± 25	81
AgGe	170.7 ± 21	49	Au_2Ho	533 ± 42	57	Dy_2	67 ± 29	82
AgHo	119.7 ± 17	50	Au_2Lu	602.1 ± 33.5	60	Er_2	71 ± 29	82
AgIn	163.2 ± 6	43	$AuSn_2$	486 ± 18	63	Eu_2	29.3 ± 16.7	48
AgLi	173.6 ± 6.3	51	$AuSn_3$	786 ± 25	63	EuRb	231.8 ± 34	48
AgMn	96 ± 12	45	Au_2Sn	542 ± 18	63	Fe_2	100 ± 21	45,62
AgNa	136.0 ± 10.5	52	Au_2Sn_2	871 ± 25	63	Ga_2	135 ± 8	78
AgSn	134 ± 21	45	Au_2Tb	582 ± 42	57	GaLi	129.3 ± 14.6	54
Al_2	149.8 ± 13.8	46	BaPd	220.0 ± 5.0	58	Gd_2	172 ± 33	82
AlAu	322.2 ± 6.3	53	BaRh	257.4 ± 25	58	Ge_2	266.4 ± 17	23
AlCu	209 ± 17	45	Bi_2	195.4 ± 5.0	17	Ge_3	639 ± 20	23–25
AlLi	172.0 ± 14.6	54	Bi_3	365 ± 13	17,67	Ge_4	999 ± 25	23–25
AlPd	250.6 ± 12.0	55	Bi_4	595 ± 8	17,18,19	Ge_5	1343 ± 42	25
$AlAu_2$	506.3 ± 25.1	56	BiGa	155 ± 17	68	Ge_6	1703 ± 54	25
Al_2Au	460.2 ± 20.9	56	BiIn	147.9 ± 4.3	69	Ge_7	2013 ± 63	25
Al_2Pd	492.4 ± 24	55	BiLi	150 ± 5	70,71	GeNi	201 ± 13	45
Au_2	221.3 ± 2.1	57	BiPb	134 ± 8	45	GePd	259 ± 17	45
Au_3	367 ± 13	44	BiSn	206.3 ± 8.4	17	Hg_2	7.5 ± 1	78
AuBa	251.1 ± 10	58	BiTl	117 ± 13	45	HgK	5.8	78

AuBe	280	45	Bi$_3$In	354.2 ± 14	69	HgLi	10.1		78
AuBi	293 ± 8.4	45	BiLi$_2$	326 ± 10	72	HgNa	5.3		78
AuCa	238	45	Bi$_2$Li	367.4 ± 7	72	HgRb	4.7		78
AuCe	355 ± 21	45	Bi$_2$Li$_2$	583.0 ± 15	72	Ho$_2$	73.5 ± 17		50
AuCo	218 ± 17	45	Bi$_2$Sn	427.6 ± 10.5	17	In$_2$	100 ± 8		45
AuCr	209 ± 17	45	Bi$_3$Sn	644.3 ± 16.7	17	InLi	86.6 ± 13.4		54
AuCs	249 ± 13	59	Ca$_2$	11.3 ± 0.05	73	IrLa	573 ± 12		83
AuCu	224.3 ± 5.1	23	Cd$_2$	8.4 ± 0.21	45	IrTh	570.7 ± 42		74
AuDy	254 ± 20	47	CdIn	134	45	IrY	452.8 ± 16		65
AuEu	238.9 ± 10.5	48	Ce$_2$	238 ± 21	45	K$_2$	53.6 ± 4.2		45
AuFe	188 ± 21	45	CeIr	570.7 ± 25	51	KLi	78.2 ± 4.2		84
AuGa	230 ± 3.8	45	CeOs	503 ± 33	75	KNa	61.0 ± 4.2		84
AuGe	270.4 ± 5.0	23,49	CePd	318.4 ± 17	76	La$_2$	243 ± 21		45
AuHo	263.4 ± 33	57	CePt	551.0 ± 25	74	LaPt	496 ± 15		85
AuLa	335 ± 21	45	CeRh	545.6 ± 25	74	LaRh	524.7 ± 16.7		76
AuLi	280.8 ± 6.5	51	CeRu	527 ± 25	75	LaY	197 ± 17		45
AuLu	328.4 ± 17	60	Co$_2$	167 ± 25	45	Li$_2$	101.0 ± 1.2		86
AuMg	243 ± 42	45	CoCu	163 ± 21	45	Li$_3$	173.6 ± 16.7		39
AuMn	188 ± 13	45	CoGe	230 ± 21	45	LiNa	86.6 ± 6.3		84
AuNa	212.1 ± 12.6	52	Cr$_2$	151 ± 21	45	LiPb	74.9 ± 8		71
AuNd	297 ± 21	45	CrCu	155 ± 25	45	LiBiPb	346.6 ± 6		72
AuNi	251 ± 21	45	CrGe	165.7 ± 29	45	Lu$_2$	138 ± 33		60
AuPb	126 ± 42	45	Cs$_2$	37.99 ± 0.9	77	LuPt	397.5 ± 34		76
AuPd	151 ± 21	45	CsHg	4.8	45	Mg$_2$	4.830 ± 0.008		45
AuPr	305 ± 21	45	CsLi	(69)	78	Mn$_2$	42 ± 29		45
AuRb	239 ± 13	59	CsK	(37)	78	Mo$_2$	404 ± 20		87
AuRh	228.9 ± 29	48	CsNa	(41)	78	MoNb	448 ± 25		88
AuSc	276.6 ± 17	61	Cu$_2$	190.2 ± 5.4	79	Na$_2$	71.1 ± 2.1		45
AuSn	251 ± 8	63	Cu$_3$	294 ± 13	44	NaRb	54.8 ± 3.8		45
AuTb	289.5 ± 33	57	CuDy	140 ± 19	79	Nb$_2$	503 ± 10		89
AuU	318 ± 29	45	CuGe	197 ± 17	45	Nd$_2$	80 ± 29		82

Table 6.1 (*Continued*)

Molecule	Dissoc. or At. Energy (KJ/mole)	Ref.	Molecule	Dissoc. or At. Energy (KJ/mole)	Ref.	Molecule	Dissoc. or At. Energy (KJ/mole)	Ref.
Ni_2	230 ± 21	45	RhTh	510 ± 21	93	Sn_5	1024 ± 25	27
Pb_2	80 ± 1	28,70	RhTl	387.0 ± 14.6	94	Sn_6	1329 ± 30	27
Pb_3	221 ± 16	28	RhU	516 ± 17	93	Sn_7	1612 ± 35	27
Pb_4	415 ± 16	28	RhV	360 ± 29	95	Tb_2	127.6 ± 25.1	57
Pd_2	105 ± 21	43	RhY	441.8 ± 10.5	92	Th_2	285 ± 21	45
Pr_2	126 ± 29	82	RuTh	587.9 ± 42	96	Ti_2	126 ± 17	45
Pt_2	358 ± 15	90,91	RuV	410 ± 29	95	Tl_2	59	45
PtTh	546.6 ± 42	74	$RhTi_2$	996 ± 42	94	Tm_2	50 ± 17	82
PtTi	394 ± 11	74	Sc_2	159 ± 21	45	U_2	218 ± 21	45
PtY	470 ± 12	90	Sm_2	50 ± 21	82	V_2	238 ± 21	45
Rb_2	45.2 ± 4.2	45	Sn_2	185.8 ± 9	27	Y_2	156 ± 21	43
Rh_2	281.6 ± 20.9	94	Sn_3	480.7 ± 17	27	Zn_2	18.4 ± 6.3	45
RhSc	440.3 ± 10.5	92	Sn_4	757.7 ± 20	27			

law'' method

$$d \ln K_P / d(1/T) = -\Delta H_T / R \qquad (6.2)$$

can be used to determine the enthalpies of formation. Ideally, one applies both second- and third-law methods and agreement in both reaction enthalpies obtained supports the reliability of the results. The specific makeup of the mass spectrometers used in these studies and the nature of the electron impact technique is considered elsewhere; however, some discussion of the contrast between electron impact and photoionization techniques will be appropriate in later sections.

The most prevalent gas-phase polymeric metal or metalloid formation in equilibrium systems is associated with the elements of Group IVA and VA. Bi_4 (18,19), Sb_4 (20), and As_4 (21) constitute major species in the vapor above their respective metallic or metalloid phases. Extensive polymer formation has also been noted in the carbon series where the predominant equilibrium vapors above the metalloids silicon and germanium are Si_3 (22) and Ge_3 (23–25). Early studies indicated the presence of mass peaks up to Sn_5 in the vapor over tin (26); however, the fraction of polymeric species in this system appears to be considerably lessened relative to the lighter members of the group. As Table 6.1 demonstrates, more refined studies (27) have now provided information on Sn_6 and Sn_7 and extended Group VA characterization to the aggregates of lead, Pb_x, $x \leq 4$ (28).

The most recent studies on polyatomic metal clusters have in large part been the province of Gingerich and coworkers (15,16) who, in addition to characterizing a number of homonuclear clusters, have been involved with the study of the majority of those heteronuclear combinations given in Table 6.1. These studies prove to be quite informative. For example, the Sn clusters $Sn_3 \rightarrow Sn_7$ (27) are found to be unusually stable, the energy required for removal of a tin atom being on the order of twice that required for the rupture of a Sn—Sn single bond. This suggests formation of doubly bonded linear chains or formation of triangular Sn_3 and polyhedra in the higher polymers with a resulting larger number of bonds. These observed characteristics are reminiscent of metal cluster complex compounds in which metallic nucleation sites are frequently found (29). The intriguing nature of these naked tin clusters focuses our attention on the need for a strong interplay among mass spectroscopy, optical spectroscopy, and molecular electronic structure theory. The lack of spectroscopic data concerning their electronic and molecular structure limits the insight into the bonding that can be gained from available atomization energies. Clearly, structural information on tin clusters will represent an important contribution to our understanding of bonding in the Group V metals. Further correlations should also be emphasized.

The substantial catalog of information in Table 6.1 has in large part been justified by the lack of detailed molecular electronic structure calculations and the need for generating representative empirical bonding models. Because the bond energies of metal molecules are among the most difficult to calculate from first principles, few detailed calculations exist and these are at present limited to examples involving metals with low atomic numbers (30). Therefore, one must rely on empirical methods for the prediction of the unknown bond energies of diatomic metals and metal clusters. The empirical models usually have a strong base in electronic structure theory, in many instances involving logical correlations with atomic and bulk metal characteristics.* The successes and merits of these empirical models have been discussed by other authors (15–17) and will not be considered in detail here;* however, we wish to emphasize again the required intertwining of experimental data from mass and optical spectroscopy and its correlation with electronic structure theory in order that bonding models be refined to the desired level. Of foremost importance is optical data on the ground and low-lying electronic states of metal clusters. Such data are not only significant to the defining of empirical models, but they also represent a necessary thread for the evaluation of the third- and second-law enthalpies defined in equations (6.1) and (6.2).

In order to evaluate "third-law" reaction enthalpies from mass spectrometric equilibrium measurements, one must determine the Gibbs free energy functions of the reactants. "Second-law" reaction enthalpies obtained at the average temperature of measurement require enthalpy functions to extrapolate to a desired reference temperature. The desired thermal functions are calculated, using statistical thermodynamics, from *known or estimated molecular parameters*. The effects of assumptions regarding the variation of assumed geometries for polyatomic metal molecules and/or the variation in the electronic structure of either diatomic or polyatomic metal molecules can be significant.

Consider the data given in Table 6.2 for the trimers of copper, silver, and gold. Although there is relatively little difference in the evaluated enthalpies when either a D_{3h} or $D_{\infty h}$ geometry is assumed, these enthalpies are notably different from that determined for an assumed C_{2v} structure. This is significant, for recent quantum chemical calculations (37) in correlation with both matrix isolation and gas-phase spectroscopic studies indicate that the C_{2v} geometry is the most probable configuration, even though the copper trimer molecule is thought to be quite fluxional.

As a result of several recent rather detailed quantum chemical calculations (38), the majority of the transition-metal molecules are thought to be characterized by a large number of low-lying electronic states. Therefore, the electronic contribution to the partition function, especially over the temperature range associated with most mass spectrometric studies will be substantial, considerably modifying determined third-law dissociation energies. It is thought that quite substantial downward corrections will be applied to the currently deduced bond energies (Table 6.1), especially for several of the middle transition-metal molecules; however, a great deal of experimental data must be obtained on the low-lying electronic states before these corrections can be applied with confidence.

The previous two examples emphasize the need to obtain a large catalog of both structural and electronic data and to correlate these data, when possible, with more

---

*Depending on the nature of the bonding in a given compound, three models are generally employed. The Pauling model and its refinements (31,32) are applicable to molecules possessing single bonds, whereas the valence bond (33,34) and atomic cell models (35,36) and their refinements are applicable to multiply bonded species.

TABLE 6.2 Effect of Variation in the Assumed Geometry for $M_3(g)$ on the Enthalpy Change, ΔH_0°, for the Reaction $M_3(g) = 3\,M(g)$ (M = Cu, Ag, Au)[a]

Molecule	Temperature Range (K)	Linear $D_{\infty h}$	Equilateral Triangle D_{3h}	Bent C_{2v}
Cu_3	1801 –	293.5	294.2	282.8
	1897	±4.7	±4.6	±4.8
Ag_3	1402 –	253.3	254.9	245.4
	1599	±5.1	±4.9	±5.1
Au_3	1780 –	367.1	366.1	354.2
	1985	±7.2	±6.9	±7.3

[a]Values are in kJ mol^{-1}.

detailed ab-initio calculations. A further example of the necessary correlation between experiment and theory can be represented by the work of Wu (39), who has used Knudsen effusion techniques to study lithium trimer, obtaining what detailed theoretical studies (40) indicate is a good estimate of the bonding and structure of this species. There are, however, some questions that arise in regard to Wu's measurement of the Li_3 ionization potential. Wu determines a value of 4.35 ± 0.2 eV. for the threshold of ionization reasoning that this measured vertical ionization energy must represent a "lower bound" since lithium trimer produced effusively is expected to have considerable vibrational and rotational excitation. As we will consider in more detail in a later section, Schumacher and coworkers (6,41) determine a value of 4.1 ± 0.1 eV, producing the trimer "cooled" in supersonic expansion and using photoionization as opposed to electron impact ionization. An explanation for this difference may come from detailed quantum chemical calculations (42) which demonstrate that the fluxional nature of neutral lithium trimer and hence a low barrier between bent and linear configurations may play an important role in determining the threshold for ionization to a very bent Li_3^+ product.

6.4 OPTICAL TECHNIQUES FOR PROBING SMALL GAS-PHASE METAL CLUSTERS

We have considered the important contribution that mass spectrometry can make to the characterization of the energetics of small metal clusters. In this section we focus primarily on methods of preparation and the application of gas-phase optical spectroscopic techniques to the elucidation of the structural and electronic properties of small metal clusters.

6.4.1 The Approach

In order to obtain the optical spectra of small metal clusters, several researchers are now attempting to couple techniques for the sampling of high-temperature, high-pressure systems and recent developments in laser technology. One wishes to develop metal cluster sources that allow not only for the production of the requisite clusters but also provide for their formation primarily with low internal temperatures ($T \leq 300K$). This requirement is necessary since the analysis of features that map energy levels, structure, and dynamic behavior may be prohibitive if not impossible at the elevated temperatures required to produce these species through vaporization of the bulk metal.

There are two general approaches currently in use to produce sufficiently cooled metal clusters. One can combine a *means* for producing metal vapor at elevated temperatures and the techniques of rarefied gas dynamics to produce an isentropic supersonically expanded "free jet" (also see following discussion) of cold metal clusters (97,98). Once formed, this jet stream of metal clusters can be simultaneously probed using a variety of laser techniques, in many cases in combination with mass spectrometry. Another approach again involves metal vaporization followed by entrainment in a flowing inert gas stream, the relatively cool gas "quenching" the metal vapor achieving large supersaturation ratios, rapid nucleation, and cluster growth (14,99). This second approach does not result in clusters that are as cold internally as those that can be obtained in supersonic expansion; however, internal temperatures approaching 80K can be achieved. At present (see also following discussion) metal vapors are created using high-temperature oven systems or through use of laser vaporization techniques. In order to understand the following discussions better, one should consider the nature of supersonic expansions.

Cooling in Isentropic Expansion

The process of high-pressure expansion to molecular flow provides a unique environment for studying nucleation, relaxation processes, and the properties of clustered species. Nozzle expansions also offer the added attraction of obtaining high-temperature molecules with very reduced internal energies. Because gases tend to nucleate during isentropic expansion from high pressures, one encounters a unique situation in which to study the condensation process itself, and, in so doing, bridges the gap between monomer formation and the formation of small metal clusters.

When a gas at high pressure ($\sim$10 to 3000 torr) is expanded isentropically into a chamber at relatively low pressure ($\sim10^{-5}$ to 10^{-2} torr), there is a cooling of its translational degrees of freedom (100). Here the translational temperature is defined by the width of the velocity distribution function, translational cooling, producing a narrowing of the velocity distribution as random kinetic energy is converted into directed mass flow. Although the entire mass of gas may be moving in one direction at high velocity, molecules in the same region of the expanding gas move at the same speed and hence the relative velocity between colliding molecules is quite low. Collisions between molecules can lead to a partial equilibration between translation and the molecular internal degrees of freedom. In effect, one has a translational bath at very low temperatures that, through collisions, acts as a refrigerant to cool the molecular vibrations and rotations. One has the potential to create an environment similar to that which characterizes studies in matrix isolation spectroscopy yet without matrix perturbations. In the course of expansion both the translational temperature

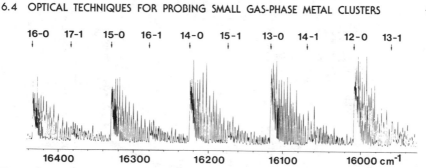

Figure 6.1 Excitation spectrum for supersonically expanded Na corresponding to the Na $A^1\Sigma_u^+ - X^1\Sigma_g^+$ transition. Bands are denoted (v', v''). Several excited state (v') levels are pumped on excitation from the lowest $v'' = 0, 1$ levels of the ground electronic state. This spectrum correlates with $T_{Vib} \cong 50K$, $T_{Rot} \cong 30K$. (From Ref. 102.)

and the density will decrease until a region of free flow is reached where there are no further collisions. At this point, the translational, vibrational, and rotational temperatures are frozen (100).

Those degrees of freedom that equilibrate most rapidly in the finite time for which they are in contact with the cold translational bath, that is, rotations, are cooled quite effectively and the final rotational temperature may be fairly close to the translational temperature. Molecular vibrations equilibrate more slowly and appear to be much more strongly influenced by the nature of the expanding metal mixtures. Figure 6.1 depicts an excitation spectrum (see also following discussion) for dimeric sodium produced in the supersonic expansion of the pure metal from an oven at 1100K; the observed vibrational and rotational temperatures are 50K and 30K, respectively (101). By comparison, Figure 6.2 contrasts the spectrum for antimony dimer obtained at an equilibrium temperature of 1175K (102) with that produced in a seeded expansion with argon (from an oven at 1225K) where the observed rotational temperature is 25K and the observed vibrational temperature is in excess of 650K! This figure demonstrates that one has the intriguing possibility of producing vibrationally hot yet rotationally cold species. In both cases, vibrational cooling trails that for rotation; however, the vibrational cooling in the pure sodium metal expansion closely follows rotation, in large part because the sodium dimer ground-electronic-state vibrational anharmonicity is quite small and the opportunity exists for several near resonant Na_2—Na_2 vibrational cooling collisions. Although the antimony dimer rotational temperature is effectively cooled through collisions with both antimony combinations and argon, Sb_2—Sb_2 collisions are considerably lessened due to the argon dilution, and hence vibrational cooling as a result of energy resonances is considerably lessened.

Cluster Formation in Isentropic Expansion

There are two approaches that one can take to expand a metal vapor supersonically, bringing the metal from a state in which it is unsaturated or barely saturated to a highly saturated mode. One can attempt to expand the pure metal vapor or carry

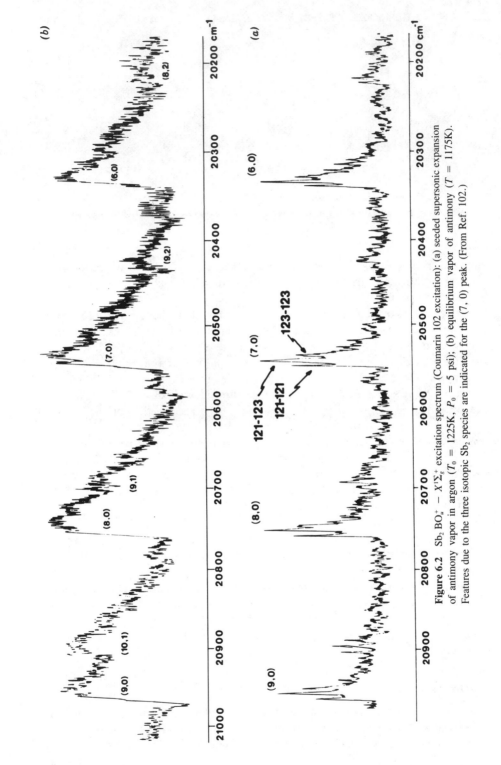

Figure 6.2 Sb$_2$ BO$_u^+$ – X$^1\Sigma_g^+$ excitation spectrum (Coumarin 102 excitation): (a) seeded supersonic expansion of antimony vapor in argon ($T_0 = 1225$K, $P_0 = 5$ psi); (b) equilibrium vapor of antimony ($T = 1175$K). Features due to the three isotopic Sb$_2$ species are indicated for the (7, 0) peak. (From Ref. 102.)

out on expansion of the metal with a noncondensible carrier gas (He, Ar, N_2, . . .). Because of the relatively low vapor pressure of most metals, the supersonic expansion of the pure metal vapor must be conducted with those metals that have the highest vapor pressures (alkalis) (5,6,9,98,102) or ovens of relatively high power must be used (103). Moderately refractory metals can be expanded at elevated temperatures to produce cooling; however, it appears that the more efficient approach involves the seeding of a metal into a carrier gas and the subsequent expansion of this mixture. Significant results have been obtained when coupling this approach with laser vaporization techniques (see also following discussion). Here, a burst of radiation from a pulsed laser is focused on a metal surface, causing intense local heating for a short time frame and creating a gas-phase metal plasma. This plasma is entrained in a carrier gas and supersonically expanded into a zone where the resulting clusters are interrogated using a variety of laser techniques (14,104).

If one desires to nucleate metals in a seeded supersonic expansion, the parameters that must be considered are the backing or stagnation pressure behind the nozzle, P_0, the stagnation temperature, T_0, the nozzle throat diameter, D, and the metal mole fraction $X_{metal} = P_m/P_0$, where P_m is the partial pressure of the metal. Cooling in the collision-dominated adiabatic expansion is controlled by the bimolecular-collision frequency that is proportional to $P_0 D$. The amount of clustering for *small* clusters is proportional to the three-body or termolecular-collision frequency (proportional to $P_0^2 D$ for a pure gas expansion), where at least two of the partners are condensible species. When a cluster grows to a sufficient size, it can continue to grow by dissipating the binding energy into its own internal degrees of freedom; however, small clusters can only grow if the binding energy is removed by a third body. Although these considerations place some constraints on the initial growth of small clusters, once the cluster size reaches the level at which energy dissipation is readily accomplished, the rate of cluster growth is expected to increase precipitously in a virtual avalanche effect (105). If one desires to increase the amount of clustering in a *pure* gaseous metal expansion, one increases the number of ternary collisions. If one wishes to optimize cooling while minimizing clustering, $P_0 D$ must be relatively large versus $P_0^2 D$, and one must keep the stagnation pressure relatively low or the nozzle diameter relatively large. If one wishes to optimize clustering with respect to binary collisional cooling, small orifice diameters and relatively large starting (backing) pressures are required. Because of the avalanche phenomena, the distribution of the energy of binding within the clusters, and the minimized effect of binary collisions, one anticipates some heating of the larger metal clusters relative to those formed in the initial ternary process.

By expanding the metal of interest in a noncondensible carrier gas, one is provided additional flexibility in that one can vary the mole fraction of the metal under study. In a mixed (seeded) expansion, binary collisions that produce cooling will occur whether the collision partners are carrier gas or metallic species; however, as we have indicated, the three-body collisions that form the smaller clusters require that at least two of the partners be condensible metals. The $P_0^2 D$ cluster formation rate is modified approximately through multiplication by the mole fraction ratio X_m^2 $(1 - X_m)$. The termolecular-collision rate for three condensibles is $P_0^2 D X_m^3$. Therefore, the total three-body cluster formation rate is $P_0^2 D X_m^2 = P_m^2 D$. The clustering rate depends on P_m, the metal vapor pressure, and D. Hence, for a given clustering rate, P_0 can be increased to enhance cooling via the binary collision rate. In a given apparatus design one would like to have available the widest possible variation of P_0 and P_m.

In an ideal experiment one would like to control both cluster size and cluster temperature. It would indeed be fortunate if one could produce a monodisperse cold cluster distribution through judicious variation of P_0 and P_m; however, this is a very difficult task. Careful and varied source design can provide a means of producing one cluster species over others and, in some cases, the various clusters produced may absorb radiation in differing spectral regions. (This is true for certain clusters of copper and silver.) By coupling optical techniques with mass spectroscopy, one can obtain much greater versatility. Many of the problems associated with multiple cluster formation are greatly alleviated through the use of mass selective optical techniques based in large part on laser-induced multiphoton ionization. Specific approaches and examples will be considered shortly.

6.4.2 Laser Techniques for Probing Cooled Metal Clusters

It is not surprising that the laser brings forth a new dimension to the characterization of cooled metal clusters. There are two basic modes by which one can excite laser-induced fluorescence and hence probe molecular quantum levels. A "fluorescence excitation spectrum" (Figs. 6.1 and 6.2) is obtained by exciting molecular fluorescence with a tunable dye laser. As the laser is tuned through an absorption feature, the molecule of interest may absorb a laser photon and be pumped to an excited electronic state. Once in this excited electronic state, the molecule may emit a photon at the same or at a different frequency from that of the pump laser. The fluorescence excitation spectrum is a plot of the total intensity of the emitted light versus the frequency of the exciting laser. In essence, one uses the emission of a photon to monitor absorption. Given unit quantum yield, the excitation spectrum contains the same information found in the absorption spectrum but is usually a far more sensitive method for obtaining this information. If a molecule is produced sufficiently cooled so that only a few vibrational and rotational levels of the ground electronic state are populated, the excitation spectrum probes the level structure of the excited electronic state.

Photoluminescence or dispersed emission spectroscopy is several orders of magnitude less sensitive than excitation spectroscopy. Here, the exciting laser is held at a specific absorption frequency and the emitted fluorescent light is dispersed to produce an emission spectrum. If, in a cold jet, we excite to a single vibronic level of an excited electronic state, the emission spectrum can be used to readily probe the level structure of the ground electronic state.

A variant of both photoluminescence and excitation spectroscopy is laser-induced atomic fluorescence spectroscopy, a technique for studying bound-free transitions. Here, the absorption of a laser photon leads to photodissociation via an excited electronic state that dissociates to products, one of which is in an electronically excited atom. The process can be represented as

$$M_x + h\nu \rightarrow M_x^* \tag{6.3}$$

$$M_x^* \rightarrow M_{x-1} + M^* \tag{6.4}$$

$$M^* \rightarrow M + h\nu' \tag{6.5}$$

The dispersed fluorescence from the individual M^* spin-orbit components is monitored as the laser is tuned and ground-state vibronic level structure is obtained. A variant of this technique involves the study of photodissociation to ground-state products (note Eq. 6.4) where a second laser (or an atomic resonance lamp) is used to excite the ground-state atom formed, the resulting dispersed fluorescence (analog of Eq. 6.5) again mapping the ground-state-level structure.

One of the most useful recent developments for the laser interrogation of supersonic jets is the technique of multiphoton ionization spectroscopy. A variant of this technique, resonant two-photon ionization (R2PI) spectroscopy, which has been used very effectively by Schumacher (6,9,107–108), Smalley (104,109–114), and coworkers, is exemplified in Figure 6.3(a). In employing multiphoton ionization spectroscopy, they tune one or more photons into resonance with an electronic transition exciting a bound state of the neutral molecule. An additional photon or photons then further excite the molecule to its ionization continuum to produce a

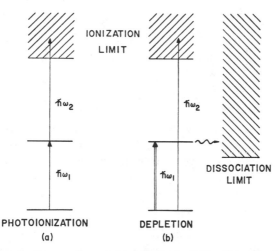

Figure 6.3   (a) Schematic diagram of resonant two-photon ionization. The molecule is pumped to an excited state by a photon energy $\hbar\omega_1$, then ionized with a single photon from a second laser ($\hbar\omega_2$). (b) To obtain a depletion spectrum, a predissociating excited level is populated by an intense laser pulse ($\hbar\omega_1$). A second probe laser ($\hbar\omega_2$) directly ionizes the remaining population in the molecular beam and provides an indication of when $\hbar\omega_1$ is tuned to a predissociating level. (From Ref. 114.)

cation. The total ion current or the mass analyzed ion current is monitored as a function of the frequency of the initial exciting laser. Provided that the second (R2PI) photoionizing photon does not lead to a dissociative ionization, there is a direct correlation between the frequency of the first exciting laser photon and a resonant transition in the parent molecule. Initial experiments using the R2PI technique employed only a single laser, the two photons used being of the same frequency; however, more recently two-color experiments employing two different lasers to excite and ionize provide a particularly powerful method for cluster identification since they allow the ionizing laser to be tuned just above threshold. This considerably heightens the possibility of observing the parent ion without extensive fragmentation although it may also be necessary to define autoionization carefully.

An alternate to the R2PI technique that is applicable to molecules undergoing predissociation is depicted in Figure 6.3(b). This technique (114) also provides a means for probing quantum levels to which the R2PI technique is blind. If a metal cluster molecule undergoes facile fragmentation due to predissociation and the predissociating transition is saturated by an intense laser field, one has an effective method for removing a particular metal cluster from the expanded cluster beam. A subsequent probe laser that directly ionizes the cluster in a single photon process can be used to monitor a depleted cluster population in the molecular beam. The photoion current for the one-photon ionization process will decrease as a first laser is scanned across the predissociating transition.

6.4.3 Overview of Small Metal Cluster Optical Spectroscopy

The groups that are pursuing the study of small metal cluster gas-phase spectroscopy employ primarily combinations or variants of those techniques described in the previous section.

Experiments Using Oven-based Technology

Schumacher and coworkers have been among the pioneers in this field (5,6,9,106–108). In extending and refining the early studies of Leckenby et al. (115), these authors have generated hot oven-pure metal supersonic nozzle beams of sodium, potassium, and Na/K molecular aggregates, studying the products of expansion with an impressive array of spectroscopies that, for the most part, have a basis in ion production and mass spectroscopy.

Particle Specific Measurement and Modeling of Alkali Cluster Ionization Potentials. Using single photon photoionization near threshold, Hermann et al. (5) have identified the clusters Na_n ($n \leq 6$), K_n ($n \leq 12$), and Na_nK_m ($n + m \leq 6$) employing a quadrupole mass filter to detect their molecular photoions. The size and range of clusters that can be identified employing the photoionization technique should be compared to that using electron impact (20 eV electrons) where only those clusters containing less than four atoms are identified. Clearly, the photoionization technique provides a much milder means of producing the cluster ions than does electron impact that is characterized by much more extensive fragmentation. As Table 6.3 demonstrates, the ionization thresholds for Na_n, $n \leq 14$, K_n, $n \leq 8$, and several of the combinations Na_nK_m ($n + m \leq 6$) have been determined. The data for the sodium clusters which are now being extended, represent the longest precisely measured contiguous series of any property as a function of cluster size. Recent measurements by Castleman and coworkers (116) indicate somewhat lower thresholds for Na_6 (3.97 ± 0.04 eV) and Na_8 (3.9 eV) but otherwise are in very good agreement with the data in Table 6.3.

As Figure 6.4 demonstrates (5,6,9), the *main* trend in the photoionization potentials of the alkali clusters as a function of size appears to be well represented if one considers the excess work function associated with a conducting curved surface as compared to an infinite flat surface. The work function of a droplet may be expressed as

$$W(R) = e\phi_w + 3/8(e^2/R) \tag{6.6}$$

TABLE 6.3 Photoionization Threshold Energies for Alkali Clusters (eV)

	Leckenby et al.[a]	Schumacher et al.[b]		Leckenby et al.[a]	Schumacher et al.[b]
Na	(5.14)[c]	(5.14)	K	(4.34)	(4.34)
Na_2	4.9 ± 0.1	4.934 ± 0.011[d]	K_2	4.0 ± 0.1	4.05 ± 0.05
	4.90 ± 0.01[e]	4.866 ± 0.014[f]			4.059 ± 0.001[g]
Na_3	3.9 ± 0.1	3.97 ± 0.05	K_3	3.4 ± 0.1	3.3 ± 0.1
Na_4	4.2 ± 0.1	4.27 ± 0.05	K_4	3.6 ± 0.1	3.6 ± 0.1
Na_5	3.9 − 4.3	4.05 ± 0.05	K_5		3.3 ± 0.1
Na_6	4.05 − 4.35	4.12 ± 0.05	K_6		—
Na_7	3.9 − 4.15	4.04 ± 0.05	K_7		3.3 ± 0.1
Na_8	4.0 ± 0.1	4.10 ± 0.05	K_8		3.4 ± 0.1
Na_9		4.0 ± 0.1	NaK	4.5 ± 0.1	4.52 ± 0.05
Na_{10}		3.9 ± 0.1	Na_2K	3.6 ± 0.1	3.7 ± 0.1
Na_{11}		3.8 ± 0.1	Na_3K		4.1 ± 0.05
Na_{12}		3.6 ± 0.1	Na_4K		4.0 ± 0.1
Na_{13}		3.6 ± 0.1	Na_5K		4.1 ± 0.1
Na_{14}		3.5 ± 0.1	NaK_2	3.4 ± 0.1	3.6 ± 0.1
			Na_2K_2	4.1 ± 0.1	4.0 ± 0.1

[a]Ref. 115.

[b]Ref. 5.

[c]In parentheses: calibration values.

[d]Twelve independent monochromator measurements.

[e]R. D. Hudson, *J. Chem. Phys.*, **43**, 1790 (1965).

[f]Determined by two-photon ionization via $A^1 \Sigma_u^+$(v' = 1–10), 10 measurements with two lasers.

[g]Determined by two-photon ionization via $B^1 \Pi_u$(v' = 1–10), 10 measurements with two lasers.

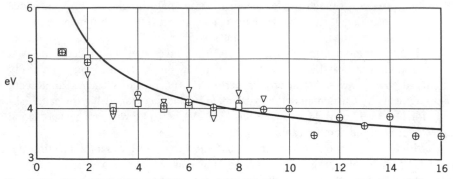

Figure 6.4 Single-photon photoionization potentials for Na$_x$. The curve corresponds to the work function of a conducting spherical drop. $\oplus$, experimental values; ∇, SCF-X_a-SW calculations; $\square$, pseudopotential ionization potentials. (From Ref. 6.)

where R is the radius of the metallic drop. The additive term results because the positive "image charge" only recedes to the center of the sphere instead of to minus infinity as characteristic of a flat surface when an electron is moved to plus infinity. Additional evidence now obtained by Kappes et al. (106) for a few very much larger clusters verifies the general trends expressed by this model and a gradual monatomic decrease to the work function of the bulk metallic phase. The radius of the metallic droplet can be approximated and interpolated between that of a diatomic metal molecule and the bulk metallic phase if one postulates that the diminution of the nearest neighbor distance between bulk and diatomic is caused exclusively by surface tension. This leads to the evaluation of an excess work function that, when subtracted from the measured Na$_n$ ionization potentials, yields a very reasonable approximation to the work function of the polycrystalline bulk metal. Deviations from this bulk value are thought to arise in some part from the electronic and structural properties of the clusters when referenced to a classical conducting metal sphere. In order to unfold the nature of these deviations, one must obtain a considerable catalog of both experimental data and detailed theory.

The data in Table 6.3 are of sufficient accuracy and resolution that they provide, in concert with theory, an indication of the complicated dynamics associated with small metal clusters and its influence on the photoionization potential. This is especially intriguing for the mixed Na/K alkali triatomics where experiment shows that the two distinct isomeric combinations of the same three atoms have virtually indistinguishable photoionization potentials (5,6,9). Theoretical calculations of sufficient accuracy on several sodium clusters indicate that the variation of geometric structure and Na—Na distance has only a small influence on the photoionization potential (6). The combination of experiment and theory (40), therefore, suggests that the alkali clusters are very floppy (fluxional) molecules (40) for which not only the Born-Oppenheimer but probably even the adiabatic approximations break down (5,6). These factors also are augmented in the heteronuclear alkali triatomics where it would appear that the nature of the transition state, with which are associated large amplitude bond-bending vibrations, is such so as to allow a rapid exchange of atoms, leading to the observation of a single photoionization potential (see also following discussion).

Although very impressive photoionization data have been obtained thus far for several of the alkali metals, the details of the threshold of photoemission, photoionization efficiency curves, and the spectrum of photoelectrons are largely unexplored. Models that relate cluster data to atomic ionization potentials and solid work functions are very promising but will require considerable fine tuning. Accurate cluster ionization potentials must be obtained for a number of other metals. Some data are available from electron impact ionization (see references cited in Section 6.3); however, these data and a portion of those obtained in the irradiation experiments are complicated by complex fragmentation processes that must also be carefully analyzed. It is hoped that many of these studies will be pursued, for the trends and models that have already emerged from the Schumacher group portend of important chemical and physical insights into this intermediate region. As we will discuss shortly, some information on ionization potentials in copper clusters has been obtained by Smalley and coworkers (113).

Application of Two-photon Ionization Spectroscopy.   An equally important contribution from the Schumacher Laboratory has been the development of particle-specific absorption spectroscopy via the two-photon ionization (TPI) technique. Important applications have been made in the study of bound-bound transitions in sodium trimer (6,9,107), the determination of extremely accurate dimer photoionization potentials (6,108), and the observation of spectral perturbations (6).

Sodium Trimer Electronic States and General Application.   In studying the spectrum of supersonically expanded cold sodium trimer, Schumacher and coworkers (6,9,107) have employed TPI spectroscopy to place an optical signature on the trimer and monitor predominantly the spectra of neutral trimer excited states. Here, they have tuned a dye laser over the excitation region while simultaneously probing the excited state population with a second ionizing argon ion laser within the ion source of a quadrupole mass spectrometer (tuned to the mass of Na_3) (107). Using this approach, a survey of the electronic absorption spectrum of the Na_3 molecule (resolution 0.4 to 0.6 cm^{-1}) encompassing almost the complete visible region, 4050 to 6900 Å, has revealed two absorption systems in the range 6100 to 6800 Å. The two-photon spectra generated are presented in Figures 6.5 and 6.6 and catalogued in Table 6.4.

The 6250 Å system (Fig. 6.5) appears to correspond primarily to a long, excited state vibronic progression with frequency spacings of ~125 cm^{-1}. This complex system may result from as many as three overlapping transitions. In the one electron picture it results from a transition that involves promotion from a strongly bonding a_1' orbital to a nonbonding e' orbital ($^2A_1'$, $^2A_2'$, $^2E' \leftarrow {}^2E'$). Therefore, a change in equilibrium bond strengths and bond lengths is consistent with a long, excited state symmetric stretch progression. The 6717 Å (Fig. 6.6) system which is approximately 20 times more intense than that at 6250 Å has been assigned to a $^2E''-{}^2E'$ transition and correlated with a short, excited state progression in conjunction with a sequence;

TABLE 6.4 Vibronic Bands Observed for Sodium Trimer Using TPI Spectroscopy

Band No.	Wavelength (Å)	Frequency (cm^{-1})	Frequency Difference
	6250 Å Band System		
1	6105.2	16379.4	$\Delta_{1,2} = 124.7$
2	6152.1	16254.7	$\Delta_{2,3} = 127.6$
3	6200.7	16127.1	$\Delta_{3,4} = 128.8$
4	6250.7	15998.3	$\Delta_{4,5} = 130.3$
5	6302.0	15868.0	$\Delta_{5,6} = 131.1$
6	6354.5	15736.9	
	6717 Å Band System		
1	6655.5	15025.0	
2	6676.8	14977.2	$\Delta_{1,2} = 47.9$
3	6717.0	14887.5	$\Delta_{1,3} = 137.6$
4	6738.0	14841.3	$\Delta_{3,4} = 46.2$
5	6757.5	14798.4	$\Delta_{4,5} = 43.0$
6	6775.8	14758.5	$\Delta_{3,6} = 129.0$
7	6798.7	14708.7	$\Delta_{6,7} = 49.8$

however, more recent laser-induced atomic fluorescence spectra and the nature of the intensity distribution in the 6717 Å system indicate that the major feature may correspond to a (0,0) band and that the bands (1) and (6) correspond to the (1,0) and (0,1) bands associated with one quanta of the symmetric stretch in the ground and excited states of the trimer. This is consistent with an expected excitation from a ground-state nonbonding e' orbital to an approximately nonbonding e'' orbital. Because both of these orbitals are nonbonding, no great change in bond length or

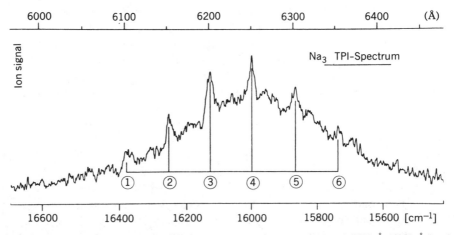

Figure 6.5 Two-photon photoionization spectrum of Na$_3$ between 6000 and 6500 Å (6250 Å band system) identifying a long, excited state vibronic progression. (From Ref. 107.)

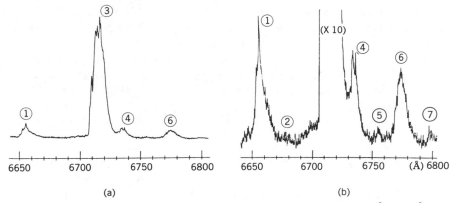

(a)   (b)

Figure 6.6 Two-photon photoionization spectrum of Na$_3$ between 6640 and 6810 Å (6717 Å band system). (From Ref. 107.)

equilibrium geometry is anticipated and the spectra should be consistent with a narrow "Frank-Condon" distribution.

The spectra in Figures 6.5 and 6.6 represent the first electronic absorption spectra of a triatomic metal cluster definitively assignable to a single species and measured under single collision conditions.* Schumacher and coworkers are now attempting to analyze the sodium trimer spectrum at a much higher resolution in order to determine definitively the structure of the ground state of the molecule. Thus far, experiments in which the trimer is probed at a resolution of 50 MHz ($\sim$0.002 cm^{-1}) have not yet produced a successful analysis, probably as a result of extensive predissociation and hence rotational line broadening. In an attempt to remedy this problem, the Schumacher group is now extending their efforts to probe excited trimer states in the near infrared where it is anticipated that predissociation will be minimized.

The TPI approach as applied to sodium trimer is generalizable in the sense that given the initial dye laser excitation, one can induce ionization using either a laser or a broad band light source within the ion source of a quadrupole or time-of-flight mass spectrometer. If one reverses the role of the two lasers, it is possible to obtain information about ionization or ion states through initial narrow band excitation to a fixed neutral *rovibronic* state of the molecule and a subsequent dye laser scan of the ionization region. Given that the stabilized power of an initial exciting dye laser is small, TPI spectroscopy will yield correct absorption intensities provided that fragmentation is absent (6). The ionization efficiency of the second laser is the bottleneck of the TPI process and, as a result, high-power density is usually necessary in order to harvest a high fraction of the excited state population for the m/e ion

*Single collision conditions are obtained at sufficiently low pressures (typically 10^{-4} to 10^{-7} torr) so that those processes under study are not contaminated by severe collisional repopulation or quenching of initially formed excited state levels. In a TPI spectral study, the initially pumped excited state of interest is ionized before undergoing subsequent collisions.

current. Using the TPI technique, one can pick out the Na_3 fluorescence at 6717 Å within the 100 times more intense vibronic progression of the Na_2 A–X transition (6,9) located in the same region. This is indeed an impressive result.

The TPI technique is also readily amenable to the detection of spectral perturbations (6,9). In contrast to the attendant weakening of spectral intensity that commonly characterizes a perturbation, they are manifest in TPI spectroscopy by an anomalously high-line intensity (in addition to frequency shift), because the perturbing state generally has a longer lifetime and hence is more efficiently ionized.

Highly Accurate Determination of Ionization Potentials—Successes and Shortcomings of the TPI Technique. In order to use TPI spectroscopy to determine very accurate dimer ionization potentials (6,108), one employs two dye lasers as outlined in Figure 6.7 and requires that the molecules under study be cooled to temperatures comparable to those characterizing the Na_2 spectrum in Figure 6.1. Rotational cooling is essential for the determination of the ionization threshold because up to approximately 1000 cm^{-1} above the ionization threshold a typical ionization efficiency curve is surprisingly rich in sharp detail (Fig. 6.8) especially if the inter-

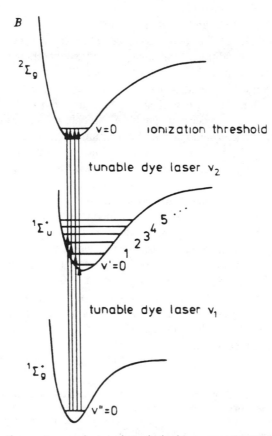

Figure 6.7 Absorption spectroscopy by two-photon ionization mass spectrometry for a dimeric species. Substates of the intermediate electronic state are investigated. (From Ref. 6.)

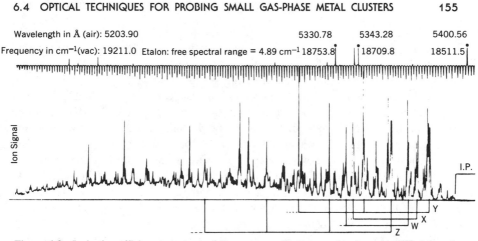

Wavelength in Å (air): 5203.90   5330.78   5343.28   5400.56

Frequency in cm⁻¹(vac): 19211.0 Etalon: free spectral range = 4.89 cm⁻¹ 18753.8   18709.8   18511.5

Figure 6.8 Ionization efficiency spectrum of Na_2 corresponding to pumping from the $B\,^1\Pi_u$ ($v' = 5$, $J'' = 0-6$) states. IP = ionization threshold. The asterisks correspond to optogalvanic Ne lines. (From Ref. 6.)

mediate state has a higher vibrational quantum number. This structured behavior resulting from autoionizing transitions portends of vastly different ionization efficiencies that may differ by more than a factor of 100 as a function of frequency. These differences are also manifest when the ionization efficiencies of various isotopic forms (e.g., $^{39}K_2$ and $^{39}K^{41}K$) (6) are compared as a function of frequency. It has been suggested that resonance photoionization via very intense molecular autoionizing transitions can be used to provide a 100-fold increase in the detection efficiency of the TPI technique (6).

The TPI technique has allowed the determination of the most precise ionization potentials obtained for Na_2, K_2, and NaK. As Figure 6.8 demonstrates for Na_2, the ionization potential is determined from the sharp onset of a photoion signal. Table 6.5 compares single PI, TPI, and electron impact measurements for Na_2, NaK, and K_2. It has been suggested that the single-photon ionization and electron impact determinations correspond to vertical ionization potentials (6), whereas the TPI technique yields an adiabatic value; however, this is clearly not definitive. Although the accuracy quoted in Table 6.5 is impressive, recent studies of autoionizing Rydberg states in Na_2 (117) that have allowed a reasonable extrapolated determination of the Na_2^+, $X^2\Sigma_g^+$ vibrational term energies indicate a Na_2 adiabatic ionization potential, IP = 39481 ± 6 cm⁻¹ (4.89414 ± 0.00074 eV). This value is some 49

TABLE 6.5 Ionization Potentials Measured for the Alkali Dimers Na_2, NaK, and K_2 (eV)

Na_2	NaK	K_2	Method[a]
4.9 ± 0.1	4.5 ± 0.1	4.0 ± 0.1	PI
4.934 ± 0.011	4.52 ± 0.05	4.05 ± 0.05	PI
4.88898 ± 0.00016	4.57 ± 0.2		EI
	4.41636 ± 0.00017	4.06075 ± 0.00015	TPI

[a]PI is a single-photon ionization; EI electron impact ionization; and TPI two-photon ionization.

cm^{-1} (0.00607 eV) above the photoionization threshold. The difference is attributable *only in part* ($\sim$20 cm^{-1}) to field ionization (lowering of the effective ionization potential due to the repeller field in the mass spectrometer ion source). The origin of the 30 cm^{-1} discrepancy remains an intriguing question.

Despite considerable effort, it has not been possible to apply the double dye laser TPI technique to determine a very precise Na_3 ionization potential (107). As Table 6.3 indicates, the Na_3 single-photon ionization potential is 3.97 eV. Using an initial dye laser operating at the wavelength (6717 Å) corresponding to the strongest Na_3 transition (Fig. 6.6), one would expect to find, as the single-photon ionization potential suggests, ion currents for ionizing wavelengths shorter than 5837 Å. In fact, no ion signal is detected when the wavelength region 5500 to 5900 Å is scanned with more than 2 watts of dye laser power. Subsequent studies of the Na_3 ionization efficiency have indicated that a two-step process via the excited state at 6717 Å requires $\sim$0.24 eV more energy to reach the ionization threshold as compared to the direct "vertical" ionization process from the ground state.

From the previous discussion, we hope that it is apparent that the determination of *precise and accurate* metal cluster ionization potentials, as well as their prediction, represents a challenging task. High accuracy will demand not only the development of new excitation schemes but also the complex intermingling of several approaches.

The Fragmentation Process—A Possible Solution. Although the two-photon technique offers a powerful means of species spectral-mass correlation, especially with the advent of two-color spectroscopy, fragmentation can represent a problem if the potential curves for the neutral and ion are shifted significantly from each other. A second photon, once absorbed, may place the molecule high on the repulsive wall of a final ionic state leading to subsequent dissociation. Delacretez et al. (118) have developed a two-photon, two-color pulsed laser time-of-flight combination to study alkali clusters. These authors have found that the excitation and ionization channels are accompanied by strong fragmentation processes and that the fragment patterns are very sensitive to the laser wavelength even when working near the ionization threshold. As noted by these authors, their results indicate that the peak intensities of cluster mass spectra cannot easily be related to the intensity distribution of the neutral cluster beam. We believe that it may be possible to alleviate this ambiguity and distinguish between direct and dissociative ionization processes.

The scattering of the fragments that result from a dissociative ionization process significantly exceeds that inherent in a direct ionization process or in the deflection of molecules (119) by the radiation pressure of laser light. If one varies the acceptance angle of the system mass spectrometer by "reproducibly" moving the mass spectrometer with respect to the fluorescence zone, it should be possible to distinguish two processes which might be exemplified by

$$Na_5 + h\nu_1 \rightarrow Na_3^* \tag{6.6}$$

$$Na_3^* + h\nu_2 \rightarrow Na_3^+ + Na_2$$

$$Na_3 + h\nu_1 \rightarrow Na_3^* \tag{6.7}$$

$$Na_3^* + h\nu_2 \rightarrow Na_3^+$$

As the mass spectrometer is moved away from the fluorescence zone or the acceptance angle of the mass spectrometer is decreased, the Na_3^+ signal from process (6.6) would be expected to drop precipitously relative to that from process (6.7). The dropoff in Na_3^+ intensity with distance of the mass spectrometer from the viewing zone should display approximately a $1/r^2$ dependence if process (6.7) is operative. The dropoff in the Na_3^+ signal will be much faster if process (6.6) is operative.

Metal Cluster Photodissociation

Bound-free Transitions in Sodium Trimer—Correlation with TPI Spectroscopy. In an effort that is complementary to the study of bound-bound transitions, Gole et al. (102,120) have developed techniques for the study of metal cluster photodissociation induced through the pumping of bound-free transitions. Their focus has been on the sodium trimer molecule. Single-photon optical pumping is used to excite from the ground electronic and low-lying states of the trimer to repulsive excited state levels with an open channel to dissociation. One monitors the electronically excited atomic states that are the products of this dissociation. Hence, one observes laser-induced atomic fluorescence, LIAF, that, in the case of sodium trimer, corresponds to the production of ^2P sodium atoms that are monitored using the well-known sodium D line. As the dye laser that induces photodissociation is frequency scanned, one records the magnitude of emission from the $^2P_{1/2}$ and $^2P_{3/2}$ components of the sodium D line. A typical LIAF spectrum obtained for Na_3 produced cold in high-purity metallic expansion is shown in Figure 6.9a. As the dye laser frequency is scanned, one produces a series of "fluctuation bands" in the intensity of both the $^2P_{1/2}$ and $^2P_{3/2}$ atomic emissions. These bands in large part reflect the vibronic structure of the ground state of sodium trimer, a preliminary analysis indicating that they correspond to a vibrationally heated yet rotationally cold distribution. A slower scan of either the $^2P_{1/2}$ or the $^2P_{3/2}$ LIAF spectrum (Fig. 6.9b) reveals a further structure which is cataloged in Table 6.6. The observed features are thought to arise from a strongly allowed ($1a_2'' \leftarrow 1a_1'$) transition that theory indicates falls at ~4800 Å (6,107).

In contrast to the $^2P_{1/2}$ LIAF spectrum, the $^2P_{3/2}$ LIAF spectrum is accompanied by sharp (vs. broad LIAF) features in close frequency coincidence with the Na_2 $B^1\Pi_u - X^1\Sigma_g^+$ excitation spectrum also shown in Figure 6.9b. These features result from the combination of a surprisingly efficient energy transfer process and a subsequent dissociation, namely,

$$Na_2^* + Na_3 \rightarrow Na_3^* + Na_2 \tag{6.8}$$

$$Na_3^* \rightarrow Na_2 + Na^* \tag{6.9}$$

$$Na^* \rightarrow Na + h\nu \ (D \ line) \tag{6.10}$$

strongly favoring the $^2P_{3/2}$ component.

The measured short-wavelength limit of the LIAF spectrum allows the determination of an upper bound to the sodium trimer bond energy, $D_0 \leq 4250 \ cm^{-1}$ $(21,200 - 16,950 \ cm^{-1})$. In addition, using the spectra in Figure 6.9, one has the

TABLE 6.6   Peak Separations Sodium Trimer Photodissociation Spectrum

Peak No.[a]	Frequency (cm⁻¹)	Δ(Separations)	
0	21133 ± 5^b		
		$113 \pm 10^b \ (\Delta_{0,1})$	
1	21020 ± 4		
		$99 \pm 4 \ (\Delta_{1,2'})$	$\Delta_{1,2} = 110 \pm 4^c$
2′	20921 ± 2		
		$112 \pm 4 \ (\Delta_{1,2''})$	
2″	20899 ± 2		
		$131 \pm 3 \ (\Delta_{2,3})$	
3	20779 ± 3		
		$133 \pm 2 \ (\Delta_{3,4})$	
4	20647 ± 3		
		$134 \pm 3 \ (\Delta_{4,5})$	
5	20513 ± 4		
		$133 \pm 8 \ (\Delta_{5,6})$	
6	20380 ± 5		

[a]From labeled peaks in Figure 6.9.

[b]Uncertainties represent standard deviations in the measurements taken from a number of experimental spectra.

[c]From average of 2′, 2″ peaks.

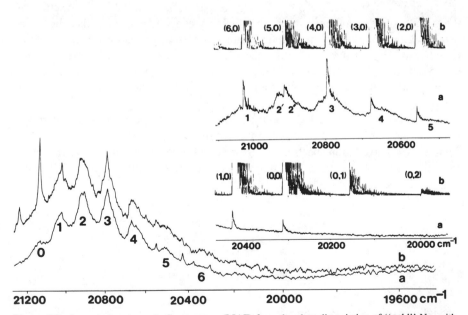

Figure 6.9   Laser-induced atomic fluorescence (LIAF) from the photodissociation of "cold" Na_3 with an Ar ion pumped (coumarin 102) dye laser, $a = {}^2P_{1/2}$, $b = {}^2P_{3/2}$ LIAF. Inset—closeup of LIAF spectrum corresponding to the ${}^2P_{3/2}$ component of the Na D lines showing sharp features which coincide with the cold Na_2 $B - X$ excitation spectrum (band heads denoted (v', v'')). See text for discussion.

potential for modeling a significant portion of the vibronic structure in the ground electronic state of the sodium trimer molecule, as well as obtaining information on the repulsive states that dissociate to excited state atoms. Aspects of the sodium trimer potential surface are discussed elsewhere (11); however, detailed quantum chemistry (121,122) indicates that the lowest energy electron configuration of the trimer, of symmetry $^2E'$, Jahn-Teller distorts to a $^2B_2$ obtuse-angled and $^2A_1$ acute-angled configuration. The $^2B_2$ configuration corresponds to the ground state (minimum on the Na$_3$ surface) and lies $\sim$0.6 kcal (200 cm^{-1}) below the $^2A_1$ acute-angled configuration which corresponds to a saddle point on the Na$_3$ potential energy surface. Another saddle point, a $^2\Sigma_u^+$ linear conformation is thought to lie some 3 kcal/mole (1050 cm^{-1}) above the $^2B_2$ conformation. Based on the location of the $^2B_2$ state and the $^2A_1$ saddle point, a barrier to pseudorotation of only 200 cm^{-1} characterizes the system.

The pseudorotation interaction in the ground state of Na$_3$ leads to a coupling of the bending and assymmetric stretching modes that will necessarily produce complicated spectral effects. However, the manifold of levels corresponding to the symmetric stretch should be only mildly affected (120). If several trimer ground-state vibronic levels are populated, one might anticipate a complicated spectral pattern from which emerge peaks associated with the symmetric stretching mode. Figure 6.9 and Table 6.6 provide support for this suggestion. The peak separations for the highest energy features (lowest levels in the ground state) are on the order of 100 to 110 cm^{-1}, and it would not be unreasonable to associate these features with significant populations in levels that result from a coupling of the bending and assymetric stretching manifolds. Of course, higher-resolution scans will be needed for definitive identification of this rather overlapped region. The 130 to 136 cm^{-1} separations for the bands 2 to 6 appear to bear a correlation with the observed features for the Na$_3$ 6717 Å system (Fig. 6.5). Based on the nature of the 6717 Å transition, one would like to suggest that the central peak corresponds to a strong origin band and that the side peaks (1) and (6) separated from the central peak by $\sim$130 to 135 cm^{-1} correspond to quanta in the excited and ground-state symmetric stretch. This indicates a ground-state symmetric stretch frequency of $\sim$134 cm^{-1} in very good agreement with that calculated by Martin and Davidson (147 $\pm$ 20 cm^{-1}) (121) and estimated by Gerber ($\sim$137 cm^{-1}) (122) based on extrapolation of detailed vibronic calculations on Li$_3$. If the symmetric stretch frequency for the ground electronic state of sodium trimer is $\sim$135 cm^{-1} and the anharmonicity of the symmetric stretch manifold is reasonably small, the features 3 to 6 in Figure 6.9 correspond, at least in part, to vibrationally hot levels of the symmetric stretch manifold. There appears to be a strong interlocking correlation between the 6717 Å system observed by Hermann et al. and the LIAF spectrum observed by Gole et al.* (42). Further higher-resolution (0.03 $\rightarrow$ 0.05 cm^{-1}) LIAF spectroscopy especially in the region 21,800 to 20,800 cm^{-1} should yield further information on the complicated vibronic coupling in the sodium trimer ground state.

---

*LIAF studies are also complementary to the study of sodium clusters in low-temperature matrices where the analog of the observed sodium trimer bound-free transition may be observed directly as a result of matrix stabilization due to the matrix cage effect (6).

Controlled Poisoning of Supersonic Expansions—Fluorescence from Hot Bands and Low-lying States of Weakly Bound Species. The production of very cold clusters requires the formation of very clean supersonic expansions. Approaching from another viewpoint, the Gole group has engineered the controlled disruption of the supersonic expansion in order to obtain "hot band" fluorescence from metal clusters. Collisional heating facilitates the possible characterization of a significantly greater component of the vibronic structure of the ground electronic state of these "weakly bound" species and offers the potential for exploring levels close to the ground-state dissociation limit. Using this collisional heating, Gole and coworkers have been able to enhance significantly the growth of a second fluctuation system, on-setting at 19,600 cm^{-1} and barely observable in the cold spectrum of Figure 6.9(a). This second fluctuation system might be associated primarily with the pumping of the $Na_3{}^2\Sigma_u^+$ linear configuration. This may occur because the sodium trimer surface is extremely flat and the trimer is very "floppy." Hence, the observed hot band fluorescence may result from the optical pumping of an averaged linear configuration. A more detailed discussion is given elsewhere (101,120).

The combination of TPI and LIAF spectra has already furnished a considerable mapping of the level structure of the sodium trimer molecule. The refinement of these studies and their extension in Na_3 and other small metal clusters should provide the grist for detailed theoretical treatment and the development of useful empirical bonding models.

Magnetic Deflection and ESR Spectroscopy of Small Alkali Metal Clusters.

Further information on the molecular electronic structure of alkali clusters can be garnered from the study of their ESR spectra. Following the very significant matrix studies of Lindsay et al. (13), Knight and coworkers (123) have attempted to study the gas-phase ESR spectra of supersonically expanded and particle-selected alkali clusters. These are very challenging experiments. At temperatures in excess of 20K, spin-rotation couplings are expected to complicate severely observed ESR spectral patterns. Knight and coworkers have succeeded in studying the magnetism of the alkali clusters of potassium and sodium and their combinations. In a beam containing clusters of potassium ranging in size between 1 and 100 atoms, magnetic deflection patterns are found to reflect three components: atoms, undeflectable clusters containing even numbers of atoms, and clusters with odd numbers of atoms with magnetic moments of approximately one Bohr magneton. In more refined studies, the Stern-Gerlach deflection profile for isolated Na_3 is characterized by one Bohr magneton peaks astride a broad central maximum (beam axis), the manifest magnetism resulting from the odd electron spin. The peaks and central maximum arise respectively from Na_3 in states with weak and strong spin-rotation couplings. Other isolated odd molecules including Na_5, K_3, Na_2K, and NaK_2 are found to show broadened profiles but Stern-Gerlach peaks are not evident.

Extensions of Oven Techniques to More Refractory Metals.

The pure metal-hot oven supersonic nozzle technique has been extended to the study of higher boiling metals including copper. In formulating these high-temperature nozzle sources, one encounters problems that are twofold. One must develop a source that maintains its structural integrity during the vaporization process. A first approach might be to consult phase diagrams (where reliable data are available); however, the conditions under which these diagrams are formulated, that is, equilibrium measurements, do not approach those conditions in a nozzle source nor do they provide information on the rate of mixing and/or alloying of various materials. One is forced to use a variety of scattered information throughout the literature. Given the development of a source that maintains its structural integrity, one must still be concerned with the physical mechanism of vaporization. It is imperative that one construct a device that maintains a stable, slightly unsaturated vapor at the backing side of the nozzle. Despite these drawbacks, appropriate devices have been constructed and cooling obtained in supersonic expansion. When pure copper vapor was expanded super-sonically from an early oven design at temperatures between 2400 and 2800K, laser-induced fluorescence demonstrated that the copper dimer produced is at vibrational and rotational temperatures that have approached 650K (103).

Using a pure copper expansion from an oven at 1800 to 2100K, Crumley et al. (124) have observed single-photon laser-induced fluorescence from copper trimer produced at rotational temperatures estimated to be between 100 and 150K. The sample spectrum shown in Figure 6.10 is dominated by what appears to be a short progression in an excited state vibronic mode. The relative intensities of the bands result primarily from the very large concentration of copper trimers produced in this experiment and the effective "Franck-Condon" factors for transition. Note that the phototube response increases (notably) with increasing frequency.

The extension of hot oven systems presents a notably more complicated problem than operating with the alkali metals; however, once developed, they have some distinct advantages in that they can produce usable cluster concentrations some three to five orders of magnitude in excess of that obtained with the very promising laser vaporization techniques that will be discussed in the following section. Using these intense continuous beams, one has been able to study metal cluster oxidation (125), and in the near future it is anticipated that they will find application in the generation

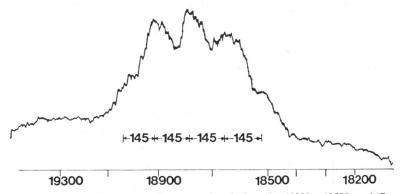

Figure 6.10 Cooled excitation spectrum for copper trimer in the region 18200 to 19500 cm^{-1} (Coumarin 7) showing a short progression in what is tentatively correlated with an excited, state vibronic mode. (From Ref. 124.)

of small cluster impregnated surfaces that may prove useful as heterogeneous catalysts (126).

Other future possibilities for oven-based systems that involve the vaporization of refractory metals may follow the lead of Riley and coworkers (127) who have reported a metal cluster beam source capable of producing continuous beams of chromium, nickel, aluminum, copper, and silver thermalized to near-room temperature. Their technique, which involves a modification of those sources used to generate very large metal aggregates (see following discussion), considers a careful balance of temperature and flow conditions as an inert gas is passed over a hot sample of the desired metal. Both low-energy electron-impact quadrupole mass spectrometry and time-of-flight analysis coupled with laser ionization have been used to assess those clusters present in the beam.

Experiments Based on the Laser Vaporization of Metals

Although the previous sections have focused on oven-based continuous source techniques for generating metal clusters, an elegant alternative to high-temperature furnaces involves vaporization of refractory materials using intense laser pulses. Two groups have been primarily responsible for the development of this inherently pulsed source technology: Bondybey and coworkers at Bell Telephone Laboratories and Smalley and coworkers at Rice University.

Nitrogen-pumped Dye Laser Fluorescence from Metal Vapors Entrained in Liquid Nitrogen-cooled Helium.   Bondybey and coworkers (14,128–133) have used the second, third and fourth harmonics of a Neodymium-YAG laser to vaporize and create a plasma of those metals of interest. The metal once vaporized by a loosely focused laser beam is entrained in helium that has been cooled to 77K (liquid nitrogen dewar) and the mixture is passed through a small orifice into an evacuated chamber. No supersonic expansion is used here and cooling is accomplished through intimate collisional contact with the cold carrier gas. The chamber into which the entrained mixture passes is usually maintained at 1 to 3 torr for spectroscopic studies but can be varied between 0.5 and 10 torr for lifetime measurements by adjusting the helium flow.

Metal clusters and metal oxides (14) that effuse from the orifice are probed by a pulsed tunable nitrogen-pumped dye laser. Dye laser pulses are suitably delayed with respect to the vaporizing laser pulse to maximize the signal of interest. Because the major variable in the system is the speed at which helium flows, this effectively represents the system clock. Typical molecular internal temperatures produced using this technique range between 80 and 120K. The cooling produced here may not approach the limits obtained for rotational or translational temperatures using very elegant supersonic expansion techniques; however, the spectra are still characterized by readily analyzable structure and the significant increase in the number of rotational levels sampled is such as to provide information on unusual features to which the very cold rotational distribution may be blind. In addition, vibrational hot bands are much more readily probed and analyzed. Because of the cooling inherent in the technique, excitation spectroscopic studies are readily accomplished; however, the use of photoluminescence spectroscopy is also feasible and will represent a future consideration.

Thus far Bondybey and coworkers have obtained electronic spectra and garnered significant lifetime data on several homonuclear metal molecules including Pb_2 (128,129), Sn_2 (129), Cu_2 (130,131), and Cr_2 (132). In addition, they have also carried out the first spectroscopic studies of the cooled heteronuclear metal molecules CuGa, CuIn (131), and SnBi (133). Data obtained for the systems studied is summarized in Table 6.7. The power of this straightforward approach and the degree of simplicity imposed on the observed spectra by virtue of their cooling is exemplified in Figure 6.11 where one observes the very clean vibrational and rotational spectra obtained for the *heavy* lead dimer molecule. The useful nature of the internal excitation range that this technique provides over very cold supersonic beams has been exemplified in studies of chromium dimer (132) where it has been possible to determine a ground-state vibrational frequency and observe an unusual modulated rotational level intensity distribution (Fig. 6.12) subsequently analyzed and assigned to a rather unique predissociative phenomena by Riley and coworkers (134).

Although only the spectra for diatomics have been reported thus far, some evidence has been obtained for the observation of Pb_3, Sn_3, and Cu_3 electronic band systems. What is very intriguing is the Cu_3 system where laser fluorescence is observed at an electronic energy in close coincidence with the data obtained in the gas phase by Crumley et al. (124) (Fig. 6.10) and Morse et al. (114) (see following discussion) and with the matrix absorption noted by Moskovits and Hulse (11). A small modification of the present technique to provide further columnation of the metal plasma effluent plus helium flow may well significantly increase the metal trimer concentration and provide an enhanced catalog of important trimer spectra. A great advantage of the technique described here is its generality and relative ease of operation. It should be applicable not only to metal clusters but also to oxides, nitrides, carbides, and many other refractory materials.

Combined Pulsed Laser Vaporization—Pulsed Supersonic Expansion Technique to Produce Ultra-Cold Metal Clusters.   Smalley and coworkers (104) have developed an elegant technique whereby cold metal clusters are produced and investigated using a combination of pulsed laser vaporization-supersonic expansion and pulsed laser ionization-time-of-flight mass spectrometry. Their approach involves an extension of the photoionization techniques developed by Schumacher and coworkers to higher boiling metals (104,109). Metal atoms (plasma) are generated by laser vaporization using the second harmonic (532 nm) of a Q-switched Nd:YAG laser. The laser impinges on a metal target that consists of a metal rod located in the throat of a pulsed supersonic nozzle assembly. The metal rod is rotated by a synchronous motor-driven micrometer screw preventing irregular laser vaporization, and thus facilitating the generation of "stable" metal cluster concentrations. The optimum laser fluence to produce a desired cluster distribution is found to vary with the properties of the metal under investigation. An intense supersonic helium gas pulse is employed to recombine and cool the laser-produced plasma. The firing of the vaporization laser is synchronized with the supersonic pulse so that it coincides

TABLE 6.7 Spectroscopic Information Obtained from Combined Laser-Vaporization Nitrogen pumped Dye Laser Induced Fluorescence (Bondybey et al.)

Molecule	Observed Transitions	Molecular Constants Obtained	Radiative Lifetimes (nanoseconds)		
Cr_2	$A^1\Sigma_u^+ - X^1\Sigma_g^+$ $\nu_{0,0} = 21751.42 \pm 0.02$	$A^1\Sigma_u^+$: $\Delta G_{1/2} = 414.8 \pm 0.5$, $B_0 = 0.2274$ cm^{-1}, $r_0 = 1.6896$ Å $X^1\Sigma_g^+$: $\Delta G_{1/2} = 452.34 \pm 0.02$, $B_0 = 0.2284$, $B_1 = 0.2246$ $B_e = 0.2303$ cm^{-1}, $r_0 = 1.68558$ Å, $r_e = 1.6788$ Å, $\alpha_e = 0.0038$ cm^{-1}			
Cu_2	$A^1\Pi_u - X^1\Sigma_g^+$ $B^1\Sigma_u - X^1\Sigma_g^+$ $B' - X\Sigma_g^+$ $\nu_{0,0} = 21848$ cm^{-1}	B': $\omega_e' \sim 221$, $\omega_e x_e' \sim 2$ cm^{-1}	$A^1\Pi_u$: $v' = 0$ 115 ± 10^a $v' = 1$ 75 ± 10 $v' = 2$ 40 ± 10 $B^1\Sigma_u$ $v' = 0\text{-}2$ 40 ± 5 B' $v' = 0$ 800 ± 100^b		

CuGa	$A - X^1\Sigma$ ($\nu_{00} \sim 25158$ cm^{-1})	$\omega'_e \sim 117$, $\omega_e x'_e \sim 0.8$ cm^{-1} $\Delta G''_{1/2} = 220 \pm 2$ cm^{-1}	57 ± 5^c
CuIn	$A - X^1\Sigma$	$\omega'_e \sim 151$, $\omega_e x'_e \sim 2.3$ cm^{-1}	40 ± 5^c
Pb$_2$	$C(O_u^+) - X(O_g^+)$ ($T_e - 15314$, $\nu_{00} = 15323$ cm^{-1}) $F(O_u^+) - X(O_g^+)$ ($T_e = 19807$, $\nu_{00} = 19832$ cm^{-1})	$X(O_g^+)$: $\omega''_e = 109.7$, $\omega_e x''_e = 0.19$ cm^{-1} $C(O_u^+)$: $\omega'_e = 129.3$, $\omega_e x'_e = 1.27$ cm^{-1} $F(O_u^+)$: $\omega'_e = 158.7$, $\omega_e x'_e = 1.16$, $\omega_e y'_e = 0.011$ cm^{-1}	1500^d 170 ± 5 ($v' = 0$)

[a]Sharp dependence of radiative lifetime on vibrational quantum level indicates the onset of predissociation, lifetime independent of pressure in range 0.5 to 15 torr.

[b]Extrapolation to zero pressure—strong pressure dependence of lifetime.

[c]Pressure range 1 to 10 torr constant lifetime.

[d]Extrapolation to zero pressure—Lifetime strongly pressure dependent.

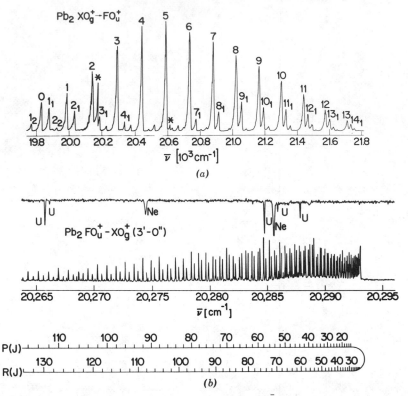

Figure 6.11   (a) Laser excitation spectrum in the region of the Pb_2 $\bar{F}(O_0^+) \rightarrow X(O_2^+)$ transition. The v' numbering is shown above the bands and the ground-state vibrational numbering is given for the hot bands as subscripts. (b) High-resolution scan showing the rotational level structure for the ($v' = 3$, $v'' = 0$) band of the $F - X$ system. (From Ref. 128.)

with the maximum pressure of helium carrier gas in the nozzle. The nozzle is pulsed by a double-solenoid mechanism with a typical backing pressure of 100 psi.

After vaporization and formation of the high-temperature plasma, the high-pressure gas (pulse) flows into a channel of variable length in which thermalizing collisions with helium atoms recombine the plasma and allow cluster formation. The contact time between the metal vapor and the helium has a profound effect on the relative concentrations of various cluster sizes in the resulting beam. Adjustments in the length of the exit channel are found to provide a convenient means of controlling the extent of cluster formation. A typical channel of length 30 mm easily can result in the production of copper clusters in the size range 1 to 29 atoms. After passing through the channel, the carrier gas and clusters are freely expanded into vacuum (10^{-5} torr) and are further cooled. Because the clustering of metal atoms requires three-body collision processes, the rate of formation of metal clusters falls off rapidly as the gas begins to expand freely; the majority of clusters are formed in the exit channel and their remaining internal energy is removed primarily in the

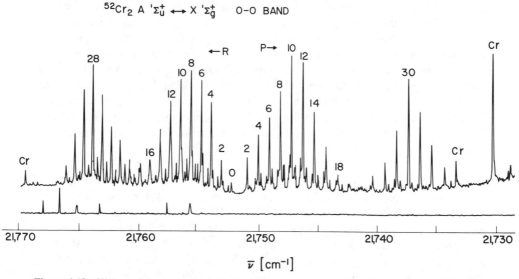

Figure 6.12 High-resolution scan of the Cr_2 O—O band. The strong labeled lines belong to $^{52}Cr_2$; the weaker lines that are apparent in the spectrum result from $^{52}Cr^{50}Cr$ and $^{54}Cr^{52}Cr$. The bottom trace corresponds to a uranium optogalvanic trace. (From Ref. 132.)

subsequent supersonic-free expansion. The rotational temperatures determined for diatomic molecules formed in this manner are routinely less than 5K.

The freely expanded cluster-helium beam passes through a skimmer into a low-pressure (10^{-6} torr) interrogation region where the cluster beam is crossed with a variety of lasers that photoionize the sample either directly or through resonant two-photon ionization (R2PI). Following photoionization, a time-of-flight mass spectrometer is used to monitor the spectral signature of a particular cluster and/or isotopic modifications of this cluster.

Using this approach, Smalley and coworkers have mapped interesting trends in the ionization potentials of copper clusters (113) and are obtaining detailed spectroscopic information on copper trimer (114). They also have obtained detailed data on several refractory transition-metal diatomics including Cu_2 (110,113), Cr_2 (111), V_2 (112), Mo_2 (104), and Ni_2 (112).

Trends in Copper Cluster Ionization Potentials. Powers et al. (113) have garnered extensive data on the ionization behavior of copper clusters ranging in size from 1 to 29 atoms. Using a number of fixed frequency outputs from an exciplex laser, these authors have obtained an approximate monitor of the threshold behavior of the photoionization cross section as a function of cluster size. The 7.9 eV photon energy of the F_2 excimer laser was found to be above the ionization potential of all clusters formed, the photoion mass spectrum showing a monotomic decrease as a function of cluster size. The 4.98 photon energy of the KrF laser was found to lie below the ionization threshold of all clusters in the range 1 to 29 atoms. The 6.4 eV ArF exciplex laser photon energy was found to be above the photoionization threshold of those odd-numbered clusters with three or more atoms, yet exceeded the threshold of the even clusters in the beam for $n \geq 8$. In the extension to clusters as large as 29 atoms, laser photoionization has produced a time-of-flight mass distribution with a pronounced even–odd alternation in cluster photoion intensity. This alternation, which appears to be much more pronounced than any such effects in the alkali metals, was first predicted by Baetzhold (135). It can be attributed to an even–odd alternation in the electronic structure of the copper clusters, the highest occupied molecular orbital (HOMO) of the even clusters with paired electron occupation being considerably more bonding than that of the open shell, odd electron clusters.

Two-photon Spectroscopy of Copper Dimer and Trimer. Resonant two-photon ionization with mass selectivity has allowed the detection of a new Cu_2 electronic band system in the region 4480 to 4660 Å (110) and five new electronic band systems in the region between 2690 and 3200 Å (113). Data for these systems are summarized in Table 6.8. The Cu_2 clusters were found to have translational, rotational, and vibrational temperatures less than 5 and 10° and between 50 and 100K, respectively. In the process of scanning the R2PI spectrum of the newly determined ultraviolet electronic states, the ionization potential of copper dimer was accurately determined to be 7.894 ± 0.015 eV.

Among the observed states for copper dimer, the C state in the range 4480 to 4660 Å is particularly intriguing. This state, which has a significant lifetime, 1.0 ± 0.2 μs, was first observed by Preuss et al. (103). A vibrational analysis has been accomplished by both the Bondybey (14) ($B' - X^1\Sigma_g^+$, Gole et al. Table 6.7) and Smalley groups (110). The Bondybey group labels this state B' for its long radiative lifetime may be indicative of a low-lying state of triplet multiplicity. The system had previously alluded all researchers studying Cu_2 because the classical high-temperature environments under which these studies were undertaken provided sufficient quenching of the upper long-lived state. Interesting facets of the molecular electronic structure of copper dimer should be revealed when this state is probed at high resolution.

TABLE 6.8   New Copper Dimer Band Systems

System	T_e (cm^{-1})	w_e (cm^{-1})	$w_e x_e$ (cm^{-1})
$C\text{-}X^1\Sigma_g^+$	21867 ±4	226 $\pm$ 3	3 $\pm$ 2
I[a]	28560 (24)	248.0 (2.7)	0.90 (0.07)
II[a]	30637 (25)	116.0 (0.26)	0.046 (0.006)
III	38782	713	
IV	35000	280	
V	37440.1 (1.9)	288.4 (2.1)	0.64 (0.44)

[a]Quoted values refer to the most abundant isotopic species, Cu_2^{63}. Values listed without estimated errors are based on measured positions of vibrational bands of uncertain assignment. Listed band origins and vibrational parameters for systems I and II are the result of an approximate deperturbation calculation. See Ref. 113.

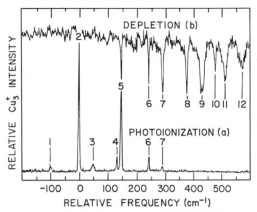

Figure 6.13 (a) Resonant two-photon ionization spectrum of $Cu_2^{63}Cu^{65}$ with strong origin band at 5397 Å; (b) depletion spectrum of $Cu_2^{63}Cu^{65}$ showing bands that predissociate rapidly relative to fluorescence.

In studying the spectrum of supersonically expanded cold copper trimer in the wavelength region 5430 to 5225 Å, Morse et al. (114) have employed three multiphoton techniques: R2PI spectroscopy, depletion spectroscopy, and the radiative repopulation of excited vibrational states. Observed spectra are depicted in Figure 6.13 and catalogued in Table 6.9. The photoionization scan obtained through the combination of a Nd:YAG-pumped dye laser followed by Nd:YAG laser ionization is dominated by an origin band at 5397 Å (feature 2) whose intensity with respect to the other features in the spectrum has been taken to indicate that a transition has

TABLE 6.9   Vibronic Band Positions Associated with Copper Trimer[a]

Band Number (Fig. 6.13)	Cu_3^{63}	$Cu_2^{63}Cu^{65}$[b]	$Cu^{63}Cu_2^{65}$[b]	Observation Technique (Sect. 6.4.2)
1	-99.2 ± 3	—	—	R2PI Spectroscopy, hot band repop.
2	0.0 ± 0.5	-0.1	-0.4	R2PI
3	46.5 ± 3.0	—	—	R2PI, hot band
4	131.3 ± 0.5	-0.7	-1.1	R2PI
5	146.4 ± 0.5	-1.1	-2.0	R2PI, depletion
6	242.9 ± 1.0	-1.4	-2.5	R2PI, depletion
7	291.7 ± 1.0	-1.5	-3.1	R2PI, depletion
8	377.9 ± 1.8	-2.0	-4.3	Depletion
9	342.0 ± 5.0	-3.5	-4.4	Depletion
10	476.6 ± 1.0	-1.7	-5.3	Depletion
11	511.6 ± 4.0	-1.7	-3.6	Depletion
12	572.8 ± 15.0	-2.2	-5.5	Depletion

[a]Relative to origin at 5397 Å (0.0 ± 0.5 in table)—all values in cm^{-1}.

[b]Relative to Cu_3^{63} band location.

occurred between two very similar electronic states (114). It has been assigned to a $^2E''\leftarrow {}^2E'$ transition of a D_{3h} molecule where both states must Jahn-Teller distort (vibrational-electronic coupling) to remove the electronic degeneracy. The sharp dropoff in spectral intensity signals the onset of a strong predissociation, at least for low-rotational quantum levels. Here, the depletion technique (note discussion in previous sections) can considerably enhance and extend the catalog of spectral information; one is able to probe predissociating levels to which R2PI spectroscopy is partially or totally blind depending on the magnitude of the predissociation. The depletion spectrum in Figure 6.13 monitors the decrease in the single photon-induced photoion signal as resonance pumping of predissociating trimer transitions depletes the population of copper trimers.

By modifying the degree of cooling inherent in their supersonic expansion, Morse et al. (114) have been able to demonstrate that bands 2, 5, 6, and 7 originate from the ground vibronic state of Cu_3, whereas bands 1, 3, and 4 correspond to hot bands associated with discrete levels 99 [1,3] or 12 cm^{-1} [4] above the ground vibronic level. If one allows radiative repopulation of ground-state levels followed by R2PI spectroscopy, it is possible to enhance slightly the intensity of those bands associated with the ground-state level at 99 cm^{-1}. Radiative repopulation apparently does not enhance the signal from band 4, apparently demonstrating the presence of a symmetry-based selection rule. The combination of bands in Figure 6.13 map primarily the vibronic structure of an excited state of copper trimer; the bands 5, 7, 9, and 12 have been assigned by Morse et al. to a progression in an excited state bending mode ($\omega_1' = 149 \pm 1$ cm^{-1}, $x_{11}' = -1.1 \pm 0.6$) (see also Fig. 6.10), whereas bands 6 and 10 are thought to be associated with a progression in the excited state symmetric stretch ($\omega_2' = 252 \pm 5$, $X_{22}' = -5 \pm 2$ cm^{-1}). Bands 8 and 11 are not readily explained by Morse et al. who suggest that they result from a third progression built on band 6, or from combination overtones. Further investigation will be needed to establish this definitively.

Despite some considerable effort, it has not yet been possible to rotationally resolve the Cu_3 spectrum. The definitive determination of the copper trimer structure will require the analysis of the rotational fine structure that cannot yet be resolved using pulsed laser techniques; however, the necessary resolution is accessible with continuous wave laser technology and the characterization of the trimer should soon be pursued by several groups.

Spectroscopy of Cooled Transition Metal Dimers.   In large part as a result of the Smalley and Bondybey groups, the spectroscopy of the transition-metal dimers is undergoing a renaissance. For example, recent investigations of the $Cr_2 A^1\Sigma_u^+ - X^1\Sigma_g^+$ transition have allowed a determination of the extremely short Cr_2 bond length and confirmed earlier widely disregarded studies by Efremov et al. (136). The Cr_2 bond length and bond dissociation energy have recently been the subject of considerable controversy focusing primarily on the nature of multiple bonding in metals. The present efforts do much to confirm the probable existence of a sextuple metal–metal bond and definitely refute recent detailed quantum chemical calculations that predict a much more weakly bound chromium dimer. Studies such as these provide

TABLE 6.10   Spectroscopic Properties of Transition Metal Diatomics

Molecule	Ground State	Equilibrium Bond Length	Dissoc. Energy
V_2	$^3\Sigma_g^-$	1.77 Å[a]	≤1.85 eV.
Cr_2	$^1\Sigma_g^+$	1.68 Å[b,c]	—
Ni_2	$^1\Gamma$ or $^3\Gamma$[a]	2.3 Å[a]	~2.07 eV.
Mo_2	$^1\Sigma_g^+$[d]	1.937 ± 0.008 Å	—

[a]Ref. 112.

[b]Refs. 111, 132.

[c]J. B. Hopkins et al. (Ref. 104) have studied the $A\,^1\Sigma_u^+ - X\,^1\Sigma_g^+$ band system extending the previous analysis of Efremov, et al. (Ref. 136b) to determine $A\,^1\Sigma_u^+$ parameters $\omega_e' = 449.0 \pm 0.2$ cm^{-1}, $\omega_e x_e' = 2.3 \pm 0.2$ cm^{-1}, $r_0' = 1.939 \pm 0.008$ Å, $\tau \sim 18 \pm 3$ ns ($v' = 0$).

[d]Ionization potential data indicate IP (Mo_2) < IP(Mo) implying a considerably stronger Mo_2^+ versus Mo_2 bond strength.

an important link in the chain of information that must be generated in order to gain a definitive understanding of the nature of bonding in the transition-metal elements. The evaluation of low-lying states in the transition-metal diatomics will also provide important input for the evaluation of mass spectral data on these species. Further initial results garnered on the transition-metal diatomics are summarized in Table 6.10.

6.5 THE FORMATION AND CHARACTERIZATION OF CLUSTERS OF INTERMEDIATE SIZE

Several groups have made significant contributions to the study of larger metal clusters that, for certain properties, have already taken on metallic character. Kappes et al. (106) have studied seeded beam expansions of high-temperature sodium vapor in helium, neon, argon, krypton, and nitrogen carrier gases. Again they employ single-photon selective photoionization mass spectroscopy. In contrast to the exponential dropoff with increasing cluster size that characterizes pure metal sodium supersonic expansions, larger abundances of higher clusters (Na_x, $x \leqslant 65$) are observed. Interestingly, a comparison of high-resolution neon and argon photoionization mass spectra obtained under very similar experimental conditions suggests that if mixed clusters do occur, they contain only one or two seed gas atoms! It appears that the spectra may be characterized by unprecedented "magic cluster numbers," that is, maxima in the photoion intensity distribution centered at $m/e = 161$, 437, and 874. Although their relative intensities are found to vary, it does not appear possible to shift the positions of the maxima by changing either the seed gas or its backing pressure. One is tempted to assign these maxima to particularly stable sodium clusters; however, it can also be argued that the growth process onsets from an extremely discontinuous abundance of smaller sodium groupings. The Schumacher group is now attempting to define clearly this growth process and measure the ionization potentials of some of the larger clusters. There is reason to believe that the temperatures associated with these larger sodium aggregates considerably exceed those of the very small cold clusters discussed previously.

Schulze et al. (137) have demonstrated that very large metal clusters (M_x, $x \geq 10,000$) can be produced by the mass aggregation technique. Recknagel (138) and coworkers have used a variation of this method to produce significant concentrations of small to intermediate size clusters of antimony, bismuth, and lead. Cluster sizes are analyzed by time-of-flight mass spectrometry. Clusters, M_x, up to $x = 650$ have been detected, whereas mass peaks up to $x = 17$ (Pb), $x = 21$ (Bi), and $x = 100$ (Sb) have been resolved. Here, metal atoms are expanded through a small orifice into 15 torr of helium enclosed in a liquid nitrogen-cooled condensation cell (in contrast to expansion into high vacuum). Clusters eventually exit this growth region through a second aperature into high vacuum where they soon pass into the time-of-flight mass spectrometer for analysis.

Recknagel et al. (138) have found that the variation of "condensation parameters" (type of inert gas, temperature of inert gas, temperature of metal oven) strongly influences the beam intensity and size distribution. A type of coalescence-growth process has been invoked by these investigators to explain their results. Cluster growth occurs only when expanding into helium and surprisingly is absent when expansion is into any of the other noble gases. In addition, clustering is found to be a stringent function of the helium temperature, slight variations preventing cluster formation entirely! Because of the many collisions occurring during passage through the growth region, the material exiting the metal vapor source (second aperature) has a thermal velocity distribution and the wide angular divergence that also characterizes a Knudsen effusion source. For this reason, it is felt that the internal excitation of the resulting metal clusters is relatively high.

The nucleation and growth of metal vapor in a cold inert gas atmosphere has been used previously by Stein and coworkers who have studied rather large aggregates of bismuth, lead, and indium. Yokozeki and Stein (139) have used a flowing-argon double orifice sampling system to produce clusters of size 2000 to 4000 atoms and characterize the clusters using electron diffraction techniques. Diffraction patterns have been obtained from metal cluster samples ranging in size from 40 to 95 Å in diameter. Analysis of the diffraction patterns reveals changes in the crystal structure from that of the bulk in the neighborhood of 50 to 60 Å in diameter. That is, the lattice parameters begin to deviate from those of the larger-size clusters. The differences are most pronounced in indium which changes from trigonal to face-centered cubic as the size decreases.

As we have noted, a major objective of many researchers working with metal clusters has been the generation of controlled size distributions. Andres and coworkers (140) have devised a source which is thought to be capable of producing intermediate size clusters of reasonably controlled size. These researchers form clusters of Cu, Ag, Au, or Ni as an aerosol supported in an inert gas which is then expanded through a sonic orifice into a vacuum region where a molecular beam is formed. The beam constituents, once deposited on a surface, are examined using electron microscopy. A controlled mean size ranging from the dimer to several thousand atoms is possible. For a mean size greater than ≈ 50, the FWHM in the cluster distribution has a constant value in diameter space of approximately twice the diameter of the monomer.

Finally, we should note the recently initiated effort by Chesnovsky and Bederson (141) to study the polarizability of large isolated metal atom clusters (20 to 1000 atoms) produced by expanding atomic vapors into a cold noble gas atmosphere followed by supersonic expansion. The deflection pattern of a given cluster determined as the result of deflection in an inhomogeneous electric field (followed by mass spectrometric detection) will be used to extract the static polarizability providing an indication of the dielectric properties of the cluster. Again, the study of molecular polarizabilities as a function of cluster size can provide important information about the evolution of the chemical bond in metallic clusters. The polarizability is sensitive to the energy gap between occupied and unoccupied orbitals in the cluster molecule and should increase with increased cluster size as a manifestation of a decrease in this energy gap. It is possible that the divergence of the polarizability at a critical cluster size will mark the onset of metallic conductivity in the isolated cluster.

6.6 METAL CLUSTER OXIDATION

Just as the study of metal clusters is in its infancy, so too is the study of their oxidation. Research efforts in this area are important, for it is thought that they will be helpful in modeling the localized phenomena, environments, and compound formation that characterize the oxidation of metallic surfaces. In the future it may be possible to assess the microscopic nature of the profound influence that "metallic" heteroatoms can have on surface oxidation when they are introduced as a fraction of a monolayer on the surface. For example, the oxidation of bismuth proceeds several orders of magnitude more rapidly (4) when this surface is contaminated with potassium. Does this result from a local bismuth-potassium interaction or does it result from a form of a long-range percolation effect along the contaminated surface? Some light can be cast on this question by studying the oxidation of mixed bismuth-potassium clusters.

Thus far, studies of metal cluster oxidation have been of two distinct types focusing on the reactions of alkali metals. Castleman and coworkers (142) have studied oxidized metal clusters using photoionization techniques. Using a coexpanding nozzle, they have formed Na_2Cl, Na_xO, and K_xO ($2 \leq x \leq 4$) from the reaction of alkali clusters and various reactant gases. The appearance potential for Na_2Cl is found to be 95.7 kcal mol^{-1} lying between the ionization potential of the metal and the electron affinity of chlorine. Appearance potentials for the alkali oxides, M_xO, are given in Table 6.11. Here, one can garner some feeling for trends in alkali oxide ionization potentials as a function of the degree of aggregation. The ionization potentials do not differ greatly from the associated metallic grouping (see Table 6.3)! In the case of sodium tetramer, however, the ionization potential does fall below that of the bare metal cluster, a finding that appears to be in accord with what has been observed as an influence of impurities on the work function of the bulk metal.

Finally, using a modified beam-gas configuration, Crumley et al. (143) have studied the chemiluminescent emission from the reaction of supersonically expanded sodium polymers (Na_n, $n \geqslant 3$) with the halogen atoms, chlorine, bromine, and iodine. The observed chemiluminescence characterizing the chlorine system is depicted in Figure 6.14; it is dominated by emission corresponding to the Na_2 $A^1\Sigma_u^+ - X^1\Sigma_g^+$, $B^1\Sigma_u - X^1\Sigma_g^+$, and $C^1\Sigma_u - X^1\Sigma_g^+$ systems, where the A, B, and C states are populated via the metathesis

$$Na_3 + X \longrightarrow Na_2^* + NaX \tag{6.11}$$

However, other features, still to be identified, may correspond to the reaction of higher sodium polymers. From the combination of the chlorine, bromine, and iodine atom reactions and an energy balance, it has been possible to determine a lower bound (2300 cm^{-1}) for the sodium trimer bond energy. Based on the results already obtained, it is possible that the generalization of reaction (6.11) can be used to excite fluorescence from high sodium polymers via the process

$$Na_x + X \longrightarrow Na_{x-1}^* + NaX$$

More recently, electron impact excitation and quadrupole mass spectrometry have been used to assess those spectral regions that will be amenable to laser probes of

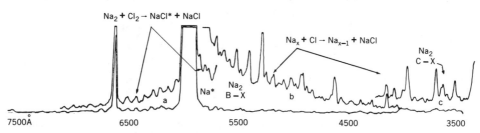

Figure 6.14 (a) Chemiluminescent spectrum (3400 to 5800 Å) associated with the processes $Na_x + Cl \rightarrow Na_{x-1} + NaCl$, where the dominant molecular spectrum corresponds to Na_2^* emission resulting primarily from the reaction of sodium trimer. The major features in the $B - X$ and $C - X$ regions correspond to the overlap of several (v', v'') emission features (e.g., for the 5278 Å feature the intensity is derived primarily from $\Delta v = v' - v'' = -9, -8$, transitions . . .). Below 4100 Å the scale for the $C - X$ region is multiplied by 4. At the long wavelength limit, the emission rises sharply as the Na D line is approached. The Na atomic emission features (Na; $3^2P - 3^2S$, $4^2P - 3^2S$), not depicted, are ~10^2 times the intensity of the molecular emission features. (b) Chemiluminescent spectra for NaCl* and Na*. The NaCl emission is tentatively attributed to the reaction $Na_2 + Cl_2 \rightarrow NaCl^* + NaCl$. The observed strong Na* fluorescence is consistent with the process $NaCl\dagger + Na \rightarrow Na^* + NaCl'\dagger$. The feature at 6600 Å corresponds to second-order emission associated with the Na $4^2P - 3^2S$ transition. In contrast, the $Na_2 + Cl$ reaction leads only to atomic emission corresponding to the Na D line and $4^2P - 3^2S$ transition. (From Ref. 143.)

TABLE 6.11 Appearance Potentials for Alkali Cluster Oxides M_xO

x	Present Measurements: M_xO	
	M = Na	M = K
1	$(6.5 \pm 0.7)^a$	—
2	5.06 ± 0.04^b	4.7 ± 0.4
3	3.90 ± 0.15	3.65 ± 0.04
4	3.95 ± 0.10	3.62 ± 0.04

[a]K. I. Peterson, P. D. Dao, and A. W. Castleman, Jr., *J. Chem. Phys.*, **79**, 777 (1983); **80**, 563 (1984).

[b]H. M. Rosenstock, K. Draxl, B. W. Steiner, and J. T. Herron, Eds., *J. Phys. Chem. Ref. Data*, **6**, Suppl. (1) 378–379 (1977).

the ground-state products of reaction. There are many other routes that these studies may soon take.

6.7 CONCLUSIONS

From the present discussion it should be clear that a number of powerful techniques and a significant catalog of spectroscopic data are now emerging in the quest of gas-phase metal cluster structure, dynamics, and reactivity. This is a very young and developing research area from which one can expect to see many exciting results in the near future. There is little doubt that we are beginning to explore a very different form of matter whose properties may soon prove to be extremely useful.

ACKNOWLEDGMENTS

It is a pleasure to acknowledge helpful discussions with Professors Ernst Schumacher, Derek Lindsay, and Martin Moskovits and Dr. Vladimir Bondybey. Special thanks also go to Mr. Robert Woodward who aided in the formulation of some of the tables. Research in Metal Cluster Chemistry at the Georgia Institute of Technology is supported by the National Science Foundation.

ADDENDUM

Since the inception of this chapter, some notable further literature has appeared; several references to recent work which poses some answers to questions raised in reference (10) can be found in Chapter 7. Here, we briefly outline some exemplary contributions both in experiment and theory (see also reference 14). An important contribution to metal cluster energetics and Table 6.1 comes from Hilbert (145), who determines the sodium trimer atomization energy as 103 ± 12 KJ/mole. There have also been refinements and an increase in our knowledge of the sodium trimer electronic states. Taking advantage of the increased cooling for the trimer produced in a very high pressure seeded beam expansion ($T_{Rot}(Na_2)$ ~7K), Delacretaz and Wöste (146) have attained a considerable increase in the resolution of vibronic features especially in the 6250 Å system (Fig. 6.5; Table 6.4). The dominant features in the 6250 Å system which form a vibronic progression with frequency spacings ~125 cm^{-1} are accompanied by a proliferation of closely spaced (10–30 cm^{-1}) vibronic groupings. A recent analysis (147) within the Jahn–Teller framework now indicates that the vibronic levels must be associated with "half integral" quantum numbers. Further band systems at wavelengths below 600 nm have now been excited using an excimer pumped dye laser, and the electronic spectrum has now been scanned to 330 nm. There are strong band systems at ~550–575 and ~475 nm and a weak unresolved system onsetting at ~525 nm. There appears to be an outstanding correlation between those spectra at wavelengths shorter than 550 nm and the sodium trimer LIAF spectrum (120).

There has also been further progress in the analysis of bound–bound transitions in copper trimer. The analysis first given by Morse et al. (114) has now been elegantly modified by Thompson et al. (148), who find that all of the excited state vibronic features for the trimer with the exception of that band labeled 10 (Fig. 6.13; Table 6.9) (two quanta of symmetric stretch (124)) can be clearly assigned in a well-founded vibronic scheme. Using laser-induced excitation spectroscopy and a trimer source producing concentrations some three to six orders of magnitude in excess of those available from laser vaporization, Crumley et al. (124) have now obtained a considerably cooler, much more clearly developed spectrum than that presented in Figure 6.10, which also extends to considerably higher frequency. Data obtained in this study allows the correlation of the recent studies by Morse et al. (114) (laser vaporization, R2PI, and depletion spectroscopies) and the recent work of Moskovits and coworkers (149,150) (resonance Raman–matrix isolation). The observed spectroscopic features are in one-to-one correspondence with the detailed excited state vibronic calculations of Thompson et al. (148) and in acceptable agreement with the "first order" ground state vibronic calculations of Moskovits (149). The generated concentration of trimer molecules is such as to allow the characterization of strongly predissociative levels lost to the R2PI technique. The extent of the spectra allow the separation of level structure associated on the one hand with the excited state symmetric stretch mode and on the other with the vibronic levels resulting from the coupling of the asymmetric stretch and bending modes. Significant changes in those features which dominate observed excitation spectra as a function of effective temperature are interpreted in terms of a change in the ground state population distribution and hence an alteration in the configuration space connecting *populated* ground state and accessed excited state levels. Observed hot-band structure implies that the pseudorotation barrier in the ground electronic state of the trimer is considerably smaller than previously suggested (114).

Work on the formation and characterization (through photoionization and mass spectrometry) of clusters of intermediate size (Section 6.5) has now been extended by both the Schumacher (151) and Knight (152) groups. The Schumacher group has extended their measurements of cluster ionization potentials and has demonstrated that the expression (6.6) is quite applicable not only to the prediction of ionization potentials for the alkali clusters but may be applicable to metal particles in general. The spherical drop model appears to have predicting power (151). The Knight (152) group has apparently found strong evidence for "magic cluster numbers" K_2, M_8 (M = Na, K), M_{20}, M_{40}, Na_{58}, and Na_{92}. These results, which have also been obtained by the Schumacher group for K_2, M_8, and M_{20}, indicate a probable higher thermodynamic stability for these particular molecular groupings. The Knight group has mapped this enhanced thermodynamic stability onto the series corresponding to the number of electrons in closed shells of an isotropic spherically symmetric potential well, the "jellium" model, finding remarkable agreement between preferred stability and shell closings. Of the first seven closed shell clusters predicted by the jellium model to be particularly stable, five are clearly observed. The application of the model is, however, the subject of some considerable controversy.

While the spherically symmetric jellium model closures have an astounding correlation with observed cluster abundances for sodium and potassium, the model appears to fall short in its description of other molecular properties and as it is applied to a wider sphere than that of the lighter alkali metals. The results obtained with the model certainly emphasize that electronic structures, especially the closing of shells, play a role comparable to that of molecular structures in the determination of stability; however, there are important refinements which must be considered in the future (153).

The subject of larger clusters and "magic numbers" has also continued to occupy the attention of Recknagel, Sattler, and coworkers (154) and the apparent "islands of stability" which characterize larger metal clusters will continue to furnish important grist for both experiment and theory.

In an effort complementary to that of Powers et al. on copper clusters (113), Rohlfing et al. (155) have obtained laser photoionization spectra over the range 4.5– 6.5 eV for iron clusters ranging in size from 2 to 25 atoms.

Finally, we should note that there has been further progress in the study of metal cluster oxidation (156). In a device which produces a metallic flow intermediate to that of a standard effusive source (atoms and small % diatomics) and laser vaporization techniques, the reaction of small silver clusters Mn, $n \geq 3$, has been monitored.

A red shift of the spectral features is associated with the increasing size of the metal cluster oxide. The spectral data combined with supplementary thermodynamic information demonstrate that AgO^* with enough energy to account for the observed chemiluminescence cannot be produced through the reaction of either Ag or Ag_2 with O_3. The smallest cluster whose reaction can yield excited states of AgO is the trimer.

REFERENCES

1. C. Kittel, *Introduction to Solid State Physics*, 4th ed., Wiley, New York, 1971, L. Solymar and D. Walsh, *Lectures on the Electrical Properties of Materials*, Oxford University Press, London, 1975; W. Harrison, *Solid State Theory*, McGraw-Hill, New York, 1970.

2. Exemplary studies: H. F. Schaeffer, III, *Acc. Chem. Res.*, **10**, 287–293 (1977); C. Bauschlicher, Jr., P. Bagus, and H. F. Schaeffer, III, *IBM J. Res. Dev.*, **22**, 213–220 (1978); A. L. Companion, D. J. Steible, and A. J. Starshak, *J. Chem. Phys.*, **49**, 3637–3640 (1968); A. L. Companion, *Chem. Phys. Lett.*, **56**, 500–502 (1978); B. T. Pickup, *Proc. R. Soc. London Ser.*, **333**, 69–87 (1973); A. Gelb, K. D. Jordan, and R. Silbey, *Chem. Phys.*, **9**, 175–182 (1975); D. W. Davies and G. del Conde, *Mol. Phys.*, **33**, 1813–1814 (1977); D. M. Lindsay, D. R. Herschbach, and A. L. Kwiram, *Mol. Phys.*, **39**, 529 (1980); P. S. Bagus, G. del Conde, and D. W. Davies, *Faraday Spectrosc. Discuss. Chem. Soc.*, **62**, 321 (1977); J. Kendrick and I. H. Hillier, *Mol. Phys.*, **33**, 635–640 (1977); R. C. Baetzold, *J. Catal.*, **29**, 129 (1973); R. C. Baetzold, *Adv. Catal.*, **25**, 1–55 (1976); W. A. Goddard, S. P. Wallach, A. K. Kappe, T. H. Upton, and C. F. Melius, *J. Vac. Sci. Technol.*, **14**, 416–418 (1977); C. Bachman, J. Demuynk, and A. Veillard, *Gazz. Chim. Ital.*, **108**, 378–391 (1978); A. B. Anderson, *J. Chem. Phys.*, **68**, 1744 (1978); A. B. Anderson, *J. Chem. Phys.*, **66**, 5108–5111 (1977); C. Bachmann, J. Demuynck, and A. Veillard, *Faraday Symp. Chem. Soc.*, **14**, 170 (1980); H. Tatewaki, E. Miyoshi, and T. Nakamura, *J. Chem. Phys.*, **76**, 5073 (1982); D. Post and E. J. Baerends, *Chem. Phys. Lett.*, **86**, 178 (1982). Further references throughout text.

3. M. Moskovits and G. A. Ozin, Cryochemistry, Wiley, New York, 1976; G. A. Ozin, *Catal. Rev. Sci. Engl.*, **16**, 191–230 (1977). Related work is discussed in the following references. A. L. Robinson, *Science*, **185**, 772–774 (1974); G. C. Demitras and E. L. Muetterties, *J. Am. Chem. Soc.*, **99**, 2796–2797 (1977); J. H. Sinfelt, *Acc. Chem. Res.*, **10**, 15–20 (1977); E. L. Muetterties, *Science*, **196**, 839–848 (1977); E. L. Muetterties, *Bull. Soc. Chim. Belg.*, **84**, 959–986 (1975); E. L. Muetterties, *Bull. Soc. Chim. Belg.*, **85**, 451–470 (1976); E. L. Muetterties, R. N. Rhodin, E. Band, C. F. Brucker, and W. R. Pretzer, *Chem. Rev.*, **79**, 91–137 (1979); E. Band and E. L. Muetterties, *Chem. Rev.*, **78**, 639–658 (1978). *Further references throughout text.*

4. T. M. Taylor, C. T. Campbell, J. W. Rogers, Jr., W. P. Ellis, and J. M. White, *Surf. Sci.*, **134**, 529 (1983).

5. A. Herrmann, E. Schumacher, and L. Wöste, *J. Chem. Phys.*, **68**, 2327 (1978).

6. E. Schumacher, W. H. Gerber, H. P. Härri, M. Hofmann, and E. Scholl, in *Metal Bonding and Interactions in High Temperature Systems*, J. L. Gole and W. C. Stwalley, Eds., A.C.S. Symposium Series, Vol. 179, 1982, pp. 83–108.

7. J. G. Fripiat, K. T. Chow, M. Boudart, J. R. Diamond, and K. H. Johnson, *J. Mol. Catal.* **1**, 59–72 (1975).

8. S. C. Richtsmeier, M. L. Hendewerk, D. A. Dixon, and J. L. Gole, *J. Phys. Chem.*, **86**, 3932 (1982); S. C. Richtsmeier, D. A. Dixon, and J. L. Gole, *ibid.*, **86**, 3937, 3942 (1982).

9. A. Herrmann, S. Leutwyler, E. Schumacher, and L. Woste, *Helv. Chim. Acta*, **61**, 453 (1978).

10. *Science,* **219**, 474, 944, 1413 (1983). (a) M. E. Geusic, M. D. Morse, and R. E. Smalley, *J. Chem. Phys.,* **82**, 5901 (1985). (b) S. C. Richtsmeier, E. K. Parks, K. Liu, L. G. Pobo, and S. J. Riley, *ibid.,* **82**, 3659 (1985). (c) K. Liu, E. K. Parks, S. C. Richtsmeier, L. G. Pobo and S. J. Riley, *ibid.,* **83**, 2882 (1985).

11. M. Moskovits and J. E. Hulse, *J. Chem. Phys.,* **66**, 3988 (1977); W. E. Koltzbucher and G. A. Ozin, *J. Am. Chem. Soc.,* **100**, 2262 (1978); W. Schulze, H. U. Becker, and H. Abe, *Chem. Phys.,* **35**, 177 (1978); T. Welker and T. P. Martin, *J. Chem. Phys.,* **70**, 5683 (1979); V. E. Bondybey and J. H. English, *J. Chem. Phys.,* **73**, 42 (1980); M. Moskovits and J. E. Hulse, *J. Chem. Phys.,* **67**, 4271 (1977).

12. W. Schulze, H. U. Becker, R. Minkwitz, and K. Manzel, *Chem. Phys. Lett.,* **55**, 59 (1978); M. Moskovits and D. P. DiLella, *J. Chem. Phys.,* **72**, 2267 (1980); K. Manzel, U. Engelhardt, H. Abe, W. Schulze, and F. W. Froben, *Chem. Phys. Lett.,* **77**, 514 (1981); D. P. DiLella, W. Limm, R. H. Lipson, M. Moskovits, and K. V. Taylor, *J. Chem. Phys.,* **77**, 5263 (1982); D. P. DiLella, K. V. Taylor, and M. Moskovits, *J. Phys. Chem.,* **87**, 524 (1983).

13. D. M. Lindsay, D. R. Herschbach, and A. L. Kwiram, *Mol. Phys.,* **32**, 1199 (1976); G. A. Thompson and D. M. Lindsay, *J. Chem. Phys.,* **74**, 959 (1981); R. Van Zee, C. A. Baumann, S. V. Bhat, and W. Weltner, Jr., *J. Chem. Phys.,* **76**, 5636 (1982); J. A. Howard, K. F. Preston, R. Sutcliffe, and B. Mile, *J. Phys. Chem.,* **87**, 536 (1983).

14. J. L. Gole, J. H. English, and V. E. Bondybey, *J. Phys. Chem.,* **86**, 2560 (1982).

15. K. A. Gingerich, *Faraday Symp. Chem. Soc.,* **14**, 109 (1980).

16. K. A. Gingerich, in *Metal Bonding and Interactions in High Temperature Systems,* J. L. Gole and W. C. Stwalley, Eds., A.C.S. Symposium Series, Vol. 179, 1982, pp. 109–124.

17. L. Brewer and J. S. Winn, *Faraday Symp. Chem. Soc.,* **14**, 127 (1980).

18. L. Rovner, A. Drowart, and J. Drowart, *Trans. Faraday Soc,* **63**, 2906 (1967).

19. F. J. Kohl, O. M. Uy, and K. D. Carlson, *J. Chem. Phys.,* **47**, 2667 (1967).

20. V. V. Illarinau and A. S. Charepanova, *Dokl. Akad. Nauk SSSR,* **133**, 1086 (1960); G. M. Rosenblatt and C. E. Birchenallis, *J. Chem. Phys.,* **35**, 788 (1961).

21. P. J. McGonigal and A. V. Grosse, *J. Phys. Chem.,* **67**, 924 (1963); F. Metzer, *Helv. Phys. Acta,* **16**, 323 (1943); L. Brewer and J. S. Kane, *J. Phys. Chem.,* **59**, 105 (1955); P. Goldfinger and M. Jeunehomme, in *Advances in Mass Spectrometry,* J. D. Waldron, Ed., Pergamon Press, New York, 1959, pp. 534–546.

22. R. E. Honig, *J. Chem. Phys.,* **22**, 1610 (1954); J. Drowart, G. DeMaria, A. J. H. Beerboom, and M. G. Ingrahm, *J. Chem. Phys.,* **45**, 822 (1966).

23. J. E. Kingcade, Jr., U. V. Choudary, and K. A. Gingerich, *Inorg. Chem.,* **18**, 3094 (1979).

24. J. Drowart, G. DeMaria, *J. Chem. Phys.* **30**, 308 (1959).

25. A. Kant and B. Strauss, *J. Chem. Phys.* **45**, 822 (1966).

26. R. E. Honig, *J. Chem. Phys.,* **21**, 573 (1953).

27. K. A. Gingerich, A. Desideri, and D. L. Cocke, *J. Chem. Phys.* **62**, 731 (1975).

28. K. A. Gingerich, D. L. Cocke, and F. Miller, *J. Chem. Phys.,* **64**, 4027 (1976).

29. B. B. King, *Prog. Inorg. Chem.,* **15**, 287 (1972); F. A. Cotton and M. H. Chisholm, *Chem. Eng. News,* (June 28, 1982), p. 40.

30. Harry Partridge, D. A. Dixon, S. P. Walch, C. W. Bauschlicher, Jr., and J. L. Gole, *J. Chem. Phys.*, **79**, 1859 (1983) and references therein.

31. L. Pauling, *The Nature of the Chemical Bond*, 3rd ed., Cornell University Press, Ithaca, NY, 1960, J. Drowart in, *Phase Stability in Metals and Alloys*, P. S. Rudman, J. Stringer and R. I. Jaffee, Eds., McGraw-Hill, New York, 1967, pp. 305–317.

32. K. A. Gingerich, *Chimia*, **26**, 619 (1972).

33. K. A. Gingerich, *Int. J. Quant. Chem. Symp.*, **12**, 489 (1978); K. A. Gingerich and S. K. Gupta, *J. Chem. Phys.*, **69**, 505 (1978).

34. L. Brewer, in *Phase Stability of Metals and Alloys*, P. S. Rudman, J. Stringer, and R. I. Jaffe, Eds., McGraw-Hill, New York, 1967, pp. 39–61, 241–249, 344–346, 560–568.

35. A. R. Miedema, R. Boom, and F. R. de Boer, *J. Less Common Metals*, **41**, 283 (1975); A. R. Miedema, *J. Less Common Metals*, **46**, 67 (1976).

36. A. R. Miedema and K. A. Gingerich, *J. Phys. B*, **12**, 2255 (1979); A. R. Miedema, *Faraday Disc. Chem. Soc.*, **14**, 136 (1980).

37. S. Richtsmeier, J. L. Gole, and D. A. Dixon *Proc. Natl. Acad. Sci.*, **77**, 5611 (1980). See also Ref. 12, DiLella et al.

38. I. Shim, J. P. Dahl, and H. Johhanson, *Int. J. Quant. Chem.*, **15**, 311 (1979); T. H. Upton and W. A. Goddard, III, *J. Am. Chem. Soc.*, **100**, 5659 (1978).

39. C. H. Wu, *J. Chem. Phys.* **65**, 3181 (1976).

40. W. H. Gerber and E. Schumacher, *J. Chem. Phys.*, **69**, 1692 (1978).

41. W. H. Gerber "Theorie des dynamischen Jahn–Teller Effekts in Li_3 und Untersuchung von. Lithium-Molekularstrahlen," Ph.D. Thesis, Bern University, 1980.

42. J. L. Gole, R. A. Eades, D. A. Dixon, and R. Childs, *J. Chem. Phys.*, **72**, 6368 (1980).

43. K. A. Gingerich, *J. Cryst. Growth*, **9**, 31 (1971).

44. K. Hilpert, K. A. Gingerich, *Ber. Bunsenges. Phys. Chem.*, **84**, 739 (1980).

45. L. V. Gurvich, G. V. Karachevstev, V. N. Kondratyev, Y. A. Lebeder, V. A. Mendredev, V. K. Potapov, and Y. S. Khodeen, *Bond Energies Ionization Potentials and Electron Affinities*, Nauka, Moscow, 1974 (in Russian).

46. C. A. Stearns and R. J. Kohl, *High Temp. Sci.*, **5**, 113 (1973).

47. K. Hilpert, *Ber. Bunsenges. Phys. Chem.*, **81**, 30:348 (1977).

48. D. L. Cocke, K. A. Gingerich, and J. Kordis, *High Temp. Sci.*, **7**, 61 (1975).

49. A. Neckel and G. Sodeck, *Monatsh. Chem.*, **103**, 367 (1972).

50. D. L. Cocke, and K. A. Gingerich, *J. Phys. Chem.* **75**, 3264 (1971).

51. A. Neubert, and K. F. Zmbov, *Trans. Faraday Soc.*, **70**, 2219 (1974).

52. V. Piacente and K. A. Gingerich, *High Temp. Sci.*, **9**, 189 (1977).

53. K. A. Gingerich and G. D. Blue, **59**, 185 (1973).

54. D. J. Guggi, A. Neubert, and K. F. Zmbov, in Proc. 4th Int. Conf. Chem. Thermodyn., M. Laffitte, Ed., Montpellier, France, August 1975, paper III/23.

55. D. L. Cocke, K. A. Gingerich, and C. Chang, *J. Chem. Soc. Faraday I*, **72**, 268 (1976).

56. K. A. Gingerich, D. L. Cocke, H. C. Finkbeiner, and C. A. Chang, *Chem. Phys Lett.*, **18**, 102 (1973).

57. J. Kordish, K. A. Gingerich, and R. J. Seyse, *J. Chem. Phys.* **61**, 5114 (1974).

58. K. A. Gingerich and U. V. Choundary, *J. Chem. Phys.* **68**, 3265 (1978).

59. B. Busse and K. G. Weil, *Ber. Bunsenges. Phys. Chem.*, **85**, 309 (1981).

60. K. A. Gingerich, *Chem. Phys. Lett,* **13**, 262 (1972).

61. K. A. Gingerich and H. C. Finkbeiner, in Proc. 9th Rare Earth Conf. Chem. Thermodyn., P. E. Field, Ed., Blacksburg, VA, October 10–14, 1971. (CONF-71101 Chemistry [TID-4500], National Information Service, Vol. 2, U.S. Dept. Comm., Springfield, VA. 22151), pp. 759–803.

62. M. Moskovits (private communication)

63. K. A. Gingerich, D. L. Cocke, and U. V. Choundary, *Inorg. Chem. Acta*, **14**, 147 (1975).

64. S. K. Gupta, M. Pelino, and K. A. Gingerich, *J. Chem. Phys.*, **70**, 2044 (1979).

65. R. Hague, M. Pelino, and K. A. Gingerich, to be published.

66. U. V. Choundary, K. Krishnan, and K. A. Gingerich, unpublished data.

67. C. L. Sullivan, J. E. Prusaczyk, and K. D. Carlson, *High Temp. Sci.*, **4**, 212 (1972).

68. V. Piacente and A. Desideri, *J. Chem. Phys.*, **57**, 2213 (1972).

69. G. Riekert, Ph.D. Thesis, Doktorarbeit Universität Stuttgart, 1980.

70. K. S. Pitzer, *J. Chem. Phys.*, **74**, 3078 (1981).

71. A. Neubert, H. R. Ihle, and K. A. Gingerich, *J. Chem. Phys.*, **73**, 1406 (1980).

72. A. Neubert, H. R. Ihle, and K. A. Gingerich, presented at 6th Int. Conf. Thermodyn., Merseburg D.D.R., August 26–29, 1980.

73. JANAF Thermochemical Tables, *J. Phys. Chem.*, Ref. Data, **7**, 793 (1978).

74. K. A. Gingerich, *J. Chem. Soc. Faraday II*, **70**, 471 (1974).

75. K. A. Gingerich and D. L. Cocke, *Inorg. Chim. Acta.*, **28**, L171 (1978).

76. D. L. Cooke, K. A. Gingerich, and J. Kordis, *High Temp. Sci.*, **5**, 474 (1973).

77. A. Kant and B. H. Strauss, *J. Chem. Phys.*, **49**, 523 (1968).

78. K. P. Huber, G. Herzberg, *Molecular Spectra and Molecular Structure IV, Constants of Diatomic Molecules*, Van Nostrand, New York, 1979.

79. K. Hilbert, *Ber. Bunsenges. Phys. Chem.*, **83**, 161 (1979).

80. V. Piacente and K. A. Gingerich, *Z. Naturforsch*, **28a**, 316 (1973).

81. J. E. Kingcade, D. C. Dufner, S. K. Gupta, and K. A. Gingerich, *High Temp. Sci.*, **10**, 213 (1978).

82. A. Kant and S. S. Lin, *Monatsh Chem.*, **103**, 757 (1972).

83. R. Hague, M. Pelino, and K. A. Gingerich, *J. Chem. Phys.*, **71**, 2929 (1978).

84. K. F. Zmbov, C. H. Wu, and H. R. Ihle, *J. Chem. Phys.*, **67**, 4603 (1977).

85. B. H. Nappi and K. A. Gingerich, *Inorg. Chem.*, **20**, 522 (1981).

86. W. C. Stwalley, *J. Chem. Phys.*, **65**, 2038 (1976).

87. S. K. Gupta, R. M. Atkins, and K. A. Gingerich, *J. Chem. Phys.*, **60**, 1958 (1974).

88. S. K. Gupta and K. A. Gingerich, *J. Chem. Phys.*, **69**, 505 (1978).

89. S. K. Gupta and K. A. Gingerich, *J. Chem. Phys.*, **70**, 5350 (1979).

90. K. A. Gingerich, D. L. Cocke, and F. Miller, *J. Chem. Phys.*, **64**, 4027 (1976).

91. S. K. Gupta, B. M. Nappi, and K. A. Gingerich, *Inorg. Chem.*, **20**, 966 (1981).

92. S. K. Gupta, M. Pelino and K. A. Gingerich, *J. Phys. Chem.*, **83**, 2335 (1979).

93. K. A. Gingerich and S. K. Gupta, *J. Chem. Phys.*, **69**, 505 (1978).

94. K. A. Gingerich and S. K. Gupta, *J. Chem. Phys.*, **69**, 505 (1978).

95. K. A. Gingerich and U. V. Choundary, unpublished data.

96. K. A. Gingerich, *Chem. Phys. Lett.*, **25**, 526 (1974).

97. For recent studies of molecular (monomeric) species formed with continuous sources, see R. E. Smalley, D. H. Levy, and L. Wharton, *Laser Focus*, November 1975; D. H. Levy, L. Wharton, and R. E. Smalley, in *Chemical and Biochemical Applications of Lasers*, C. B. Moore, Ed., Vol. II, Academic, New York, 1977, p. 1; L. Wharton, D. Auerbach, D. Levy, and R. Smalley, in *Advances in Laser Chemistry*, A. H. Zewail, Ed., Springer Series in Chemical Physics, Springer, New York, 1978. For recent studies of van der Waal's complexes, see, for example, W. Klemperer, *Ber. Bunsenges. Phys. Chem.*, **78**, 128 (1972); *J. Chem. Soc. Faraday Discuss.*, **62**, 179 (1977) and references therein. See also Benjamin J. Blaney and George E. Ewing, *Ann. Rev. Phys. Chem.*, **27**, 553 (1976) and references therein.

98. For the initial demonstration of cooling using laser fluorescence, see M. P. Sinha, A. Schultz, and R. N. Zare, *J. Chem. Phys.*, **50**, 549 (1973). NO_2 work: R. E. Smalley, B. L. Ramakrishna, D. H. Levy, and L. Wharton, *J. Chem. Phys.*, **61**, 4363 (1974); *ibid.*, **63**, 4977 (1975). S-tetrazine work: R. E. Smalley, L. Wharton, D. Levy, and D. W. Chandler, *J. Mol. Spectr.*, **66**, 375 (1977); *J. Chem. Phys.*, **68**, 2487 (1978).

99. K. Sattler, J. Mühlbach, and E. Recknagel, *Phys. Rev. Lett.*, **45**, 821 (1980) and references therein.

100. R. E. Smalley, L. Wharton, and D. H. Levy, *Acc. Chem. Res.*, **10**, 139 (1977).

101. J. L. Gole and G. J. Green, in *Metal Bonding and Interactions in High Temperature Systems*, J. L. Gole and W. C. Stwalley, Eds., A.C.S. Symposium Series, Vol. 179, 1982, pp. 125–152.

102. G. J. Green, Ph.D. Thesis, Georgia Institute of Technology, 1982.

103. D. R. Preuss, S. A. Pace, and J. L. Gole, *J. Chem. Phys.*, **71**, 3553 (1979).

104. J. B. Hopkins, P. R. R. Langridge-Smith, M. D. Morse, and R. E. Smalley, *J. Chem. Phys.*, **78**, 1627 (1983). R. L. Whetten, D. M. Cox, D. J. Trevor, and A. Kaldor, *J. Chem. Phys.*, **89**, 566 (1985); *Phys. Rev. Lett.*, **54**, 1494 (1985); D. J. Trevor, R. L. Whetten, D. M. Cox, and A. Kaldor, *J. Am. Chem. Soc.*, **107**, 518 (1985); E. A. Rohlfing, D. M. Cox, and A. Kaldor, *J. Chem. Phys.*, **81**, 3322 (1984).

105. R. E. Smalley, private communication.

106. M. M. Kappes, R. W. Kunz, and E. Schumacher, *Chem. Phys. Lett.*, **91**, 413 (1982).

107. A. Hermann, M. Hofmann, S. Leutwyler, E. Schumacher, and L. Wöste, *Chem. Phys. Lett.*, **62**, 216 (1979).

108. S. Leutwyler, A. Hermann, L. Wöste, and E. Schumacher, *Chem. Phys.*, **48**, 253 (1980); S. Leutwyler, M. Hofmann, H. P. Härri, and E. Schumacher, *Chem. Phys. Lett.*, **77**, 257 (1981).

109. T. G. Dietz, M. A. Duncan, D. E. Powers, and R. E. Smalley, *J. Chem. Phys.*, **74**, 6511 (1981).

110. D. E. Powers, S. G. Hansen, M. E. Geusic, A. C. Puiu, J. B. Hopkins, T. G. Dietz, M. A. Duncan, P. R. R. Langridge-Smith, and R. E. Smalley, *J. Phys. Chem.*, **86**, 2556 (1982).

111. D. L. Michalopoulos, M. E. Geusic, S. G. Hanson, D. E. Powers, and R. E. Smalley, *J. Phys. Chem.*, **86**, 3914 (1982).

112. P. R. R. Langridge-Smith, M. D. Morse, G. P. Hansen, R. E. Smalley and A. J. Merer, *ibid.*, **80**, 593 (1984).

113. D. E. Powers, S. G. Hansen, M. E. Geusic, D. L. Michalopoulos, and R. E. Smalley, *J. Chem. Phys.*, **78**, 2866 (1983).

114. M. D. Morse, J. B. Hopkins, P. R. R. Langridge-Smith, and R. E. Smalley, *J. Chem. Phys.*, **79**, 5316 (1983).

115. E. J. Robbins, R. E. Leckenby, and P. Willis, *Adv. Phys.*, **16**, 739 (1967); P. J. Foster, R. E. Leckenby, and E. J. Robbins, *J. Phys.*, **B2**, 478 (1969).

116. K. I. Peterson, P. D. Dao, R. W. Farley, and A. W. Castleman, Jr., *J. Chem. Phys.*, **80**, 1780 (1984).

117. S. Leutwyler, T. Heinis, M. Jungev, H. P. Härri, and E. Schumacher, *J. Chem. Phys.*, **76**, 4290 (1982).

118. G. Delachretaz, J. D. Ganiere, R. Monot, and L. Wöste, *Appl. Phys.*, **B29**, 55 (1982).

119. A. Hermann, S. Leutwyler, L. Wöste, and E. Schumacher, *Chem. Phys. Lett.*, **62**, 444 (1979).

120. J. L. Gole, G. J. Green, S. A. Pace, and D. R. Preuss, *J. Chem. Phys.*, **76**, 2247 (1982).

121. R. L. Martin and E. R. Davidson, *Mol. Phys.*, **35**, 1713 (1978).

122. W. Gerber private communication and Ref. 41.

123. W. D. Knight, R. Monot, E. R. Dietz, and A. R. George, *Phys. Rev. Lett.*, **40**, 1324 (1978). W. A. de Heer, A. R. George, W. H. Gerber, and W. D. Knight (unpublished).

124. W. H. Crumley and J. L. Gole, *J. Phys. Chem.* (in press, 1986).

125. R. B. Woodward, W. H. Crumley, and J. L. Gole, *J. Phys. Chem.* (in press, 1986).

126. See Chapters by Gates and Brenner in this book.

127. S. J. Riley, E. K. Parks, C. R. Mao, L. G. Pobo, and S. Wexler, *J. Phys. Chem.*, **86**, 3911 (1982).

128. V. E. Bondybey and J. H. English, *J. Chem. Phys.*, **74**, 6978 (1981).

129. V. E. Bondybey and J. H. English, *J. Chem. Phys.*, **76**, 2165 (1982).

130. V. E. Bondybey, G. P. Schwartz, and J. H. English, *J. Chem. Phys.*, **78**, 11 (1983).

131. V. E. Bondybey *J. Chem. Phys.*, **77**, 3771 (1982).

132. V. E. Bondybey and J. H. English, *Chem. Phys. Lett.*, **94**, 443 (1983).

133. J. H. English and V. E. Bondybey, *J. Chem. Phys.*, **79**, 4746 (1983).

134. S. J. Riley, E. K. Parks, L. E. Pobo, and S. Wexler, *J. Chem. Phys.*, **79**, 2577 (1983).

135. R. C. Baetzhold, *J. Chem. Phys.*, **55**, 4363 (1971); R. C. Baetzhold and R. E. Mack, *J. Chem. Phys.*, **62**, 1513 (1975); R. C. Baetzhold, *J. Chem. Phys.*, **55**, 4355 (1971); R. C. Baetzhold and R. E. Mack, *Inorg. Chem.*, **14**, 686 (1975).

136. Yu. M. Efremov, A. N. Samoilova, and L. V. Gurvich, *Opt. Spectros.*, **36**, 381 (1974).

137. H. Abe, K.-P. Charlé, B. Tesche, and W. Schulze, *Chem. Phys.*, **68**, 137 (1982).

138. J. Mühlback, E. Recknagel, and K. Sattler, *Surf. Sci.*, **106**, 188 (1981); K. Sattler, J. Mühlbach, O. Echt, P. Pfau, and E. Recknagel, *Phys. Rev. Lett.*, **47**, 160 (1981).

139. A. Yokazeki and G. D. Stein, *J. Appl. Phys.*, **49**, 2224 (1978).

140. R. S. Bowles, J. J. Kolstad, J. M. Calo, and R. P. Andres, *Surf. Sci.*, **106**, 117 (1981).

141. B. Bederson, private communication.

142. K. I. Peterson, P. D. Dao, and A. W. Castleman, Jr., *J. Chem. Phys.*, **79**, 777 (1983); P. D. Dao, K. I. Peterson, and A. W. Castleman, Jr., *ibid.* **80**, 563 (1984).

143. W. H. Crumley, J. L. Gole, and D. A. Dixon, *J. Chem. Phys.*, **76**, 6439 (1982).

144. J. L. Martins, R. Carr, and J. Buttet, *J. Chem. Phys.*, **78**, 5646 (1983); J. L. Martins, J. Buttet, and R. Carr, *Phys. Rev.*, **B31**, 1804 (1985); K. Lee, J. Callaway, K. Kwong, R. Tang, and A. Ziegler, *ibid.*, **31**, 1796 (1985); G. Paccioni and J. Koutecky, *J. Chem. Phys.*, **81**, 3588 (1984). P. Fantucci, J. Koutecky, and G. Pacchioni, *ibid.*, **80**, 325 (1985); M. Chou, A. Cleland, and M. Cohen, *Solid State Comm.*, **52**, 645 (1984).

145. K. Hilbert, *Ber. Bunsenges. Phys. Chem.*, **88**, 260 (1984).

146. G. Delacretaz and L. Wöste, *Surf. Sci.*, **156**, 770 (1985).

147. R. Whetten, G. Delacretaz, and L. Wöste, private communication.

148. T. C. Thompson, D. G. Truhlar, and C. A. Mead, *J. Chem. Phys.*, **82**, 2392 (1985).

149. M. Moskovits, *Chem. Phys. Lett.*, **118**, 111 (1985).

150. D. P. DiLella, K. V. Taylor, and M. Moskovits, *J. Phys. Chem.*, **87**, 524 (1983).

151. M. Kappes and E. Schumacher in *Electronic and Atomic Collisions,* J. Eichler, I. Hertel and N. Stolterfoht, eds., Elsevier, Amsterdam (1985). M. M. Kappes, M. Schär, P. Radi, and E. Schumacher, *On the Manifestation of Electronic Structure Effects in Metal Clusters,* in press.

152. W. D. Knight, K. Clemenger, W. deHeer, W. Saunders, M. Chou, and M. Cohen, *Phys. Rev. Lett.*, **52**, 2141 (1984); W. D. Knight, K. Clemenger, W. A. deHeer, and W. A. Saunders, *Solid State Comm.*, **53**, 445 (1985).

153. J. L. Gole and W. C. Stwalley, *Characterization of Alkali Metal Aggregation from Atom to Bulk,* Advances in Atomic and Molecular Physics, in press.

154. K. Sattler, *Surf. Sci.*, **156**, 292 (1985); H. J. Novinsky, R. P. Pflaum, P. Pfau, K. Sattler, and E. Recknagel, *ibid.*, **156**, 165 (1985); P. Pfau, K. Sattler, R. Pflaum, and E. Recknagel, *Phys. Lett.*, **104A**, 262 (1984); E. Recknagel, Proc. 9th Int. Conf. Atomic Physics, R. S. vanDyck, Jr., and E. N. Fortson, eds., Seattle, 1984, p. 153; O. Echt, *J. Phys.*, **C18**, L663 (1985).

155. E. A. Rohlfing, D. M. Cox, A. Kaldor, and K. H. Johnson, *J. Chem. Phys.*, **81**, 3846 (1984).

156. J. L. Gole, R. Woodward, J. S. Hayden, and D. A. Dixon, *J. Phys. Chem.*, **89**, 4905 (1985).

7

APPLICATION OF MATRIX
ISOLATION TO THE STUDY
OF METAL CLUSTERS
WITH A POSTSCRIPT ON
THE REACTIVITY OF CLUSTERS
IN SUPERSONIC BEAMS

MARTIN MOSKOVITS
Department of Chemistry
University of Toronto
Toronto, Ontario, Canada

Matrix isolation spectroscopy is one of the first techniques used to study metal clusters. For the purposes of this chapter, metal clusters will be defined in a manner so as to exclude the stable species such as those discussed in Chapter 3 but will include "naked" species such as Cu_3, Sc_{13}, and so on and those with an incomplete complement of ligands such as Cu_3CO, $Ni_2(CO)_2$, and the like.

A pioneering study of metal clusters by matrix isolation is due to Blyholder and Tanaka (1) who applied the technique to the study of the interaction of CO with small nickel particles. Since then, dozens of papers have appeared that report on the use of various spectroscopic techniques to determine the electronic ground and excited state structure, vibrational frequencies, and reactivities of small metal particles isolated in solid matrices. Among the techniques used are absorption and emission uv-visible spectroscopy, infrared, Raman and resonance Raman spectroscopies, ESR, magnetic circular dichroism, and Mössbauer, and a few examples of NMR (2) and secondary ion mass spectroscopy (3).

In this chapter we deal almost exclusively with experimental results obtained via matrix isolation. At the end of the chapter a small section on the reactivity of gas-phase clusters will also be included.

7.1 TECHNIQUE

Matrix isolation refers to the technique whereby unstable molecular, atomic, or ionic species are trapped in a solid inert environment such as the rare gas solids. These compositions can then be studied spectroscopically at leisure. Several options are available during such an experiment. For example, the guest molecule may be irradiated to promote photochemistry or photoisomerization, the matrix may be warmed slightly to allow diffusion to bring together reagents that have been trapped in the same matrix or the matrix may be ablated by ion bombardment to analyze its content by SIMS. Clearly, most matrix experiments must be conducted at low temperatures (4 to 100K).

An often unspoken assumption of matrix isolation is that, at least in their ground states, the trapped molecules interact with the matrix host sufficiently little to allow them to retain much of their gas-phase properties, such as geometry, electronic ground state, and vibrational structure, intact. Experience has shown this to be true in most instances. So, for example, Jacox (4) has recently shown that the vibrational frequencies of most matrix-isolated species are within 2% of their gas-phase counterparts. This implies that the structures of the guests are likewise only slightly perturbed by the matrix. Excited electronic states of guest species, by contrast, may be significantly modified by the matrix and in some cases an electronically excited matrix-isolated atom may treat rare gas atoms as ligands (5). Thus, in general, electronic absorption and emission bands of matrix-isolated species will differ markedly from their gas-phase analogs regarding both frequency and intensity.

Matrix-isolated clusters are normally formed by co-condensing metal vapor with rare gas onto a surface cooled sufficiently to freeze out the gas. The metal vapor is usually generated either by heating a filament made of the desired metal electrically or heating a refractory metal (Ta, W, Mo) filament, basket, or knudsen effusion cell carrying or containing the desired metal. Since a well-founded technology for vaporizing metal exists (6), it will not be discussed in greater detail here. With the exception of a few metals such as bismuth and antimony these vaporization techniques produce largely monatomic metal vapor, hence the metal clusters that arise in the matrix result from aggregation of the metal atoms by diffusion in the as yet uncongealed layer of the matrix. Clusters containing many metal atoms are thereby formed presumably by a stepwise addition of atoms. By judiciously varying the metal to a gas atom ratio (usually in the range 10^{-3} to 10^{-1}), one may generate distributions of clusters of varying average size.

Likewise, if the matrix gas is doped with a potentially reactive molecule like CO, one may make small cluster carbonyls that may be construed to be models for CO adsorbed on metals.

Several simple calculations simulating this cluster growth have been reported (7,8). In one of these the authors (8) begin with the network of reactions shown in Figure 7.1. The reactions are assumed to be diffusion-limited, with only the metal atom and ligand molecule diffusing to any appreciable extent. Only two rate constants are needed, one describing the motion of a metal atom k_M, and the other representing the ligand molecule k_L. All other species are assumed to be immobile and all other reactions are ignored. The temperature inhomogeneity of the "reaction zone" is

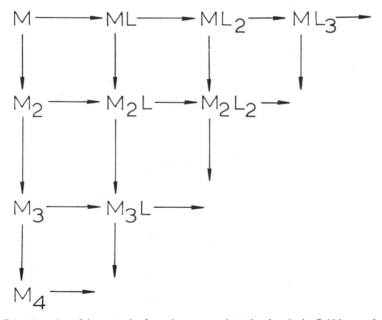

Figure 7.1   A portion of the network of reactions assumed to take place in the fluid layers of a matrix immediately prior to solidification.

also neglected. Instead, they assume that temperature-averaged rate constants may be substituted into the ordinary differential equations governing rates of reaction. This maneuver is, in fact, shown to be correct for the reaction $2M \rightarrow M_2$ and its rectitude is assumed by extrapolation for the more complex set of simultaneous reactions shown in Figure 7.1.

The equations describing this reaction scheme are:

$$\frac{d[L]}{dt} = -k_L[L]\left\{\sum_{ij} A_{ij}\right\}$$

$$\frac{d[M]}{dt} = k_M A_{11}\left\{\sum_{ij}(1 + \delta_{i1}\,\delta_{1j})A_{ij}\right\}$$

$$\frac{dA_{ij}}{dt} = k_M A_{11}\left\{A_{i-1,j} - A_{ij}\right\} + k_L[L]\left\{A_{i,j-1} - A_{ij}\right\}$$

where A_{ij} is the concentration of the species M_iL_{j-1} and terms such as $A_{0,j}$ and $A_{i,0}$ are zero. The $\delta_{i1}\delta_{1j}$ term in the preceding equation arises from the fact that in forming M_2 in the reaction $2M \rightarrow M_2$ the rate constant must be doubled.

In solving these equations, one assumes that the reactions shown in Figure 7.1

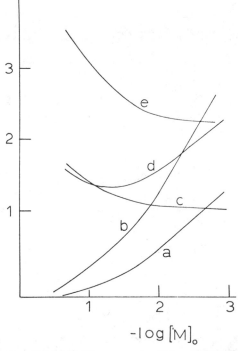

Figure 7.2 A plot of $-\log_{10}([M_xL_y]/[ML])$ versus $-\log[M]_0$ calculated by the technique described in the text. a, b, c, d, and e refer, respectively, to M_2L, M_3L, ML_2, M_2L_2, and ML_3.

evolve to time τ_q and then suddenly stop. This very rough approximation essentially replaces rate constants that presumably decrease over an infinite span of time as the products progressively cool, with average, unchanging rate constants applicable over a finite time.

The preceding equations were solved numerically by stepwise integration. A sample result is shown in Figure 7.2 using $k_M\tau_q = k_L\tau_q = 50$ (mol fraction^{-1}) and an initial ligand concentration of 0.004. The constants k_L and k_M were shown always to occur as products with τ_q, hence, the products are taken as single parameters. The figure shows log $[M_xL_y]/[ML]$ plotted against log (metal/argon ratio). Attention should be focused on the region below 1% metal where the log-log plot appears linear, implying that an equation of the form

$$[M_xL_y]/[ML] = (\text{metal ratio})^m$$

is obeyed.

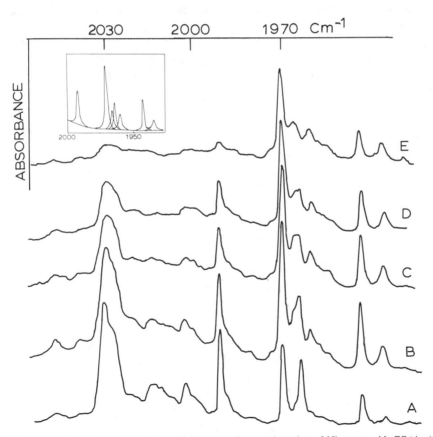

Figure 7.3 Portions of the spectra of the products of co-condensation of Ni atoms with CO/Ar (1/250). *A* through *E* refer to increasing total metal concentration. Inset is a curve-resolved version of spectrum. *B*.

This resulting simple expression has been used as a rationale for identifying the size of the cluster giving rise to a given spectral feature. So, for example, in a series of experiments in which nickel was co-condensed with CO/Ar mixtures (9), the infrared spectra shown in Figure 7.3 were obtained. In the series only the metal/argon ratio was varied, hence, by plotting the quantity log $[A_x/A_{NiCO}]$ as a function of log [total metal concentration] (where A_x and A_{NiCO} are the absorbance of a band of unknown origin and a bond of NiCO, respectively), one may discern from the slope the nuclearity of the cluster that is responsible for the absorption A_x. This is shown in Figure 7.4, from which infrared frequencies carried by Ni_2CO, Ni_3CO, $Ni(CO)_2$, and $Ni_2(CO)_2$ are identified.

Great care must be exercised in using this technique to identify the sizes of clusters in matrices. This is especially true of matrices containing high metal doping and thus larger clusters. In this concentration domain the simple rule fails and spectral overlaps abound. It is, in fact, safe to say that cluster identification in matrices remains a very tricky, difficult, and imperfect task.

One way to avoid this would be to preform the clusters in the gas phase, mass select them, and then matrix isolate them, hoping that fragmentation does not occur either during mass analysis or upon impact on the matrix. Although a number of gas-phase cluster sources have been reported, this technique has only been applied sparingly (10). For example, gas-phase sodium atoms, diatomics, and triatomic formed by effusion from a nozzle were separated in a Stern-Gerlach magnet and

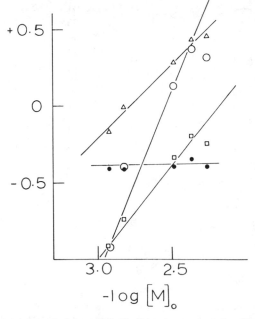

Figure 7.4 Experimental data taken from Ref. (9). Triangles, open circles, filled circles, and squares refer to Ni_2CO, Ni_3CO, $Ni_2(CO)_2$, and $Ni(CO)_2$, respectively.

the Na_2 fraction was matrix-isolated (10). The resulting visible absorption spectrum was almost free from sodium atoms. Likewise, the so-called gas-aggregation technique has been used to generate silver clusters in the A to 200 Å range which were subsequently matrix-isolated (11).

In the latter technique, metal is vaporized into a low-pressure atmosphere of cooled noble gas, such as argon, where clusters form in three-body collisions. The products then pass through a hole in the reaction chamber to form an effusive cluster source that can be collected on a cold surface. It is difficult to make silver clusters of diameters less than 10 Å in this way. This size corresponds roughly to 50 silver atoms.

7.2   SPECTROSCOPIC TECHNIQUES

7.2.1   Uv-visible Absorption Spectroscopy

Uv-visible absorption spectra yield information about electronic transitions. A typical example of a matrix *uv*-visible spectrum involving metal clusters is shown in Figure 7.5 which shows results obtained with silver in krypton. At low metal loading the spectrum is dominated by absorptions due to silver atoms. With an eightfold increase in the metal/krypton ratio numerous new bands appear that are attributed to silver aggregates Ag_2—Ag_6. The assignments, especially of the higher clusters, must by no means be construed as firm. Aside from the aforementioned difficulties in assignment, one often finds several bands belonging to a cluster of a given size that has been isolated in different sites within the solid matrix. The relative abundances of the molecule in the various sites may differ according to the temperature and thermal history of the matrix. In some cases one can convert clusters from one site to another reversibly by alternately irradiating into the appropriate absorption (12). Such a reversible process has been observed for isolated Ag_3 and interpreted as photoisomerization (13a). It is, however, more likely that the Ag_3 cluster does not change shape, but rather, that the matrix surrounding it is changed somewhat in structure upon irradiation. Occasionally, coupling with matrix vibrations can affect even the ground-state vibrational spectrum. This is wonderfully shown by Bechthold et al. (13b) for Ag_2 isolated in solid Xe. When excited with Kr^+ laser radiation one sees a progression due to Ag_2, as well as progressions built on combinations of the Ag_2 fundamental and a xenon lattice vibration.

The spectra of Figure 7.5 are quite representative of cluster absorption spectra. With a few exceptions the bands are broad and rather featureless. No definite structural information may be obtained from them. When vibrational fine structure is present as in the case of Mn_2 (14), Fe_2 (14), Ni_2 (15), and Pt_2 (16), it is, of course, characteristic of the excited state rather than of the ground state. The main thrust of uv-visible absorption spectroscopy is in demonstrating that a cluster is present in the matrix and indicating at what frequency one must excite when laser-induced emission is attempted. Recently, matrix measurements on clusters have been extended to the vacuum ultraviolet (17) by using a synchrotron source. In that

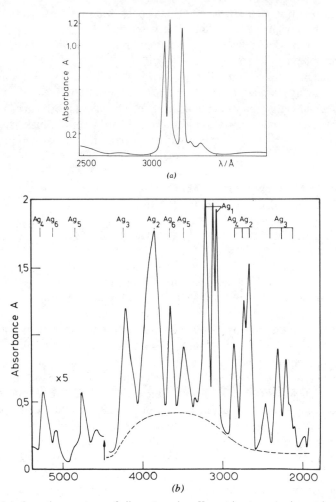

Figure 7.5 (*a*) absorption spectrum of silver atoms in a Kr matrix at an atomic metal concentration below 0.1%; (*b*) absorption spectrum of silver atoms and molecules in a Kr matrix at an atomic concentration of 0.8% prepared at 29K. (Reproduced by permission, from W. Schulze, H. V. Becker, and H. Abe, *Chem. Phys.*, **35**, 177 (1978).)

frequency domain, interesting new questions regarding the nature of Rydberg states of metal atoms and clusters arise since the electron in those states will be delocalized into the region of the matrix atoms.

7.2.2 Magnetic Circular Dichroism (MCD)

Considerably more information may be obtained from matrix uv-visible absorption spectra when it is combined with MCD. In this technique the matrix is placed between the poles of a magnet and the circularly polarized light used to record the spectrum is passed through the sample coaxially with the magnetic field. Electronic states of the cluster that have nonzero spin or space angular momenta will be shifted in energy and, in particular, degenerate states with nonzero spin will be split into spin-orbit terms. Because the selection rule allowing transitions to these multiplets depends on the sense (right or left) of circular polarization of the incident light, the spectrum recorded with right circularly polarized light will in those cases be shifted in frequency from that recorded with its left-handed complement. Hence, a difference spectrum $\Delta A = A_L - A_R$, (where L and R stand for right and left) will contain positive features, negative features, couplets, and multiplets according to the nature of the states involved in the transition. With degenerate ground states one has the further effect of a temperature-dependent MCD spectrum as a result of the varying population of the spin-orbit states that have been split out of the ground state.

Just how powerful this technique may be in the hands of experts is illustrated by the work of Schatz (18), Vala (19), and Grinter (20). Miller et al. (18), for example, have reported MCD spectra of magnesium, calcium, and strontium clusters. Figure 7.6 shows their spectrum for matrix-isolated magnesium atoms and clusters. A band centered near 27,000 cm^{-1} and assigned to Mg$_2$ shows only a very weak MCD spectrum. It is therefore assigned to a $^1\Sigma_g^+ \rightarrow {}^1\Sigma_u^+$ transition. This contrasts with the Mg$_2$ band near 38,000 which shows an intense couplet. It is assigned on that basis to a $^1\Sigma_{g^+} \rightarrow {}^1\Pi_u$ transition. Information is not only obtained from the type of MCD spectral features observed but also from quantitative analysis of these spectral features. The powerful deductions based on these is illustrated by the analysis surrounding the band at 31,700 cm^{-1} assigned to Mg$_3$ (Fig. 7.6). The

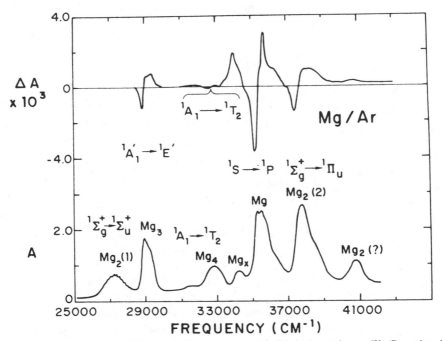

Figure 7.6 Absorption and MCD spectra ($\Delta A = A_L - A_R$) for Mg in Ar matrix at $\approx$6K. (Reproduced by permission, Ref. 18.)

couplet is unambiguous evidence for a degenerate excited state; possible only for linear or D_{3h} geometries. This restriction, together with arguments regarding the intensity of the MCD couplet relative to the integrated absorbance of the band, further restricts the options to D_{3h} (18).

The use of MCD, coupled with varying sample temperature, is illustrated by the work of Rivoal et al. (19) on Mn_2. The MCD spectrum shows a striking temperature dependence. This prompted the authors to confirm the conclusion of Van Zee et al. (21), based on an ESR study, that Mn_2 is an antiferromagnetically coupled diatomic, the temperature dependence being due to varying the population in the low-lying spin states makes up the ground-state manifold.

MCD has been applied mainly to nickel, palladium (22), and groups IB (20) and IIA (18) metal molecules.

7.2.3 Mössbauer Spectroscopy

Mössbauer spectroscopy is a form of γ-ray absorption spectroscopy that is applicable to atoms bound in solids, which absorb without recoiling and hence have little doppler shift from the frequency of the emitter. The technique is restricted to atoms possessing nuclides with appropriate lifetimes. Primary among these are [57]Fe and [117]Sn; however, matrix studies with [151]Eu have also been reported (23).

Mössbauer spectra yield three parameters from which useful information may be gleaned. These are the isomer shift, the electric quadrupole interaction, and the magnetic hyperfine interaction.

The isomer shift (IS) is proportional to the difference in the electronic charge density at the nucleus of the absorber and the source; that is,

$$IS = \tfrac{2}{3}\pi Ze^2 \Delta(r^2)[|\psi(o)|_A^2 - |\psi(o)|_S^2]$$

where $\Delta(r^2)$ is the change in the mean square radius of the nucleus on going from the ground to its excited state and $|\psi(o)|^2$ is the electronic probability density at the nucleus for the absorber and source (according to the subscript). Because only s electrons penetrate the nucleus to any extent, the IS is a sensitive measure of the s electron density.

The quadrupole shift is proportional to the nuclear electric quadrupole moment and the electric field gradient at the nucleus.

Using these parameters, McNab et al. (24) showed, for example, that the effective electron configuration about an iron atom in Fe_2 is approximately $3d^64s^{1.47}$. The decrease from its single atom value of $3d^64s^2$ is attributed to the involvement of s electrons in the FeFe bond. Likewise, Montano (25) has shown how the nuclear electric quadrupole splitting and magnetic hyperfine interaction may be used to narrow down the ground-state representation in Fe_2. Using an external magnetic field, he determined that the magnetic hyperfine interaction was large. This implies that the ground state of Fe_2 is a high spin state. Likewise, the electric quadrupole splitting implies a negative field gradient, eliminating Δ_g and Δ_u (25) as possible candidates for the representation of the ground state. However, the expected magnitude of the electric quadrupole splitting for a Σ state is double that for a Π state. The observed magnitude was closer to that expected for Σ than for Π. Accordingly a $^7\Sigma$ ground state was postulated.

Although this remarkable bit of deduction might imply that Mössbauer is a powerful technique for determining the electronic structure of clusters, it is, in fact, rather inconclusive in most cases, much like absorption spectroscopy. (One should, in fact, consider even the Fe_2 result as not entirely conclusive.) Nevertheless, it has been used to detect a large number of species of the form FeM with $M =$ Cr, Mn, Co, Ni, Cu, Pt (26). With ^{151}Eu the diatomic (Eu_2) and multimers, possibly Eu_3—Eu_5 have been observed using Mössbauer (23). Multimers such as Fe_3, Fe_3Pt, Fe_2Cr, Fe_2Cr_2, and $FeCr_3$ (27) have also been reported. In the case of these and other multimers the assignments must be considered to be tentative. Mössbauer has also been used to study tin (28) and antimony (29) clusters.

7.2.4 EXAFS and XANES

Two very promising related techniques for cluster structure determination are extended X-ray absorption fine structure (EXAFS) and X-ray absorption near edge structure (XANES). In EXAFS and XANES the low-intensity interference structure observed near X-ray absorption edges is analyzed to yield an atom–atom correlation function from which atom–atom distances are determined. Using this technique, researchers determined that the internuclear separation in Fe_2 was 1.87 ± 0.13 Å in argon (30) and 2.02 ± 0.02 Å in neon matrices. With matrices richer in iron, iron–iron distances that were intermediate between those of Fe_2 and bulk iron were determined for clusters of unknown nuclearity.

EXAFS has also been used to follow the progress with the size of the lattice parameter in small silver microcrystallites made by the gas-aggregation technique (31). Remarkably good EXAFS spectra were obtained from which it was determined that the lattice parameter decreased with decreasing cluster size, for crystallites ranging in size between 20 and 100 Å. The results were adequately explained in terms of striction due to surface tension.

7.2.5 Electron Spin Resonance Spectroscopy

ESR has been perhaps the most powerful technique in determining the geometrical and electronic structures of matrix-isolated clusters. Measurements are performed on a cooled sapphire or metal rod placed between the poles of a magnet. Electron spin resonances are then detected by sweeping the magnetic field in order to achieve resonance between a microwave frequency applied to the sample simultaneously with the magnetic field and the energy interval between Zeeman levels split out of degenerate spin-orbit states by the magnetic field.

The measurements yield two major parameters, the g tensor and a tensor. Both of these are isotropic in fluid systems where orientation is random. In matrices anisotropic g and a tensors are often encountered. The g values play roughly the role of the chemical shift in NMR, determining the magnetic field strength at which the "center" of the ESR spectrum occurs. It may therefore be used as a rough chemical indicator. For cluster studies it is the a tensor that is the most revealing. It measures the magnitude of electron spin/nuclear spin hyperfine interactions. The size of the a value is proportional to the Fermi contact term $|\psi(o)|^2$ mentioned earlier in the context on Mössbauer spectroscopy. This term is a measure of the "spin density" of the unpaired electrons "upon" a given atom. Since the hyperfine interaction splits the ESR spectrum into multiplets, one may use the information as an indication of the number of atoms in the cluster (from the number of hyperfine lines) and as a clue to geometry (from the spin-density distribution among the atoms in a cluster). As an example, a molecule such as M_3, where M is an alkali atom, may have a D_{3h} geometry, a linear geometry, and two types of C_{2v} geometries. In one, the apical angle is greater than 60° (obtuse); in the other, the angle is less than 60° (acute). The bonding in alkali trimers is due largely to s electrons. In order to discover the geometry of the M_3 molecule, one has to determine the distribution of the spin of the odd electron among the three atoms. Clearly, in a molecule with D_{3h} geometry the spin will be shared equally by the three atoms. In a C_{2v} molecule the unpaired electron can occupy either an a_1 state in which the spin density is concentrated mainly on the apical atom or a b_2 state which has spin density only on the terminal atoms. A simple molecular orbital argument indicates that for the obtuse geometry the b_2 state lies below the a_1 state, and vice versa for the acute geometry. Knowing the s-electron spin distribution leads directly, therefore, to the structure of M_3-type molecules.

ESR does have some drawbacks. It is, of course, only sensitive to paramagnetic species. So, for example, it will not detect molecules such as Na_2, Na_4, Na_6, and so on. Depending on which species one is seeking, this limitation may be beneficial in that it simplifies the spectrum of a matrix containing a range of clusters. It is also necessary occasionally to work with isotopically pure metals in cases where the naturally occurring element has several isotopes each with different nuclear spin. The jumble of lines that would occur otherwise would make interpretation difficult. Finally, it is not always so easy to interpret ESR spectra as has been implied earlier. Occasionally, more than one size cluster may result in the same number of ESR lines. With very large clusters one may not be sure that one has observed all the lines in a hyperfine set. Other complications in interpretation might also arise as a result of superhyperfine interactions, with nuclear spins belonging to the matrix host atoms, large g and a tensor anisotropies, quadrupole terms, and large zero-field splittings.

Matrix ESR studies of metal clusters are due mainly to three groups: Weltner (21), Lindsay (32), and Howard and Mile (33). The alkali trimers (Li_3, Na_3, and K_3) were originally studied by Lindsay and coworkers (34). Li_3 is found to be a pseudorotating (dynamic Jahn-Teller) molecule in all the matrices studied. In argon Li_3 was formed only after photolyzing the species assigned to Li_7. In both Na- and K-containing matrices two triatomic species were detected, one ascribed to an obtuse C_{2v} molecule, the other to a pseudorotating triatomic. The sensitivity of the ESR spectrum to geometrical changes is wonderfully illustrated by the spectra of the two forms of $^{39}K_3$. The obtuse triatomic carries a spectrum consisting of a septet of

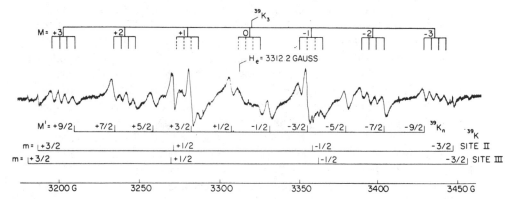

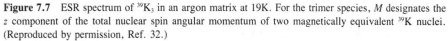

Figure 7.7 ESR spectrum of $^{39}K_3$ in an argon matrix at 19K. For the trimer species, M designates the z component of the total nuclear spin angular momentum of two magnetically equivalent ^{39}K nuclei. (Reproduced by permission, Ref. 32.)

quartets, Figure 7.7. The septet is due to the spin density on two magnetically equivalent potassium nuclei, whereas the quartet arises from a third, nonequivalent nucleus. The pseudorotating K_3 molecule has three equivalent nuclei. Its ESR spectrum consists of 10 unsplit lines of intensity ratio $1:3:6:10:12:12:10:6:3:1$ (Figure 7.8). The line widths of some of these lines are temperature dependent, yielding estimates of the correlation time that, in turn, is related to the inverse hopping rate of the pseudorotating K_3 molecule among three equivalent obtuse forms (34). Howard et al. (35) obtained similar results for Li_3 to those of Lindsay et al.

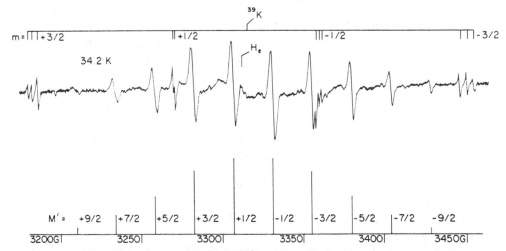

Figure 7.8 ESR spectrum of pseudorotating $^{39}K_3$ in an Ar matrix at 34K. The stick spectrum shows the predicted intensity distribution for three equivalent ^{39}K nuclei. Lines belonging to ^{39}K atoms are also indicated. (Reproduced by permission Ref. 34.)

in a solid adamantane. In that medium the molecule was stable for several hours even at room temperature.

Howard et al. were also the first to report the ESR spectra of Cu_3 (33), Ag_3 (36), and Cu_2Ag (36) in solid adamantane. In both cases, obtuse, non-pseudorotating (i.e., $^2B_2$) molecules were inferred. This contrasts with the findings of Kernisat et al. (37) who report Ag_3 to be an acute (i.e., $^2A_1$) triatomic in solid N_2. The molecules Cu_5 and Ag_5 have also been reported. The former is described as a distorted trigonal bipyramidal molecule with 60% of the unpaired $4s$ spin population on two of the equitorial copper atoms. (The molecules are best described as obtuse angle triatomics interacting with two equivalent atoms and below the molecular plane.) Despite the elegance of the interpretation, the structure and indeed the assignment of the ESR spectra reported in Ref. (38) must be considered tentative.

ESR has also been used to investigate transition-metal clusters such as Mn_2, Mn_5 (21), Sc_2 (39), and Sc_3 (40). A remarkable ESR spectrum with more than 60 hyperfine lines has also been reported (40) and assigned to Sc_{13} isolated in neon matrices. Its structure was postulated to be icosahedral.

Because ^{45}Sc has a nuclear spin of $7/2$ one expects to see for Sc_2 an ESR spectrum consisting of 15 lines with an intensity variation of $1:2:3:4:5:6:7:8:7:6:5:4:3:1$. In the spectrum actually observed (39) each of these lines is further split by higher-order interactions into submultiplets of equal intensity, making 64 lines in all. The authors were able to simulate the observed spectrum (including two sets of perpendicular fine-structure lines) only by assuming the molecule to possess a quintet ground state, hence $^5\Sigma$. Likewise, Mn_2 was unequivocally shown to be an antiferromagnetically coupled molecule (21). Upon changing the temperature, one observed that the molecule populated successively, states of total spin 0, 1, 2, 3 separated by the Landé interval spacing J, $2J$, $3J$ from which J was determined to be -9 ± 3 cm^{-1}. By assuming the anisotropic exchange to be due entirely to magnetic dipole interactions (which vary as r^{-3}), the authors estimated the internuclear separation in Mn_2 to be 3.4 Å.

ESR results are also reported for Y_3 (40) and interpreted as being a bent triatomic (probably $^2B_2$) in contrast with Sc_3 (40) which was determined to have an equilateral triangular form. Mn_5 (21) was tentatively interpreted to be a planar pentagonal molecule.

Occasionally, the nonobservance of an ESR spectrum of a cluster whose presence in the matrix is indicated by optical data may be a telling result. In the most banal case, it indicates a non-paramagnetic ground state. In certain other cases, for example, for La_3 (40) it may indicate that the molecule is linear with a ground state of nonzero, orbital angular momentum leading to a highly broadened ESR spectrum due to the large g-tensor anisotropy.

The nonobservance of Cu_3 in solid argon by ESR is somewhat more puzzling. It may indicate that in that medium the molecule is a dynamic Jahn-Teller molecule; this too leading to extreme broadening, in contrast with the result in the adamantane matrix (33). Something akin to this has been observed with pseudorotating Na_3 (41) in whose ESR spectrum some hyperfine lines are broadened to the point of disappearance while others are observed.

7.2.6 Fluorescence, Raman, and Resonance Raman

Two techniques that have been successful in providing structural information on matrix-isolated metal clusters are laser-induced fluorescence and resonance Raman. For diatomics these methods can yield vibrational constants, whereas for polyatomics structural information is possible of either a general sort as in cases when more than one fundamental is observed or a more precise variety when isotopic fine structure can be resolved in the vibrational lines. Isotopic fine structure, when it is present, is also of great use in confirming or, at least, corroborating the assignment of a fluorescence or resonance Raman progression.

The technique is simple. The matrix is irradiated with a laser line and the luminescence excited is dispersed spectroscopically. Alternatively, a part of or the total fluorescence is collected as a function of the frequency of the exciting laser. This is called an excitation spectrum and its vibrational structure is characteristic of the upper state; that is, the information it provides is related to that obtained in an absorption spectrum.

A beautiful example of fluorescence is due to Ahmed and Nixon (42). The authors used one of the uv lines of an Ar^+ laser to excite Ni_2 isolated in solid argon, obtaining the spectrum shown in Figure 7.9. The isotopic fine structure due to the three species $^{58}Ni_2$, $^{58}Ni^{60}Ni$, and $^{60}Ni_2$ is plainly visible and confirms the assignment. A possible danger in interpreting fluorescence spectra is that one cannot identify the two states connected by the transition. So, for example, the fluorescence may represent transitions between two excited states, the lower of which may erroneously be assumed to be the ground state. Such a state of affairs indeed prevailed with some of the alkaline earth dimers and with the molecules Pb_2. In the latter case it was matrix data (43) that clarified the situation. In the case of Bi_2 it was shown that the wrong carrier had been identified and the published data ascribed to Bi_2 actually resulted from a spectrum of Bi_4 (44).

The problem of ensuring that the lower state involved in an emission is the ground

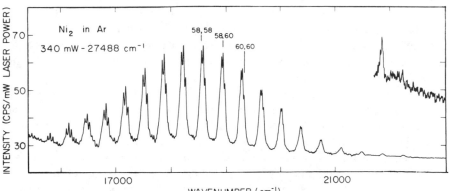

Figure 7.9 The $A \rightarrow X$ emission system for Ni_2 in Ar matrix at 10K, excited by 363.8 nm Ar^+ laser line. Isotopic fine structure is indicated. (Reproduced by permission Ref. 42.)

state is resolved in most cases where Raman or resonance Raman emissions are observed. These spectra are distinguished from (relaxed) fluorescence by using various exciting frequencies and noting that the positions of the spectra *relative* to the frequency of the exciting radiation remains constant. By contrast, in fluorescence the *absolute* frequencies of the emissions are constant.

Raman emissions have been reported for Zn_2, Cd_2 (45), Ag_2, Ag_3 (46), and Cr_3 (47), while resonance Raman has been used to study Sc_2, Sc_3 (48), Mn_2 (48), Ti_2, V_2 (49), Cr_2, Cr_3 (47), Fe_2 (50), Ni_3 (51), Cu_3 (52), Li_2, Li_3 (53), and possibly Li_4 (53). In addition, Raman and resonance Raman emissions have been reported for Ga_2, In_2, and Tl_2 (54). These refer, of course, to matrix studies. In some cases gas-phase studies followed and in a few cases preceded the matrix work. Fluorescence studies have yielded ground-state vibrational constants for the group IIA dimers (55), Sn_2 (56), Pb_2 (43), Sb_2 Sb_4 (57), Bi_2, Bi_4 (44), and Mo_2 (58).

In all cases but two the gas-phase and matrix vibrational results agreed to within one or two wave numbers. The exceptions were the group IIa dimers and Cr_2. With Mg_2 and Ca_2 the matrix frequencies were greater than their gas-phase counterparts. This was interpreted by the authors (55) as being due to the matrix cage effect. In the matrix the long-range portion of the effective interatomic potential is steepened as a result of interactions with the matrix cage. This has negligible effect on the harmonic frequency of the molecule for deep potentials and for short-bonded molecules. It may, however, be important for long-bonded molecules characterized by a shallow interatomic potential as in Mg_2 and Ca_2. Dichromium is a strongly bound molecule, and hence for it this effect is expected to be minor. In this instance, however, the authors (59) argued on the basis of the inordinately large anharmonicity observed for this molecule and the observation of a long-lived state of Cr_2 that the potential describing this molecule is somehow unusual perhaps due to a second minimum in the ground-state potential at about 3 Å. Resonance Raman has also produced evidence that the molecule Cu_3 is a dynamic Jahn-Teller species (52) akin to its alkali trimer brethren. The spectrum consisted of a series of progressions each built on a different vibronic state of Cu_3, but in each the vibrational spacing corresponded to the symmetric vibration of Cu_3, its only proper vibration. The same sort of observation was made for Li_3 isolated in solid xenon (53). In that case, however, another form of Li_3 was observed (Fig. 7.10) which according to the isotopic fine structure (Fig. 7.11) was not pseudorotating but was a rigid triatomic molecule with an apical angle of approximately 50°. In solid krypton only the rigid form of Li_3 was observed (53).

Although it was stated earlier that resonance Raman ensures ground-state data, the possibility exists that the same laser radiation that excites the resonance Raman emissions also pumps the molecule into a long-lived electronically excited state, which then acts as the lower state in a resonance Raman process. Such observations have been made with V_2 (49) and Pb_2 (60). It is normally distinguishable from true ground-state resonance Raman by the fact that the emission intensity of the excited state variety depends quadratically on laser fluence, at least before saturation occurs. Upon saturation the fluence dependence reverts to linear, and excited state resonance Raman cannot be distinguished from that originating in the ground state. With the

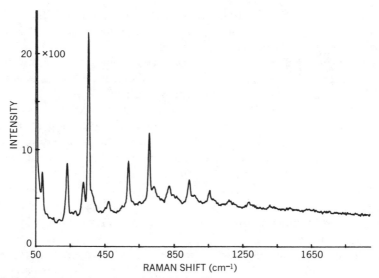

Figure 7.10 Spectrum attributed to "rigid" Li₃ in solid Xe matrix excited with 457.9 nm Ar⁺ laser light.

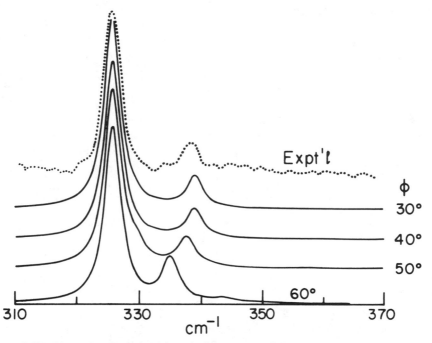

Figure 7.11 Observed and calculated isotopic fine structure of the symmetric stretch fundamental attributed to "rigid" Li₃ isolated in solid Kr matrix.

large number of low-lying electronically excited states that transition-metal molecules possess, the possibility that even a resonance Raman spectrum is not characteristic of the ground state cannot be totally discounted.

Aside from data relating to the electronic and geometrical structure of metal diatomics and clusters, the type of information produced by the techniques discussed earlier has already allowed some more global insight to be acquired regarding transition-metal diatomics. One notes, for example, on moving across the series (Table 7.1) that the values of the metal–metal stretching force constants increase to a maximum for V_2, decrease again to a low point for Mn_2, increasing to another (lower) maximum for Ni_2, and decreasing again thereafter. This behavior is in remarkable agreement with the cohesion energy of the respective bulk metals, also shown in Table 7.1, which lists the bond-dissociation energies reported for the diatomics. This implies that a large measure of pairwise bonding, undoubtedly due largely to the d electrons, is maintained in the bulk metal similar to that which exists in the diatomics.

The nature of the bonding in the diatomics that is discussed most often in the language of hydrogenic wavefunctions and that which exists in the bulk metals that is normally couched in terms of band theory can indeed be reconciled. Gelatt et al. (61) have decomposed the cohesive energy (ΔE_c) of metals into five contributions, three attractive and two repulsive. The first three are the d band, s-d hybridization, and conduction band; the next two are atomic preparation and renormalization. The last may be construed in some ways as an artifact of their calculational method whereby they renormalize the wavefunction of the constituent atoms so that the

TABLE 7.1　Selected Observed and Calculated Properties of Metal Diatomics and Bulk Metals

Metal	k_e (mdyn/Å)[a]	ΔE_c (eV)[b]	D^0 (eV)[c]	$\Delta E_c / k_e$
K	0.10	0.941	0.56	9.4
Ca	0.074	1.825	0.12 ± 0.01	24.7
Sc	0.76	3.93	1.3	5.2
Ti	2.35	4.855	1.3 ± 0.18	2.1
V	4.34	5.30	2.47 ± 0.22	1.2
Cr	2.80	4.18	1.56 ± 0.22	1.5
Mn	0.25	2.98	0.44 ± 0.3	11.9
Fe	1.48	4.29	1.30 ± 0.22	2.9
Co	1.46	4.387	1.73 ± 0.22	3.0
Ni	2.48	4.435	2.38 ± 0.22	1.8
Cu	1.30	3.50	1.97 ± 0.06	2.7
Zn	0.12	1.35	0.19 ± 0.07	11.3

[a]See references associated with Table 7.2.

[b]C. Kittel, *Introduction to Solid State Physics*, 3rd ed., Wiley, New York, 1967.

[c]The bond dissociation energy of the diatomics were taken from K. A. Gingerich, Ref. 86 for all the species with the exception of that of Sc_2 which was taken from Ref. 85.

squares of their amplitudes integrate to unity over the Wigner-Seitz cell rather than over all space. This results in an almost constant positive contribution to ΔE_c. The atomic preparation term is the energy required to promote the constituent atom from its normal $d^n s^2$ state to the $d^{n+1}s$ state that correlates best with the orbitals in the bulk metal. (The chromium atom, although already in a $d^5 s$ ground state still requires a sizable atomic preparation correction that promotes the atom from its ground term to the center of all the multiplets associated with the $d^5 s$ configuration.) The d-band contribution reflects two properties: (1) the breadth of the d band which is a monotonic function of the energy by which bonding states derived largely from d electrons are separated from their antibonding counterparts, this quantity increasing as one moves leftward across the transition metals, and (2) the degree to which the d band is filled. Filling increases from left to right in the periodic table. The two effects cause the energy contribution of the d electrons to ΔE_c to increase slowly as one goes from scandium to chromium and then to decrease rapidly thereafter. At chromium one has complete filling of bonding states (and therefore the greatest amount of cohesion due to d electrons) while maintaining empty antibonding states. The s-d hybridization term, again almost a constant, expresses the apparent gain in energy when one hybridizes the conduction electrons and the d band. And finally, the conduction band expresses, in addition to the energy separation (i.e., the width of the band) between the bonding and antibonding orbitals of mainly s (or perhaps s-p) character and its degree of filling, the major contributions from effects that cannot be easily expressed as pairwise interactions.

Most of these energy contributions have their counterparts in the diatomics. The d bandwidth correlates well with the energy separation between bonding and antibonding counterparts of σ, π, and δ levels derived from the d electrons. The conduction band width also corresponds, albeit much more approximately, to the energy separation between the s-derived σ_g and σ_u orbitals. The atomic preparation has also been long recognized (62) to be required in discussing the nature of the bonding in transition-metal complexes.

This atomic preparation term is more important and more subtle than it would seem at first glance. In Mn_2, for example, one expects the d-orbital separation to be large (61); hence, on the basis of d bonding alone one expects Mn_2 to be a very strongly bonded molecule. Unfortunately, manganese also has the highest atomic preparation term of all the first row transition-metal atoms which offsets the stabilization due to the d manifold.

The degree to which the bonding in the bulk metals may be understood in terms of what exists in diatomics may be surmised by considering the ratio of the cohesive energy of the bulk metal to the harmonic force constant of the analogous diatomic as a function of the position in the periodic table (Table 7.1). This ratio is seen to be constant (within rather broad limits) for Ti, V, Cr, Fe, Co, Ni and Cu. Sc lies a little outside this limit, whereas the metals K, Ca, Mn, and Zn possess ratios of $\Delta E_c / k_e$, which are substantially higher than those of the aforementioned set. What this indicates is that in the "strongly bonded" metals such as Ti, V, and so on, pairwise bonding is the dominant contribution to the cohesive energy, whereas in metals such as Ca, Mn, and Zn the cohesive energy, although rather weak, is

TABLE 7.2 Selected Ground-State Constants of Homonuclear Metal Clusters[a]

	r_e/(Å)	ω_e/(cm^{-1})	$\omega_e X_e$/(cm^{-1})	D_e/(eV)	Ground-State
Diatomics					
[7]Li$_2^b$	2.6729	351.43	2.610	1.07	$^1\Sigma_g^+$
[7]Be$_2$ (82)	2.45	275.8	26.0	0.098	$^1\Sigma_g^+$
[11]B$_2$	1.59	1051.3	9.35	3.08	$^3\Sigma_g^-$
[23]Na$_2$	3.0788	159.124	0.7254	0.730	$^1\Sigma_g^+$
[24]Mg$_2$ (gas)	3.890	51.12	1.645	0.0526	$^1\Sigma_g^+$
Ar matrix (55)	—	90.8	0.60	0.10–0.4	
[27]Al$_2$	2.466	350.01	2.022	1.57	$^3\Sigma_g^-$
[39]K$_2$	3.9051	92.021	0.2829	0.520	$^1\Sigma_g^+$
[40]Ca$_2$ (gas) (83)	4.277	65.07	1.09	0.135	$^1\Sigma_g^+$
Ar matrix	—	81.7 (55) 74.0 (84)	0.52 (55) 0.61 (84)	0.40	
[45]Sc$_2$	—	239.9 (48)	0.93 (48)	1.3 (85)	$^5\Sigma$ (39)
[48]Ti$_2$	—	407.9 (49)	1.08 (49)	1.3 (86)	—
[51]V$_2$	1.76 (87)	537.5 (49)	4.3 (49)	1.85 (87)	$^3\Sigma_g^-$ (87)
[52]Cr$_2$ (gas) (88)	1.6888	$\Delta G_{\frac{1}{2}} = 452.34$	—	1.56 (86)	$^1\Sigma_g^-$
matrix (47)	—	427.5	15.75	—	
[55]Mn$_2$ matrix	3.4 (21)	124.7 (48)	0.24 (48)	0.44 (86)	$^1\Sigma_g^+$
[56]Fe$_2$ matrix	1.87–2.02 (30)	300.26 (50)	1.45 (50)	1.3 (86) 1.2 (89)	—
[59]Co$_2$	—	290 (90)	—	1.7 (86)	—
[58]Ni$_2$	2.20 (91)	380.9 (42)	1.08 (42)	2.07 (91)	$^1\Gamma g$ or $^3\Gamma u$
[63]Cu$_2$	2.2197	264.55	1.025	2.05	$^1\Sigma_g^+$

	r_e	ω_e			Ground state
^{64}Zn$_2$	—	80 (45)	—	0.19 (86)	$^1\Sigma_g^+$
^{69}Ga$_2$	—	180 (54)	1 (54)	—	$^3\Sigma_u^-$
^{85}Rb$_2$	—	57.31	0.105	0.49	$^1\Sigma_g^+$
^{88}Sr$_2$ (gas)	—	39.6 (92)	—	0.14 (92)	$^1\Sigma_g^+$
(Ar matrix)		~47 (55)			
^{89}Y$_2$	—	—	—	1.6 (86)	—
^{93}Nb$_2$ (Kr matrix)	—	421 (93)	1.4 (93)	5.2 (86)	—
^{98}Mo$_2$ (gas)	1.939 (94)	477.1 (94)	—	4.2 (94)	$^1\Sigma_g^+$
matrix		476 (58)			
^{103}Rh$_2$	—	—	—	2.9 (86)	—
^{106}Pd$_2$	—	—	—	1.1 (86)	—
^{107,109}Ag$_2$	2.482	192.4	0.643	1.66	$^1\Sigma_g^+$
^{114}Cd$_2$	—	~58 (45)	—	0.09 (86)	$^1\Sigma_g^+$
^{115}In$_2$	—	118 (54)	0.8 (54)	1.0 (86)	—
^{120}Sn$_2$	—	118 (56)	0.5 (56)	1.9 (86)	$^1\Sigma_g^+$
^{121}Sb$_2$ (gas)	2.34	269.9	0.55	3.1	$^1\Sigma_g^+$
Ne matrix		270.4 (57)	0.57 (57)		
^{133}Cs$_2$	4.47	42.022	0.0823	0.397	$^1\Sigma_g^+$
^{197}Au$_2$	2.4719	190.9	0.420	2.31	$^1\Sigma_g^+$
^{202}Hg$_2$	—	36 (95)	—	0.08 (86)	$^1\Sigma_g^+$
^{205}Tl$_2$	—	80 (54)	0.5 (54)	0.6 (86)	—
^{208}Pb$_2$ (gas) (96)	2.93	109.7	0.19	0.85	0_g^+
Ar matrix (43) (60)		112.2	0.35	—	
^{209}Bi$_2$ (Kr matrix)	—	172.7 (97)	0.34 (97)	2.0 (86)	$^1\Sigma_g^+$
^{210}Po$_2$	—	155.715	0.3353	—	—

TABLE 7.2 (Continued)

	$r_e/(\text{Å})$	$\omega_e/(\text{cm}^{-1})$	$\omega_e X_e/(\text{cm}^{-1})$	$D_e/(\text{eV})$	Ground-State

Triatomics

$^7\text{Li}_3$; $^2\text{E}'$ geometry (dynamic Jahn Teller)

Atomization energy: 1.8 eV (98)

Symmetric stretching frequency = 310 (53) (Xe matrix)

A rigid form was found in solid Kr and in another site of solid Xe (53)

With frequencies

	ν_1	ν_2
	353	231 (Xe)
	351	247 (Kr)

$^{45}\text{Sc}_3$; (48) Geometry close to equilateral triangle

ν_1	ν_2	ν_3
247	145	151

$^{52}\text{Cr}_3$; (47) Geometry: bent triatomic

ν_1	ν_2	ν_3
308	123	226

$^{58}\text{Ni}_3$; Geometry: bent triatomic

ν_1 (51)

232

$^{63}\text{Cu}_3$; $^2\text{E}'$ Geometry: dynamic Jahn Teller in Ar matrix (52)

bent triatomic in adamantane matrix (33)

Atomization energy: 3.05 eV (99)

Symmetric vibration 354 (52)

Y$_3$: $^2B_2$ (40)

^{107}Ag$_3$: $^2A_1$ (37) $^2B_2$ (36) geometry: bent triatomic

Atomization energy: 2.62 eV (99)

ν_1
120.5 (46)

Au$_3$:

Atomization energy: 3.80 eV (100) (99)

Tetratomics

Be$_4$: Geometry T_d

$\nu_1(a_1)$	$\nu_2(e)$	$\nu_3(t_2)$	
151	—	—	
149.7	89.8	120.4	[Ar matrix (44)] [Ne matrix (44)]

Sb$_4$: Geometry T_d

$\nu_1(a_1)$	$\nu_2(e)$	$\nu_3(t_2)$	
241	(140)	(179)	[in Ne matrix (57)]

[a]This table draws heavily on the compilation of Weltner and Van Zee (81). Only numbers based on experimental data are included. Speculative assignments are marked with a (?).

[b]When no reference is given, data were obtained either from ref. 81 or from K. P. Huber and G. Herzberg, *Constants of Diatomic Molecules*, Van Nostrand, New York, 1979.

unusually strong as compared to what one expects on the basis of the strength of the bond in the analogous diatomic. This clearly points to the presence of interactions in the bulk metal which cannot be understood in terms of two-body forces alone. These interactions in fact dominate the cohesive energy of the more weakly bound metals and accounts for roughly half the cohesive energy of V, the most strongly bound among the metals considered. A compendium of properties of homonuclear metal clusters is presented in Table 7.2.

7.3 METAL CLUSTERS WITH LIGANDS

By doping the matrix gas with an appropriate ligand such as CO, N_2, or C_2H_4, one may generate species of the form M_xL_y. These may be construed to be small cluster analogs of adsorbed molecules. More ambitious possibilities exist wherein more than one ligand is attached to the same cluster, for example, N_2 and H_2 on an iron cluster in order to model coadsorbed species that may be intermediates in a catalytic process such as the reduction of N_2 to ammonia.

An even more intriguing possibility exists that once the species are coadsorbed, one might actually see a reaction between two molecules attached to the same matrix-isolated cluster. A clean example of such an event has not yet been reported, although it has been shown in one case based on a uv-visible absorption study (63) and in another on an infrared study (64) that iron atoms do not react with N_2, whereas Fe_2 and Fe_3 bond to N_2, strongly reducing its vibrational frequency by more than 300 cm^{-1} in one case (from 2331 cm^{-1}). Likewise, $Ag(CO)_3$ was found to react to form $Ag_2(CO)_6$ (65) upon diffusion in the matrix. The dimeric species then decomposes further to the metal and perhaps weakly adsorbed carbonyl.

Despite their obvious mimicry of molecules adsorbed on metals, matrix studies of ligands bonded to very small matrix-isolated metal clusters have been relatively few. The systems Ni/CO (9), Cu/CO (66), Pd/CO (63), and Ni/N_2 (67), have been studied more or less systematically. Ag/pyridine (68) was reported in the context of clarifying the role of the chemisorptive bond in producing the so-called surface-enhanced Raman effect. Fe_2 is found to interact with CH_4 (69) to form a weak complex which upon uv-irradiation does not rearrange to a methyl iron hydride complex, unlike the Fe atom that does. More recently, Klabunde and coworkers (70) have studied the reaction of a number of metal systems (Al, Ga, In, Mg) with CH_3Br in an attempt to form Grignard intermediates of the form CH_3M_xBr. Among these Mg was remarkable in that its atoms did not seem to be very reactive toward methyl bromide, whereas Mg_2 and Mg_3 performed the desired oxidative addition. The results of an ab initro calculation (71) supported this observation, with the finding that RMg_2X is more stable than $RMgX$.

Hints of cluster size-dependent reactivity based on reactions of organics with low-temperature metal slurries preceded these examples (72); however, the precise sizes of the clusters involved were not known.

The studies involving CO and metal clusters indicated that the infrared spectrum of the carbonyl approaches that of CO adsorbed on bulk metal for relatively small clusters. With copper (66), for example (Fig. 7.12), one finds that although the CO stretching frequency of CuCO is rather far from that of copper adsorbed on poly-crystalline copper, by the time Cu_4CO is reached the CO stretching frequency lies in the same neighborhood as that for the adsorbed CO. Upon prolonged warm-up large enough matrix-isolated copper clusters were produced so that one was able to ascertain that the neighboring CO molecules were more or less parallel to each other much like CO's bonded to a (locally) flat surface. This conclusion was deduced from the relative infrared band intensities produced in a mixed $^{12}CO/^{13}CO$ experiment. It had been previously shown that laterally coupled CO molecules produce the observed intensity pattern only when they are parallel (73). With palladium, however, small palladium clusters produced some species that were rather different

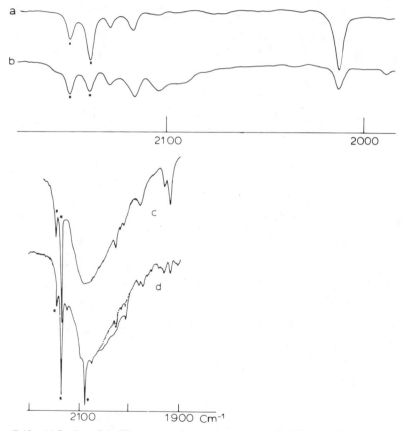

Figure 7.12 (a) Portion of the IR spectrum of a matrix containing Cu/CO/Ar in the ratio 1/10/4000; (b) the spectrum of the same matrix after warmup to 35K; (c) the spectrum obtained with a matrix approximately 50 times richer in Cu (asterisks mark uncoordinated CO); (d) same as (c) but with an equimolar mixture of ^{12}CO and ^{13}CO (dagger marks uncoordinated ^{13}CO). The dotted line is the low-frequency edge of spectrum c drawn to emphasize the small change in the spectrum of CO "adsorbed" on large Cu clusters brought about by the addition of ^{13}CO. Note frequency scale contraction in (c) and (d).

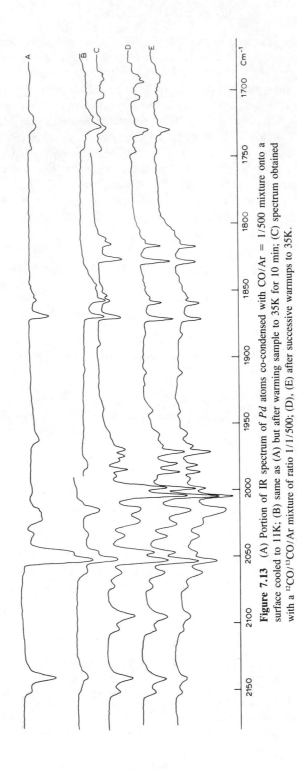

Figure 7.13 (A) Portion of IR spectrum of *Pd* atoms co-condensed with CO/Ar = 1/500 mixture onto a surface cooled to 11K; (B) same as (A) but after warming sample to 35K for 10 min; (C) spectrum obtained with a $^{12}CO/^{13}CO/Ar$ mixture of ratio 1/1/500; (D), (E) after successive warmups to 35K.

210

from those encountered in CO adsorbed on bulk palladium (Fig. 7.13). For example, an absorption at approximately 1725 cm^{-1} was ascribed to Pd$_3$CO in which the ligand triply bridged (μ^3) the tripalladium cluster. Upon warm-up, large matrix-isolated palladium cluster carbonyls were formed whose spectra were remarkably similar to those obtained with CO adsorbed upon silica-supported palladium (large) clusters. No trace of the very low-frequency CO vibrations remained in this limit.

In all these studies, cluster sizes were obtained from metal concentration studies as outlined earlier, whereas CO stoichiometry was obtained from matrices containing a mixture of $^{12}CO/^{13}CO$. The goal was to distinguish species such as M_x/CO from those to which more than one CO is bonded.

Similarly, Smardzewski and coworkers (68) showed that the ring-breathing vibration of pyridine bonded to silver approaches that were observed for pyridine adsorbed on bulk silver even for Ag$_2$py. In this case, the consequence of this result is more complicated since the aim of the authors was to show that on the bulk silver surfaces, with which comparison was made, the *py* molecule was bonded to small surface clusters.

Another example (67): When nickel is deposited into pure N$_2$, a second band was observed at 2216 cm^{-1} in addition to that at 2174 cm^{-1} previously observed and assigned to Ni(N$_2$)$_4$ (Fig. 7.14). The new band was assigned to a species Ni$_2$(N$_2$)$_x$ on the basis of a metal concentration study. This frequency is quite close to that reported for N$_2$ adsorbed on bulk nickel (74) and differed from that of an analogous dinickel carbonyl species by the absence of bridge bonded N$_2$ ligands.

Although thought provoking, these studies also demonstrate the difficulties involved in studying cluster effects on bonding and reactivity. In the first instance, the generation and identification of clusters is difficult. In the second, the low temperature and the presence of the solid "solvent" impedes reactions. The spectacular results obtained in this regard by the cluster beam groups limits further the territory in which matrix isolation can make a unique contribution. For this reason we will summarize some recent results obtained with metal cluster beams in the domain of metal cluster reactivity.

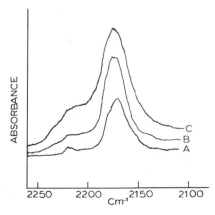

Figure 7.14 Portion of the IR spectrum obtained from a N$_2$ matrix containing Ni; (*a*), (*b*), and (*c*) refer to matrices of progressively higher Ni concentration.

7.4   REACTIVITY OF CLUSTERS IN BEAMS

The experimental technique used in the following studies are very similar although several independent groups are involved. The technique is described in Chapter 6. Briefly, metal is vaporized by focusing the second harmonic of a Nd:YAG laser on it. The metal vapor is entrained in helium vapor which is pulsed across the metal ablation region in synchronism with the laser pulse. The metal/helium mixture is then forced through a nozzle where adiabatic expansion cools the metal clusters formed to several degrees Kelvin (rotational) and several tens of degrees Kelvin (vibrational). A reagent may then be introduced downstream of the nozzle or directly in the helium stream. The cluster and cluster product distribution can then be measured after (usually ArF laser) photoionization by a suitable mass spectroscopic technique.

Several groups have studied the gas-phase reaction of iron clusters with H_2 and D_2 (75–77). The adsorptive bond between the metal and molecular H_2 is sufficiently weak that only H_2 which adsorbs dissociatively on the clusters will be represented in their mass spectrum. The extent of reaction may then be gauged either from the growth of new iron cluster hydride species or by noting the relative depletion of a given cluster size in the mass spectrum of the cluster beam. Richtsmeier et al. (75) find, for example (Fig. 7.15), that the reaction rate for H_2 chemisorption is not a monotonic function of the iron cluster size. In particular, Fe_{10}, Fe_{11}, and Fe_{12} are unusually reactive as compared to their immediate neighbors, whereas Fe_{17} is inordinately inert to H_2 chemisorption. Beyond Fe_{25} the H_2 chemisorption rate remains, more or less, constant. Similar results were reported by Geusic et al. (77) and by Whetten et al. (76). The latter group showed that an intriguing anticorrelation existed

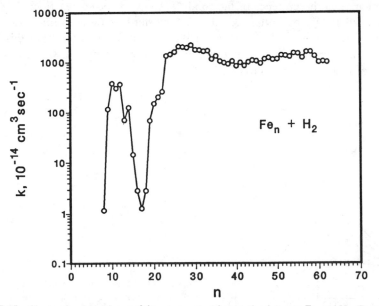

Figure 7.15   Cluster size dependence of the rate constant for reaction between Fe_x and H_2. (Reproduced by permission Ref. 75.)

between the ionization potential of the iron cluster and its relative reactivity towards H_2 (Fig. 7.16). The implication is that the dissociation of H_2 on a cluster surface is preceded by a harpooning mechanism in which H_2^- is formed, at least fleetingly. Clusters in the Fe_{15} to Fe_{18} range are characterized by relatively high-ionization potentials, hence low reactivity. Such a simple explanation does not, however, seem adequate to account for other systems. Moreover, Fe_x show a remarkable D_2 isotope effect which is as yet not understood. The Exxon group is by far the most prolific to date in this area.

Recently, this group has reported cluster size dependence in a variety of systems. Iron clusters in the range Fe_2 to Fe_{15} are found to react with O_2 to form a product of the form Fe_xO_2 and with H_2S to yield a cluster sulfide Fe_xS, devoid of hydrogen (78). The reactivity increases with cluster size but less rapidly than $x^{2/3}$ as one might have expected naïvely on the basis of the cross section for a sphere containing x Fe atoms. Significantly, the iron atom did not react with either O_2 or H_2S, recalling its previously discussed inertness in reactions with N_2. Intriguingly, no reaction was found with CH_4 and iron clusters of any size.

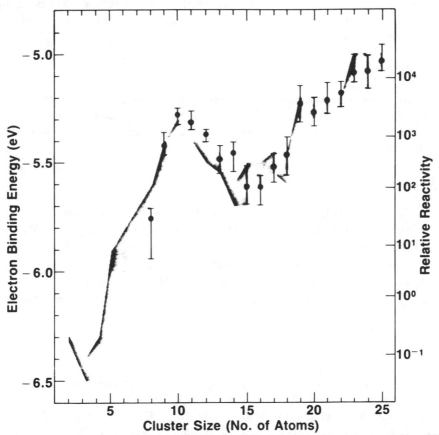

Figure 7.16 Comparison of measured ionization thresholds (left scale) with relative reactivities of Fe clusters (right scale). The grey band indicates the uncertainty in ionization threshold while the vertical bars mark the uncertainty in relative reactivity. (Reproduced by permission Ref. 76.)

A most thought-provoking set of reactions are reported with platinum clusters in the range Pt_2 to Pt_8 and four hydrocarbons: benzene, cyclohexane, hexane, and 2,3-dimethylbutane (79). With benzene Pt_2 and Pt_3 form complexes, the most prominent of which was $Pt_2\phi_2$, although mono-, di-, tri- and tetrabenzene complexes are also observed. With clusters in the range Pt_4 to Pt_8 the benzene begins to dehydrogenate, with the greatest hydrogen loss suffered by $Pt_8\phi_3$.

With cyclohexane, dehydrogenation occurs upon clusters in the entire range Pt_2 to Pt_8. The degree of dehydrogenation increases with cluster size. For the larger clusters the C:H ratio approaches unity, suggesting that the cyclohexane has been converted to benzene, a well-known reaction on supported platinum catalysts.

The Exxon group has also reported unusual distributions with carbon clusters (80). These do not fit within the terms of reference of this book save for the fact that upon soaking the graphite rod, which served as a carbon atom source, in KOH, mixed carbon/potassium clusters were generated of stoichiometry K_2C_{2n}, and the carbon cluster distribution was radically altered from that obtained in the absence of potassium. Potassium is a well-known promoter of catalysts used in hydrocarbon reactions such as Fischer-Tropsch. The relationship between the cluster beam result and the role of potassium in catalysis is, at present, obscure. It may not, however, remain so for long.

ADDENDUM

Recently Bordybey (private communication) has called into question the assignment of the progression shown in Fig. 7.9. It is claimed that the carrier is not Ni_2 but rather Se_2 which was presumably present in the metal atom source from a prior experiment.

REFERENCES

1. G. Blyholder and M. Tanaka, *J. Colloid Interfac. Sci.*, **37**, 753 (1971).

2. D. White, Abstract, 3rd Int. Conf. Matrix Isolated Species, Nottingham, 1981.

3. H. T. Jonkman and J. Michl, *J. Chem. Soc. Chem. Comm.*, 751 (1978).

4. M. E. Jacox, *J. Phys. Chem.*, Ref. Data., **13**, (1984).

5. M. Moskovits and J. E. Hulse, *J. Chem. Phys.*, **67**, 4271 (1977).

6. P. L. Timms, in M. Moskovits and G. Ozin, Eds., *Cryochemistry*, Wiley, New York, 1976.

7. R. Niedermayer, *Angev. Chem. Int.*, **14**, 212 (1975).

8. M. Moskovits and J. E. Hulse, *J. Chem. Soc. Faraday II*, **73**, 471 (1977).

9. J. E. Hulse and M. Moskovits, *Surf. Sci.*, **57**, 125 (1976).

10. W. Schulze, E. Schumacher, and S. Leutwyler, private communication; M. Hofman, S. Leutwyler, and W. Schulze, *Chem. Phys.*, **40**, 145 (1979).

11. H. Abe, W. Schulze, and B. Tesche, *Chem. Phys.*, **47**, 95 (1980).

12. L. C. Balling, M. D. Harvey, and J. J. Wright, *J. Chem. Phys.*, **70**, 2404 (1979).

13. (a) U. Kettler, P. S. Bechthold, and W. Krasser, *Surf. Sci.*, **156**, 867 (1985). (b) P. S. Bechthold, U. Kettler, and W. Krasser, *Solid State Commun.*, **52**, 347 (1984).

14. T. C. DeVore, A. Ewing, H. F. Franzen, and V. Calder, *Chem. Phys. Lett.*, **35**, 78 (1975).

15. M. Moskovits and J. E. Hulse, *J. Chem. Phys.*, **66**, 3988 (1977).

16. K. Jansson and R. Scullman, *J. Mol. Spectrosc.*, **61**, 299 (1976).

17. D. M. Kolb, H. H. Rotermund, W. Schrittenlacher, and W. Schroeder, *J. Chem. Phys.*, **80**, 695 (1984).

18. J. C. Miller, R. L. Mowery, E. R. Krausz, S. M. Jacobs, H. W. Kim, P. N. Schatz, and L. Andrews, *J. Chem. Phys.*, **74**, 6349 (1981).

19. J. C. Rivoal, J. Shakhs Emampour, K. J. Zeringue, and M. Vala, *Chem. Phys. Lett.*, **92**, 313 (1982).

20. R. Grinter and D. R. Stern, *J. Chem. Soc. Faraday II*, **79**, 1011 (1983).

21. R. J. Van Zee, C. A. Baumann, and W. Weltner Jr., *J. Chem. Phys.*, **74**, 6977 (1981); C. A. Baumann, R. J. Van Zee, S. V. Bhat, and W. Weltner, Jr., *ibid.*, **78**, 190 (1983).

22. R. Grinter and D. Stern, *J. Chem. Soc. Faraday II*, **79**, 1011 (1983).

23. P. A. Montano, *J. Phys.*, **C15**, 565 (1982).

24. T. K. McNab, H. Micklitz, and P. Barrett, *Phys. Rev.*, **B4**, 3787 (1971).

25. P. A. Montano, *Faraday Symp. Chem. Soc.*, **14**, 79 (1980).

26. W. Dyson and P. A. Montano, *J. Am. Chem. Soc.*, **100**, 7439 (1978).

27. W. Dyson and P. A. Montano, *Phys. Rev.*, **B20**, 3619 (1979); *Solid State Commun.*, **33**, 191 (1980); S. Shamai, M. Pasternak, and H. Micklitz, *Phys. Rev.*, **B26**, 3031 (1982).

28. A. Bos and A. T. Howe, *J. Chem. Soc. Faraday II*, **70**, 440 (1974).

29. H. M. Nagarathna, H-J Choi, and P. Montano, *J. Chem. Soc. Faraday I*, **78**, 923 (1982).

30. P. A. Montano and G. K. Shenoy, *Solid State Commun.*, **35**, 53 (1980); H. Purdum, P. A. Montano, G. K. Shenoy, and T. Morrison, *Phys. Rev.*, **B25**, 4412 (1982).

31. P. A. Montano, W. Schulze, B. Tesche, G. K. Shenoy, and T. I. Morrison, *Phys. Rev.*, **B30**, 672 (1984).

32. G. A. Thompson and D. M. Lindsay, *J. Chem. Phys.*, **74**, 959 (1981).

33. J. A. Howard, K. F. Preston, R. Sutcliffe, and B. Mile, *J. Phys. Chem.*, **87**, 536 (1983).

34. D. M. Lindsay, D. Garland, F. Tischler, and G. A. Thompson, in J. L. Gole and W. C. Stwalley, Eds., ACS symposium 179, *Metal Bonding and Interactions in High Temperature Systems*, 1982, Ch. 7; D. A. Garland and D. M. Lindsay, *J. Chem. Phys.*, **78**, 2813 (1983); **80**, 4761 (1984); **78**, 5946 (1983).

35. J. A. Howard, R. Sutcliffe, and B. Mile, *Chem. Phys. Lett.*, **112**, 84 (1984).

36. J. A. Howard, K. K. Preston, and B. Mile, *J. Am. Chem. Soc.*, **103**, 6226 (1981); J. A. Howard, R. Sutcliffe, and B. Mile, *ibid.*, **105**, 1394 (1983).

37. K. Kernisant, G. A. Thompson, and D. M. Lindsay, *J. Chem. Phys.*, **82**, 4739 (1985).

38. J. A. Howard, R. Sutcliffe, J. S. Tse, and B. Mile, *Chem. Phys. Lett.*, **94**, 561 (1983); *J. Phys. Chem.*, **87**, 2268 (1983).

39. L. B. Knight, Jr., R. J. Van Zee, and W. Weltner, Jr., *Chem. Phys. Lett.*, **94**, 296 (1983).

40. L. B. Knight, Jr., R. W. Woodward, R. J. Van Zee, and W. Weltner, Jr., *J. Chem. Phys.*, **79**, 5820 (1983).

41. D. M. Lindsay and G. A. Thompson, *J. Chem. Phys.*, **74**, 1114 (1982).

42. F. Ahmed and E. R. Nixon, *J. Chem. Phys.*, **71**, 3547 (1979).

43. V. E. Bondybey and J. H. English, *J. Chem. Phys.*, **67**, 3405 (1977).

44. V. E. Bondybey and J. H. English, *J. Chem. Phys.*, **73**, 42 (1980).

45. A. Givan and A. Loewenschuss, *Chem. Phys. Lett.*, **62**, 592 (1979).

46. W. Schulze and H. V. Becker, *Chem. Phys. Lett.*, **35**, 177 (1978).

47. D. P. DiLella, W. Limm, R. H. Lipson, M. Moskovits, and K. V. Taylor, *J. Chem. Phys.*, **77**, 5263 (1982).

48. M. Moskovits, D. P. DiLella, and W. Limm, *J. Chem. Phys.*, **80**, 626 (1984).

49. C. Cossé, M. Fouassier, T. Mejean, M. Tranquille, D. P. DiLella, and M. Moskovits, *J. Chem. Phys.*, **73**, 6076 (1980).

50. M. Moskovits and D. P. DiLella, *J. Chem. Phys.*, **73**, 4917 (1980).

51. M. Moskovits and D. P. DiLella, *J. Chem. Phys.*, **72**, 2267 (1980).

52. D. P. DiLella, K. V. Taylor, and M. Moskovits, *J. Phys. Chem.*, **87**, 524 (1983).

53. M. Moskovits, W. Limm, and T. Mejean, in B. Pullman, et al. Eds., *Dynamics on Surfaces*, Reidel, Amsterdam, 1984. p. 437.

54. F. W. Froben, W. Schulze, and U. Kloss, *Chem. Phys. Lett.*, **99**, 500 (1983).

55. J. C. Miller and L. Andrews, *Appl. Spect. Rev.*, **16**, 1 (1980).

56. V. E. Bondybey and J. H. English, *J. Mol. Spect.*, **84**, 388 (1980).

57. V. E. Bondybey, G. P. Schwartz, and J. E. Griffiths, *J. Mol. Spect.*, **84**, 328 (1981).

58. M. J. Pellin, T. Foosnaes, and D. M. Gruen, *J. Chem. Phys.*, **74**, 5547 (1981).

59. M. Moskovits, W. Limm, and T. Mejean, *J. Chem. Phys.*, **82**, 4875 (1985).

60. H. Sontag, B. Eberle, and R. Weber, *Chem. Phys.*, **80**, 279 (1983).

61. C. D. Gelatt, Jr., H. Ehrenreich, and R. E. Watson, *Phys. Rev.*, **B15**, 1613 (1977).

62. H. A. Skinner and F. Summer, *J. Inorg. Nucl. Chem.*, **4**, 245 (1957).

63. M. Moskovits, *Acc. Chem. Res.*, **12**, 229 (1979).

64. P. H. Barrett and P. A. Montano, *J. Chem. Soc. Faraday II*, **73**, 378 (1977).

65. D. McIntosh, M. Moskovits, and G. A. Ozin, *Inorg. Chem.*, **15**, 1669 (1976).

66. M. Moskovits and J. E. Hulse, *J. Phys. Chem.*, **81**, 2004 (1977).

67. J. E. Hulse, M.Sc. Thesis, University of Toronto.

68. D. E. Tevault and R. R. Smardzewski, *J. Chem. Phys.*, **77**, 2221 (1982).

69. Z. H. Kafafi, R. H. Hauge, and J. L. Margrave, *J. Am. Chem. Soc.*, in press.

70. Y. Imizu and K. J. Klabunde, *Inorg. Chem.*, **23**, 3602 (1984).

71. P. G. Jasieu and C. E. Dykstra, *J. Am. Chem. Soc.*, **105**, 2089 (1983).

72. T. O. Murdock and K. J. Klabunde, *J. Org. Chem.*, **41**, 1076 (1976); K. J. Klabunde, T. Groshens, M. Brezinski, and W. Kennelly, *J. Am. Chem. Soc.*, **100**, 4437 (1978).

73. M. Moskovits and J. E. Hulse, *Surf. Sci.*, **78**, 397 (1978).

74. R. P. Eischens, *Acc. Chem. Res.*, **5**, 75 (1972).

75. S. C. Richtsmeier, E. K. Parks, K. Liu, L. G. Pobo, and S. J. Riley, *J. Chem. Phys.*, **82**, 3659 (1985).

76. R. L. Whetten, D. M. Cox, D. J. Trevor, and A. Kaldor, *Phys. Rev. Lett.*, **54**, 1494 (1985).

77. M. E. Geusic, M. D. Morse, and R. E. Smalley, *J. Chem. Phys.*, **82**, 590 (1985).

78. R. L. Whetten, D. M. Cox, D. J. Trevor, and A. Kaldor, *J. Chem. Phys.*, **89**, 566 (1985).

79. D. J. Trevor, R. L. Whetten, D. M. Cox, and A. Kaldor, *J. Am. Chem. Soc.*, **107**, 518 (1985).

80. E. A. Rohlfing, D. M. Cox, and A. Kaldor, *J. Chem. Phys.*, **81**, 3322 (1984).

81. W. Weltner, Jr., and R. J. Van Zee, *Ann. Rev. Phys. Chem.*, **35**, 291 (1984).

82. V. E. Bondybey, *Chem. Phys. Lett.*, **109**, 436 (1984).

83. W. J. Balfour and R. F. Whittlock, *Can. J. Phys.*, **53**, 472 (1975); C. R. Vidal, *J. Chem. Phys.*, **72**, 1864 (1980).

84. V. E. Bondybey and C. Albiston, *J. Chem. Phys.*, **68**, 3172 (1978).

85. G. Vertraegen, S. Smoes, and J. Drowart, *J. Chem. Phys.*, **40**, 239 (1964).

86. K. A. Gingerich, *Faraday Discuss.*, **14**, 109 (1980) (compilation).

87. P. R. R. Langridge-Smith, M. D. Morse, G. P. Hansen, R. E. Smalley, and A. J. Jerer, *J. Chem. Phys.*, **80**, 593 (1984).

88. S. J. Riley, E. K. Parks, L. G. Pobo, and S. Wexler, *J. Chem. Phys.*, **79**, 2577 (1983). D. L. Michalopoulos, M. E. Geusic, S. G. Hansen, D. E. Powers, and R. E. Smalley, *J. Phys. Chem.*, **86**, 3914 (1982); V. E. Bondybey and J. H. English, *Chem. Phys. Lett.*, **94**, 443 (1983).

89. M. Moskovits and D. P. DiLella, in J. L. Gole and W. C. Stwalley, Eds., ACS symposium Vol. 179, *Metal Bonding and Interactions in High Temperature Systems*, 1982.

90. M. Moskovits, D. P. DiLella, and A. Loewenschuss, unpublished.

91. M. D. Morse, G. P. Hansen, P. R. R. Langridge-Smith, L-S. Zheng, M. E. Geusic, D. L. Michalopoulos, and R. E. Smalley, *J. Chem. Phys.*, **80**, 5400 (1984).

92. T. Bergman and P. F. Liao, *J. Chem. Phys.*, **72**, 886 (1980).

93. M. Moskovits and W. Limm, *J. Chem. Phys.*, in press.

94. J. B. Hopkins, P. R. R. Langridge-Smith, M. D. Morse, and R. E. Smalley, *J. Chem. Phys.*, **78**, 1627 (1983).

95. L. R. Epstein and M. D. Powers, *J. Phys. Chem.*, **57**, 336 (1953).

96. V. E. Bondybey and J. H. English, *J. Chem. Phys.*, **76**, 2165 (1982).

97. K. Manzel, U. Engelhardt, H. Abe, W. Schulze and F. W. Froben, *Chem. Phys. Lett.*, **77**, 514 (1981).

98. C. H. Wu, *J. Chem. Phys.*, **65**, 3181 (1976).

99. K. Hilpert and K. A. Gingerich, *Ber. Bunsenges, Phys. Chem.*, **84**, 739 (1980).

100. A. Kent, S-S. Lin, and B. Strauss, *J. Chem. Phys.*, **49**, 1983 (1968).

8

METAL CLUSTERS IN ZEOLITES

PIERRE GALLEZOT
Institut de Recherches sur la Catalyse
C.N.R.S.
Villeurbanne, France

8.1   INTRODUCTION

During the last decade increasing attention has been paid to metal clusters hosted in zeolites. Because of their small and homogeneous sizes, they are well-suited materials for the study of the atomic and electronic structure of highly divided metals. Because of the high fraction of surface atoms and because their environment can be easily modified, they are unique materials for the study of metal-support and metal-adsorbate interactions. Their properties are in many respects intermediate between those of metals and those of molecular clusters; therefore, they are interesting species to be studied with the scope of bridging the gap between metal physics and molecular metal cluster chemistry. In applying these properties to catalysis, the first aim was to use metal clusters as catalysts for CO conversion reactions to synthesize hydrocarbons used as substitutes for petroleum-derived products. The second aim was to prepare catalytic systems combining the advantages of homogeneous and heterogeneous catalysts, that is, a good activity and selectivity coupled with a high stability and the ease of handling and recovery.

The crystal structure, adsorption, ion exchange, and catalytic properties of zeolites are described in the books of Breck (1), Rabo (2), and Barrer (3) and in the proceedings of the international conferences on zeolites (4–9). A complete review of metals in zeolites has been given by Minachev and Isakov (10) and three review papers on noble metals have been published by Gallezot (10–13).

Metal clusters in zeolites include a large number of chemical species. They comprise dimers as well as aggregates containing several tens of atoms. The metal atoms can be naked or bonded to various kinds of ligands. Their oxidation state might change to some extent. For the sake of clarity, distinction will be made in this paper among metal aggregates, molecular metal clusters, and ionic clusters on the following bases:

1. Metal aggregates are small metal crystals where the bond lengths are essentially similar to those of bulk metal although they might be slightly elongated or contracted because of incomplete coordination or bonding with adsorbates. The atoms are formally zerovalent but their electronic structure can be modified by intrinsic size effects or environment effects.

2. Molecular metal clusters are polynuclear metal complexes with the same molecular frame and metal–metal bonds as in the corresponding molecular crystals of a known structure. They are solvated in zeolite cages and interact

to a variable extent with framework and extraframework atoms. They can lose reversibly part of their ligands.

3. Ionic clusters include incompletely reduced metal atoms. The metal–metal distances are intermediates between those observed in oxides and in metals. Their stability is due to weak ionic bonding with framework or extraframework oxygen atoms.

The distinction between these three types of metal clusters is meaningful only if they can be prepared and characterized unambiguously. Therefore, the preparation and characterization of the metal clusters will be emphasized, as well as the modification of structure in the course of treatments, adsorptions, and catalytic reactions. The extent of these modifications can be such that the metal clusters are completely transformed. Of special interest are the transformations of metal aggregates into cations (redispersion) and of the cations into molecular metal clusters (in situ cluster synthesis). Finally, the catalytic properties will be examined in relation to the physical properties of the metal clusters. The different aspects of intrazeolitic metal cluster chemistry will be examined through a limited number of experimental investigations in which a good characterization of entrapped metal clusters has been achieved.

8.2   PREPARATION AND CHARACTERIZATION OF METAL CLUSTERS

Preparation of metal clusters encaged in the porous frameworks of zeolites is a prerequisite step to obtain materials of interest. Metal aggregates larger than the cages or pores but still occluded in the zeolite crystals will also be considered. They can be hosted in preexisting holes or can grow by destroying the cage walls. Metal clusters on the external surface of the zeolite crystal will not be considered. They are not stable enough because they can easily sinter or be leached away in the course of treatments or catalytic reactions.

8.2.1   Preparation and Characterization of Metal Aggregates

Metal aggregates are usually prepared by reduction of ion-exchanged zeolites but can also be obtained by thermal decomposition of entrapped metal carbonyls. Reduction of zeolites impregnated with metal salts gives a heterogeneous dispersion.

Reduction of Ion-exchanged Zeolites

Ion exchanges should result in a uniform repartition of cations in the zeolite crystals. In this respect the pH of the exchange solution should be adjusted to avoid precipitation of metal hydroxide on the external surface. Ion exchange with metal ammine cations could also prevent the formation of a hydroxide phase.

The preparation of metal aggregates requires very precise treatment conditions to avoid either incomplete reduction or agglomeration of atoms into large particles. There are no treatment conditions generally applicable to all metal zeolites because the cation reducibility depends on both its reduction potential and its coordination and accessibility. The ease of migration and agglomeration of the reduced atoms depends on the metal vapor pressure, the presence of residual molecules that can favor the transport of the atoms in the form of volatile complexes, and the configuration of the zeolite pores. Thus, metal aggregates can be more easily trapped in cages with restricted aperture (faujasite-type zeolite) than in cylindrical pores (mordenite, L). Large metal concentration does not preclude the preparation of small aggregates because of the large inner surface, the aggregates remain far apart in distant cages even for metal concentration as large as 10 wt %.

Noble metals are usually loaded in Y zeolite by exchange of metal ammine cations such as $[Ru^{III}(NH_3)_6]^{3+}$, $[Rh^{III}(NH_3)_5Cl]^{2+}$, $[Pd^{II}(NH_3)_4]^{2+}$, $[Ir^{III}(NH_3)_5Cl]^{2+}$, and $[Pt^{II}(NH_3)_4]^{2+}$. They are easily reduced by hydrogen but should be pretreated to eliminate the NH_3 molecules because reduction in the presence of evolving ammonia could lead to metal agglomeration. Table 8.1 gives the treatment conditions required to obtain encaged noble metal aggregates in Y-type zeolite. Note that pretreatment in oxygen is a prerequisite step except for $[Ru(NH_3)_6]Y$ where heating in O_2 results in the formation of bulk oxide. The treatments should be carried out in flowing O_2 with a low heating rate up to the temperature range indicated in Table 8.1, otherwise heterogeneous samples are obtained. Pretreatment in O_2 results in the formation of encaged oxide-like species in the case of Rh^{III} and Ir^{III} (26), whereas Pd^{2+} and Pt^{2+} cations appear in PdY and PtY zeolites (11). Details on the size and location of the noble metal aggregates can be found in the papers referenced in Table 8.1 and in review papers (11,12,27). Note that the aggregates in the supercages are within the size range 1 ± 0.3 nm which means that the number of atoms can range from a

TABLE 8.1 Preparation of Noble Metal Aggregates in Y Zeolites^a

Zeolites	Treatments (atm., temp./K)	Cage	Refs.
	vac., 300 → 623 : H_2, 623	Supercage^b	14
$[Ru(NH_3)_6]Y$	He, 300 → 673 : H_2, 673	Supercage	15,16
	vac., 300 : H_2, 533	Holes^c	14
$[Rh(NH_3)_5Cl]Y$	O_2, 300 → 623 : H_2, 473–873	Supercage	17,18
$[Pd(NH_3)_4]Y$	O_2, 300 → 400 : H_2, 290–400	Supercage^d	19
	O_2, 300 → 570–870 : H_2, 290–500	Sodalite cage	19–21
$[Ir(NH_3)_5Cl]Y$	O_2, 300 → 480–550 : H_2, 500–900	Supercage	22
$[Pt(NH_3)_4]Y$	O_2, 300 → 580–630 : H_2, 470–670	Supercage	23,24,25
	O_2, 300 → 750–870 : H_2, 370–570	Sodalite cage^e	23,24,25

^{a}1 nm aggregates in supercages or atoms in sodalite cages.

^{b}20% of atoms in sodalite cage.

^{c}2 to 4 nm occluded particles.

^{d}Aggregates filling adjacent supercages.

^{e}25% of 2 to 3 nm occluded particles.

few units to a few tens. The morphology and number of atoms in platinum aggregates are discussed in Section 8.3.2.

Cations of the first transition row are less easily reduced because of their lower reduction potential and their location in sites of high stabilization energy. Therefore, stronger reducing agents and modification treatments prior to reduction leading to a better reducibility should be used. Sodium vapor has been used to reduce FeY zeolite (28) but partial sintering could not be avoided. The presence of small amounts of noble metals dissociating hydrogen favors the reducibility of Ni^{2+} cation (29). The highest degree of reduction and the smallest nickel aggregates have been obtained by reducing NiCaX with atomic hydrogen at room temperature (30,31). The blocking of hidden sites (e.g., SI sites in faujasite-type zeolite) by exchange of Ca^{2+} cations forces the transition cations to occupy less stable and more accessible sites. Thus, a complete reduction of Ni^{2+} ions yielding 1 nm nickel aggregates was made possible using a NiCaX zeolite (31). Exchange with rare earth ions (Ce^{3+}, La^{3+}) is also beneficial to obtain small aggregates. The role of these high electrostatic field cations is multiple: They increase the accessibility and reducibility of the cations and hamper the migration of the reduced atoms. Thus, nickel aggregates smaller than 1 nm were characterized by magnetic methods on reduced NiCeX zeolites (32). The low crystal field stabilization of Ni^{2+} cations in mordenite allows an easier reduction than when in faujasites (33). Complexing the cations with extraframework ligands eases the formation of small aggregates. Thus, room temperature H_2 reduction of Ni^{2+} in mordenite after complexation with CO has been reported (33). The increased reducibility of Ni^{2+} in the presence of NH_3 (34) can be attributed to a better accessibility of Ni^{2+} ions since NH_3 can induce migration of the Ni^{2+} to the supercages (35). The precipitation of hydroxide-like species inside the cages also favors the formation of small aggregates (36,37).

A new preparation method has been described by Scherzer and Fort (38). Treatments of Fe^{II}, Co^{III}, Ni^{II}, $Cu^{II}Y$ zeolites with water solutions of alkaliferrocyanide lead to the precipitation inside zeolite crystals of insoluble metal ferrocyanides which can be reduced by H_2 to yield monometallic or bimetallic particles. The same preparative method has been used by Iton et al. (39) to obtain Fe, Fe—Ru, and Fe—Co aggregates in modified ZSM-5 zeolites. However, in both cases data on the aggregate sizes are lacking.

To sum up, the preparation of metal aggregates of the first transition row is a difficult task. Homogeneous preparation of 1 nm aggregates can only be achieved by using all the treatment conditions favoring the reduction of cations at low temperature, for example as in the NiCaX reduction (30) where exchange by Ca^{2+} cations, pretreatment with CO, and reduction by atomic hydrogen were combined.

Decomposition of Encaged Metal Carbonyls

Metal compounds small enough to enter the zeolite pores can be adsorbed and thermally decomposed to obtain entrapped metal atoms. There are early reports in the patent literature on the use of metal carbonyls, acetylacetonate, halogenides, and alkyl-derivatives (40): since then, mainly carbonyls were used. The adsorption and the intermediate states of decomposition of metal carbonyls are discussed later.

The thermal decomposition should be carried out by rapid heating in an inert or reducing atmosphere to avoid back diffusion of the molecules toward the external surface. Upon decomposition the metal atoms can be oxidized by reaction with the hydroxyl groups of the zeolite. Thus, oxidation of Mo^0 atoms produced by $Mo(CO)_6$ decomposition in HY zeolite was proved by H_2 formation, uv spectroscopy, and decrease of the IR bands corresponding to OH groups (41). However, when $Mo(CO)_6$ is decomposed at 573K in a PtNaHY zeolite containing 1 nm platinum aggregates most of the molybdenum atoms are deposited on the surface of the platinum aggregates producing a slight increase of the aggregate size and masking the chemisorptive properties of platinum (42). Therefore, $Mo(CO)_6$ decomposition on metal zeolite could be used to prepare bimetallic aggregates. The decomposition of $Fe_3(CO)_{12}$ in HY zeolite also leads to iron cations (43,44), whereas decomposition in NaY at 473K yields active Fischer-Tropsch catalysts which provides indirect evidence for the formation of iron aggregates (44). Nagy et al. (45) have studied the decomposition of $Fe(CO)_5$ in HY zeolite. Clustering into aggregates occurs via thermal or photochemical decomposition; because of a stronger interaction between iron and support atoms, the photochemical decomposition leads to smaller aggregates which cannot be detected by electron microscopy.

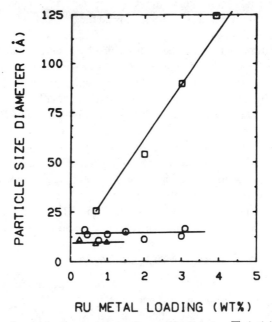

Figure 8.1 Effect of metal loading on average Ru particle diameter. □, incipient wetness; ○, ion exchange; △, vapor impregnation. (Reproduced with permission from Ref. 46.)

Gallezot et al. (41) have shown that the complete decomposition of $Re_2(CO)_{10}$ and $Ru_3(CO)_{12}$ in Y zeolite gives rhenium and ruthenium aggregates. Further characterization of ruthenium aggregates produced by this technique has been described (44,45). Chen et al. (46) have shown that ruthenium aggregates issued from $Ru_3(CO)_{12}$ decomposition in Y zeolite are smaller (1 nm) than those prepared by ion exchange and reduction (1 to 1.5 nm) or by impregnation. This is clearly illustrated in Figure 8.1 where the aggregate diameter is plotted as a function of the metal loading for different conditions of preparation.

Nazar et al. (47) have used bis-(toluene)Fe^0 (or Co^0) to prepare metal aggregates. The complexes were prepared by metal evaporation into a slurry of dehydrated zeolite and toluene at low temperature. The organometallic molecules thus formed enter the zeolite and decompose upon warming to room temperature into metal aggregates fitting into the supercages.

To sum up, vapors or solutions of metal carbonyls or of any other neutral organometallic complexes can be adsorbed and decomposed in zeolite to prepare metal aggregates. This technique should be developed because metal atoms released at low or moderate temperature are expected to give smaller aggregates than by ion exchange-reduction technique. However, only large pore zeolites can be loaded this way because of the large molecular size of most organometallic complexes.

8.2.2 Preparation of Molecular Metal Clusters

The simplest way to introduce polynuclear metal complexes into zeolite is to adsorb them from the vapor or liquid phase. This technique is obviously restricted to a small number of molecules small enough to enter the zeolite pores. The other way is to perform in situ syntheses of molecular cluster in cages larger than the pore opening.

Grafting of Polynuclear Carbonyls

Metal carbonyls are usually sublimated in vacuum so that the vapor can enter the zeolite previously dehydrated. Thus, $4Re_2(CO)_{10}$ and $3Ru_3(CO)_{12}$ have been introduced per unit cell in the HY zeolite at 383 and 393K, respectively, without loss of CO (41). The location of the molecules in the zeolite cages has been determined by crystal structure analysis, and the interaction with the framework has been probed by IR spectroscopy. The linear $Re_2(CO)_{10}$ molecule stretches across the supercage; it is slightly distorted because of the hydrogen bonding of the CO ligands with the hydroxyl groups of the framework. At 573K the Re—Re bond is broken and $Re(CO)_3$ species are directly bonded to the 12-membered oxygen ring of the cage aperture. The $Ru_3(CO)_{12}$ molecule is adsorbed in the vicinity of the supercage aperture (41). In NaY, the structure of the $Ru_3(CO)_{12}$ molecule is not disrupted but the original symmetry of the molecule is perturbed (48).

The intermediate states of decomposition of $Fe_3(CO)_{12}$ and $Co_2(CO)_8$ in Y zeolite have been studied in detail (49). The encaged $Fe_3(CO)_{12}$ molecules are bonded to

Na^+ or H^+ *via* a bridging CO ligand. Decarbonylation of $Fe_3(CO)_{12}$ proceeds by steps involving the reversible dismutation process. Decarbonylation of $Co_2(CO)_8$ also proceeds via a multistep route involving $Co_4(CO)_{12}$ species. Bein et al. (50) have detected $Fe_3(CO)_{12}$ as an intermediate in the decarbonylation of $Fe(CO)_5$ in HY zeolites.

Bulky carbonyls deposited on the external zeolite surface can to some extent be adsorbed in the zeolite framework even though they cannot enter the pores. Thus, Gelin et al. (51) have shown by IR spectroscopy that $Rh_6(CO)_{16}$ sublimed at 353K onto HY zeolite can be decarbonylated by heating in O_2, H_2, or vacuum at 373K. The partially stripped rhodium carbonyl species can enter the zeolite pores and be recarbonylated in situ by CO at 373K to give polynuclear rhodium carbonyls, possibly $Rh_6(CO)_{16}$. This process was effective only for low metal loading (0.5 wt % Rh); otherwise, fragmentation of $Rh_6(CO)_{16}$ and agglomeration into metal particles occurs on the external surface.

In Situ Synthesis of Molecular Clusters

Transition-metal complexes in zeolites can be synthesized by adding ligands to cations previously introduced by ion exchange: the zeolite framework acts as solvent, anion, and eventually additional ligand (52). Polynuclear metal complexes can also be synthesized in large pore zeolites. Mantovani et al. (53) reported that entrapped polynuclear rhodium carbonyl can be formed in Y zeolite upon reduction of $[Rh(NH_3)_6]Y$ by a CO:H_2 mixture (80 atm, 403K). The IR spectrum given in Figure 8.2 exhibits three bands at 2095 (vs), 2080 (sh), and 2060 (w) cm^{-1} corresponding to terminal CO ligands and a band at 1765 cm^{-1} corresponding to bridging CO. Unambiguous identification of the polynuclear carbonyl was not possible because the CO frequencies shift with respect to those in $Rh_4(CO)_{12}$ or $Rh_6(CO)_{16}$ due to steric and charge density effects imposed by the zeolite matrix. Since low loss of rhodium occurred during catalytic reaction, the possibility that the rhodium carbonyl was deposited on the external surface was excluded. Lefebvre et al. (54) have shown that $Rh^{III}Y$ zeolite reacts with CO to give $Rh^I(CO)_2$ species which then react with a mixture of CO:H_2 at 293K to give polynuclear carbonyls. From IR and ^{13}C NMR spectroscopies, it was concluded that $Rh_4(CO)_{12}$ molecules are formed according to the overall reaction

$$4[Rh^I(CO)_2]^+ + 4CO + 2H_2 \rightarrow Rh_4(CO)_{12} + 4H^+$$

This cluster can be transformed slightly above room temperature and in the presence of water to give a new polynuclear carbonyl whose spectrum is similar to that found by Mantovani et al. (53) (Fig. 8.2) or to that observed after in situ recarbonylation of $Rh_6(CO)_{16}$ on NaY (51); the spectrum was assigned to encaged $Rh_6(CO)_{16}$ (54,55). The zeolite $Rh^{III}Y$, treated under similar conditions, has been studied by the radial electron distribution method which gives interatomic distances that are present in the samples (56). Figure 8.3 shows that the radial distribution exhibits a large peak at 2.77 Å which corresponds to the Rh—Rh distances similar to those observed in $Rh_6(CO)_{16}$.

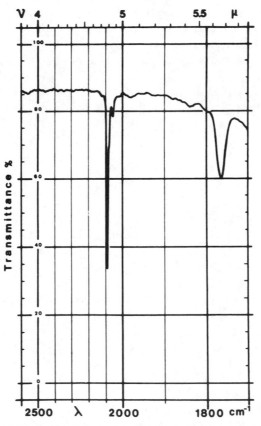

Figure 8.2 Infrared spectrum of Rh-exchanged zeolite after heating at 403K under 80 atm of CO:H$_2$. Absorption bands of CO at 2095 (vs), 2080 (sh) 2060 (w), and 1765 (s) cm^{-1}. (Reproduced with permission from Ref. 53.)

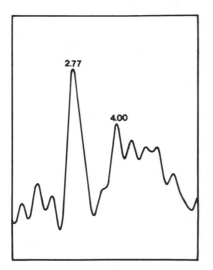

Figure 8.3 Distribution of interatomic distances in RhY zeolite activated in O$_2$ at 570K and treated under CO:H$_2$O at 300K. The main peak at 2.77 Å corresponds to Rh—Rh distances similar to those in Rh$_6$(CO)$_{16}$.

227

Gelin et al. (55) have prepared polynuclear iridium carbonyls by treating with $CO:H_2O$ or $CO:H_2$ mixtures, encaged $[Ir^I(CO)_3]$ species previously obtained by carbonylation of Ir^{III} in the zeolite cages (57). The structure of the clusters was identified by the combined use of IR, mass spectrometry, and magic angle spinning ^{13}C NMR.

These studies clearly demonstrate that in situ synthesis of molecular clusters within zeolite cages is possible under very mild conditions. Catalytic applications based on these homogeneous complexes hosted in zeolites will be discussed in Section 8.5.1.

8.2.3 Preparation of Ionic Metal Clusters

Early reports described the formation of Na_4^{3+} ionic clusters in NaY zeolite exposed to sodium vapors (58,59), but most of the subsequent investigations were conducted on charged silver clusters. Kim and Seff (60,61) interpreted the crystal structure of strongly dehydrated AgA zeolite in terms of octahedral Ag_6 metal clusters interacting with Ag^+ cations in sodalite cages. Gellens et al. (62) reinvestigated the crystal structure of AgA zeolites dehydrated and exchanged to a variable extent, and concluded that linear Ag_3^{2+} clusters are formed with an Ag^0 atom between two Ag^+ cations. The Ag^0—Ag^+ distances range between 2.85 and 3.00 Å; they are slightly larger than in silver metal (2.89 Å). The number of Ag_3^{2+} species increases with dehydration temperature; at most four clusters can be formed in the cages. Figure 8.4 gives the coordination of Ag_3^{2+} clusters in A zeolite. The formation of clusters is due to an autoreductive process involving framework oxygen (63). Linear Ag_3 clusters were also characterized by crystal structure analysis in faujasite-type zeolites with different Si/Al ratios (64). Dehydration in vacuum leads to an autoreductive process resulting in simultaneous occupancy of hexagonal prisms (SI) and two adjacent SI′ sites in sodalite cages. This arrangement was described in terms of linear Ag_3 clusters with Ag(I)—Ag(I′) distances in the range 3.0 to 3.16 Å. Clearly, these distances are closer to those observed in oxides than in silver metal. The mechanism of Ag reduction by hydrogen in AgY and Ag mordenite has been studied by Beyer et al. (65). Highly dispersed silver aggregates are formed at low temperature, the presence of charged clusters was postulated. Ozin et al. (66,67) have characterized by optical reflectance spectroscopy the presence of low nuclearity Ag_n^{q+} ($n = 5, 6, 7, \ldots, 13$) clusters in AgY zeolites of low-exchange level dehydrated at 773K. In totally Ag-exchanged zeolites larger clusters are formed at the expense of Ag_n^{q+}.

Bergeret and Gallezot (20) have shown that in PdY zeolite activated at 870K in O_2, 12 Pd^{2+} cations/unit cell occupy SI′ sites. They are interacting with 3 framework oxygen atoms, O(3), at 2.07 Å. After hydrogen reduction at 420K, 8.8 palladium atoms are still located on SI′ sites but are more distant from the anions (Pd(I′)-O(3) = 2.74 Å). There are also 1.8 palladium atoms in II′ sites at 2.75 Å from O(2) anions. Since there are 10.6 palladium atoms in 8 sodalite cages, a few number of dimers are present. These palladium atoms bear a positive charge since they still

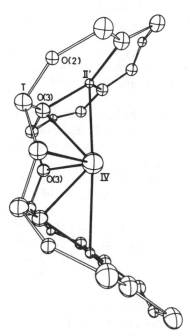

Figure 8.4 Representation of the coordination of the Ag_3 cluster in dehydrated Na, Ag—A zeolite. (Reproduced with permission from Ref. 64.)

interact weakly with framework anions. Furthermore Pd^+ species were detected by ESR (68) and the binding energy of Pd3d 5/2 level measured by XPS was shifted by $+1.4$ eV with respect to palladium metal (69). Similarly, few dimers of electropositive platinum atoms could be present in the sodalite cages of PtY zeolite outgasssed at 870K in O_2 and reduced at 570K in H_2 (24).

To sum up, charged metal clusters especially dimers or trimers can be formed by an autoreductive process in vacuum or by reduction at low temperature under hydrogen. The atoms are still interacting with framework oxygen because they are positively charged. The reduction is not complete because low-valence states are stabilized by the interaction with the zeolite framework.

8.3 MODIFICATION OF THE ATOMIC STRUCTURE OF METAL CLUSTERS

So far, there are few well-documented examples of structure modifications concerning encaged molecular metal clusters (see, however, earlier discussion on the transformation of $Rh_4(CO)_{12}$ into $Rh_6(CO)_{16}$). Therefore, this section will deal with structure modification of metal aggregates, namely, (1) transformation into smaller aggregates or into atoms and ions (redispersion) and (2) modification of the atomic structure (morphology, atom packing, interatomic distances) induced by the aggregate environment.

8.3.1   Redispersion of Metal Aggregates

Examples of redispersion into smaller particles of metals supported on zeolites can be found in the patent literature (70). These processes are based on the formation, migration, and decomposition of mobile metal compounds involving treatments by O_2, Cl, S, NH_3, and water vapor.

Redispersion of palladium aggregates into Pd^{2+} cations by heating in oxygen is possible provided the aggregates are in the zeolite cages. Thus, palladium atoms in sodalite cages of Y zeolite can be reoxidized within the temperature range 450 to 780K into Pd^{2+} cations occupying SI′ sites (19,21). Bergeret et al. (19) have also shown by X-ray techniques that 1 nm palladium aggregates filling adjacent super-cages are reoxidized in O_2 at 470K into Pd^{2+} cations. Larger particles occluded in the zeolite crystals or on the external surface are transformed into PdO. The same behavior was observed on CuY zeolite. Herman et al. (71) concluded from ESR and gas uptake measurements that encaged copper aggregates are reoxidized into Cu^{2+} ions in O_2 at 673K but larger particles are transformed into CuO. On the other hand, Beyer et al. (65) showed by X-ray diffraction that even 200 Å silver particles on the external surface can be reoxidized at 620K into Ag^+ in the zeolite framework. Silver aggregates formed at low-reduction temperatures can be reoxidized at 363K. Oxidative redispersion of metal into cations by O_2 treatment is probably limited to metal cations such as Ag^+ and Pd^{2+} which have a high-stabilization energy when they are bonded to the zeolite framework. On the contrary, ruthenium is oxidized into RuO_2 whatever its state of dispersion (14).

Oxidative redispersion of palladium aggregates by NO is easier than with O_2. Thus, 2 nm palladium aggregates occluded in PdY zeolite crystals are oxidized into PdO by O_2 treatment, whereas adsorption of NO at room temperature can redisperse the palladium atoms into cations. This process has been studied in detail by Che et al. (72) with the combined use of X-ray diffraction, ESR, IR, and mass spectroscopy. The overall reaction stoichiometry reads

$$2(Z{-}0)^-, 2H^+ + Pd^0 + 2NO \rightarrow N_2O + H_2O + 2(Z{-}O)^-, Pd^{2+}$$

The metal atoms are oxidized and complexed by NO before migrating toward the zeolite cages. Similarly, Garbowski et al. (33) reported that nickel particles as large as 8 nm supported on mordenite can be reoxidized into cations by treating the zeolite with NO at 473K, the disappearance of the metal was monitored by magnetic measurements, and the appearance of Ni^{2+} cations by uv spectroscopy.

Another interesting example of oxidative redispersion has been demonstrated by the combined use of infrared spectroscopy and radial electron distribution (56). Rhodium aggregates smaller than 1 nm encaged in Y zeolite have been prepared by treating [Rh(NH$_3$)$_5$Cl]Y in O_2 at 623K, then in H_2 at 473K. The distribution of interatomic distances is given in Figure 8.5, curve a. All the peaks correspond to the interatomic distances of face-centered cubic (fcc) rhodium. After contacting the zeolite with CO at 300K for 24 h, all the Rh—Rh distances characteristic of the metal disappear (curve b). The IR study shows that Rh^I(CO)$_2$ species are formed.

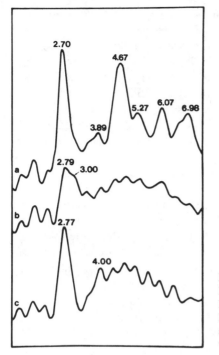

Figure 8.5 Distribution of the interatomic distances in RhY activated at 570K in O_2 and reduced at 470K in H_2—curve *a*, after reduction, distances correspond to fcc Rh aggregates—curve *b*, after treatment of the previous sample under CO—curve *c*, after treatment of the previous sample under CO:H_2O at 300K.

Therefore, the rhodium aggregates have been completely disrupted, and rhodium atoms are oxidized and complexed by CO. Upon further treatment at 300K with CO:H_2O mixtures, the $Rh^I(CO)_2$ species migrate and rearrange into polynuclear carbonyls. At the end of the process the formation of $Rh_6(CO)_6$ clusters is shown by the appearance of the interatomic distance at 2.77 Å (curve *c*) and by an IR spectrum similar to that given in Figure 8.2. This study shows that metal aggregates can be transformed into molecular metal clusters under very mild conditions.

8.3.2 Structure and Morphology of Metal Aggregates

Two methods can be used to determine the microstructure of supported metal aggregates or molecular metal clusters—the EXAFS method based on X-ray absorption spectroscopy measurements and radial electron distribution (RED) method based on X-ray scattering measurements. The relative advantages of these methods have been compared in a recent review (73). EXAFS probes mainly the first coordination sphere around the absorbing atom, whereas RED gives the complete set of interatomic distances present in the cluster. These distances do not need any correction unlike those determined by EXAFS which should be corrected for phase shift using a model compound. Most of the RED investigations discussed hereafter have been performed on 1 ± 0.3 nm platinum aggregates in Y zeolite prepared as described in Table 8.1 (74–79).

Figure 8.6 gives the distribution of distances corresponding to adsorbate-free platinum aggregates (curve a) and to H_2-covered aggregates (curve b). The latter have the normal fcc structure and interatomic distances of bulk platinum (74), whereas the structure of the bare aggregates is distorted and contracted (75,76). Similar conclusions have been obtained by EXAFS (80). The distances are relaxed in H_2-covered aggregates because dissociated hydrogen fulfills the coordination of surface atoms. However, the fact that the relaxed distances are the same as in bulk platinum is probably coincidental. Thus, recent results on iridium aggregates (81) show that the distances are relaxed upon H_2 adsorption but they remain smaller than in bulk iridium.

Before considering the effect of other adsorbates on the aggregate structure, let us examine the information given by RED on the average number of atoms and on the morphology of the platinum aggregates. The experimental distribution of distances was compared with the distributions calculated from the coordinates of atoms in model aggregates of various shapes and sizes (79). The distributions corresponding to spherical polyhedra containing 25 to 45 atoms are in good agreement with the experimental distribution, whereas plate-like shapes (rafts) can be readily discarded. The distribution corresponding to a truncated fcc tetrahedron containing 40 atoms (Fig. 8.7, curve a) fits well with the experimental distribution (curve b). Its largest and smallest dimensions (1.27 and 0.96 nm, respectively) are compatible with the supercage diameter 1.3 nm and with the 1 ± 0.3 nm size measured by SAXS (24) or TEM (25). It has the same symmetry as the supercage, each of the truncated corners pointing toward a cage aperture. Therefore, the aggregate morphology seems related to that of the cage by a three-dimensional epitaxy.

Fraissard et al. (107) have designed a novel method to count the average number of atoms per aggregate. It is based on the measurements of the NMR chemical shift of ^{129}Xe adsorbed in zeolite which depends on the chemical coverage of the aggregate

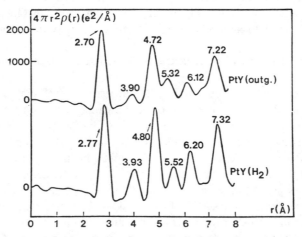

Figure 8.6 Distribution of distances in 1 nm Pt aggregates in PtY—curve a, naked aggregates—curve b, H_2-covered aggregates.

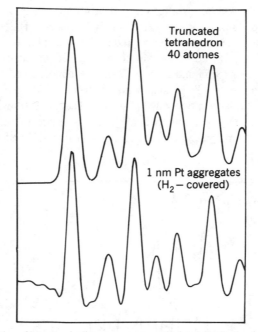

Figure 8.7 Distribution of interatomic distances curve *a*, calculated distribution for a 40 atoms truncated fcc tetrahedron—curve *b*, experimental distribution corresponding to H_2-covered 1 nm Pt aggregates.

surface. Thus, the NMR spectrum of xenon adsorbed on PtNaY zeolite exhibits well-characterized shifts δa, δb, δc which have been associated with a coverage of 0, 2, and 4 H atoms, respectively, per aggregate. The number of H_2 molecules adsorbed as line *b* reaches its maximum (lines *a* and *c* absent) is therefore equal to the number of aggregates in the sample. Then, knowing this number by H_2 uptake measurements and the metal concentration, the number of atoms per aggregate is deduced easily. It was found that this number ranges between 4 and 8 depending on the samples, whereas TEM and X-ray methods (24,25) point to an average size of 1 ± 0.3 nm corresponding to 20 to 50 atoms in agreement with RED estimation.

The adsorption of CO on the bare aggregates produces a 50% decrease in the intensity of the Pt—Pt peaks (76). This was interpreted in terms of the displacement disorder of the platinum atoms induced by Pt—CO bonding. Some distances are elongated and the others contracted around a mean distance close to the normal distance in bulk platinum. Similar contraction and elongation of metal–metal bond lengths are also observed in carbonyl clusters. Thus, in $[Pt_{18}(CO)_{24}]^{4-}$ the nearest neighbor distances are in the range 2.52 to 2.95 Å around an average distance of 2.79 Å (82). This shows that the structure of platinum aggregates covered by CO bears some relationships to that of platinum carbonyl clusters. Dissociative hydrocarbon adsorption resulting in the formation of Pt—C bonds also induces a displacement disorder of the platinum atoms (77).

Corrosive adsorbates such as oxygen and sulfur produce larger structural disorder.

Thus, the Pt—Pt peaks vanish after oxygen adsorption at 300K (74) and a new weak peak corresponding to Pt—O—Pt bondings appears. The adsorption of H_2S induces a decrease of the Pt—Pt peaks, indicating a displacement disorder of the platinum atoms (Fig. 8.8, curve 2). Moreover, the second peak at 3.92 Å is split into two components at 3.55 and 3.95 Å (78). The appearance of these new peaks means that ordered Pt—S—Pt arrangements are formed; more specifically, it gives evidence for a sulfur adsorption on C_{4v} sites of facets (100) with a preferential bonding with 2 of the 4 platinum atoms. The adsorption of SO_2 on H_2-covered aggregate (curve 4) produces a larger displacement disorder and the adsorption of

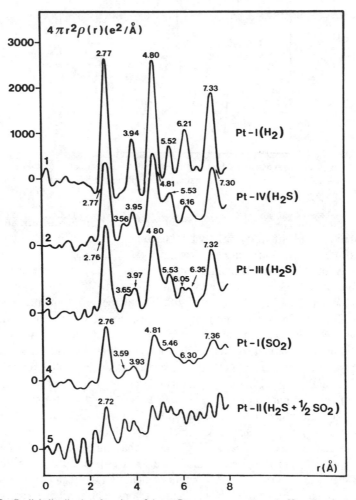

Figure 8.8 Radial distribution function of 1 nm Pt aggregates—curve 1, H_2-covered—curve 2, adsorption of H_2S on bare aggregates—curve 3, adsorption of H_2S on H_2-covered aggregates—curve 4, adsorption of SO_2 on H_2-covered aggregates—curve 5, adsorption of (H_2S + $\frac{1}{2}$ SO_2) on H_2-covered aggregates.

a mixture of H_2S and SO_2 yielding elemental sulfur leads to a complete disorder (curve 5). The increasing disorder in the series $H_2S < SO_2 < S$ is probably due to a larger sulfur coverage and to a smaller hydrogen coverage.

The structure of 1 nm platinum aggregates has been studied in the course of benzene hydrogenation at 300K (75) and n-butane hydrogenolysis at 570K (79). During C_6H_6 hydrogenation with a stoichiometric $C_6H_6 + 3H_2$ mixture the surface is covered by both associatively bonded C_6H_6 molecules and dissociated hydrogen since the structure is intermediate between that under pure hydrogen and that under pure benzene. The respective coverages in H_2 and C_6H_6 do not change in excess of C_6H_6 with respect to the stoichiometry, which means that adsorbed hydrogen is not displaced by benzene. At 570K under pure H_2 the hydrogen coverage is half that at 300K. During the conversion of n-butane under a pressure ratio $H_2/nC_4 = 10$, the coverage increases both in hydrogen and in hydrocarbon but at least half the aggregate surface is still covered with dissociated hydrogen. Therefore, in the course of C_6H_6 hydrogenation and nC_4 hydrogenolysis the hydrocarbon molecules do not displace markedly adsorbed hydrogen, since the H_2 coverage remains at least 50%. This unexpected result could be due to the particular configuration of the aggregates in zeolite cages. Because the aggregate fills the cage, about 50% of the aggregate surface is not accessible to hydrocarbon but instead remains available for hydrocarbon chemisorption.

To sum up, the atomic structure of metal aggregates encaged in zeolite is very flexible in response to adsorbates. The extent of atom displacement induced by surface ligands depends on their binding energy and coverage. The ease of metal oxidation is also an important factor. Thus, CO adsorption on platinum aggregates produces a limited displacement disorder which is reversible upon CO desorption (76), whereas CO adsorption disrupts the rhodium aggregates because of an oxidation and complexation of rhodium atoms into $Rh^I(CO)_2$ species (56).

8.4 MODIFICATIONS OF THE ELECTRONIC STRUCTURE OF METAL CLUSTERS

Modifications of the XPS-binding energies are expected when the aggregate size decreases because of relaxation effects (final states effects) and because of intrinsic change in the electronic structure (initial states effects) (83). The hybridization of d states with empty s and p states above the Fermi level reduces the d-electron occupation. Therefore, the area of the L edge peak (white line) on X-ray absorption spectra should be higher because the probability of $p \rightarrow d$ transitions increases with the number of empty d states. Furthermore, the electronic structure is modified by the interaction with the aggregate environment including other metal atoms, support atoms, and chemisorbed molecules. Thus, one can expect a combined effect of the aggregate size, the electron-donor or electron-acceptor properties of the zeolite, and the nature and coverage of adsorbates. It is difficult to evaluate the relative importance of intrinsic size effect and environment effect on the electronic structure. These effects can be studied with experiments in which the environment is modified without changing the size, and vice versa.

XPS measurements performed on Y zeolites containing encaged metal aggregates indicate that binding energies are positively shifted whatever the nature of the metal: ruthenium (16,84), rhodium (85), palladium (69), platinum (69,86), rhenium (84). The positive shifts have been interpreted in terms of metal electron deficiency due to electron transfer to the support (69,84,86). In an XPS study of PtLaY and PtNaY zeolites, Foger and Anderson (86) have decomposed the experimental profile of the 4f $\frac{5}{2}$ and 4f $\frac{7}{2}$ lines into three components. The presence of high-energy components above 71.1 eV was attributed to the platinum electron deficiency which was larger in PtLaY than in PtNaY. Since lanthanum ions were loaded in the reduced PtNaY zeolite, the aggregates have the same size in both catalysts, and the larger electron deficiency in PtLaY is well due to a modification of the aggregate environment.

Gallezot et al. (42,87,90) have studied the effect of the chemical environment on the electronic structure of 1 nm platinum aggregates in Y zeolites. The displacement of the L_{III} X-ray absorption edge and the relative area of the near edge peak are given in Table 8.2. The edge shift and white line area are larger when the support acidity increases and in the presence of multivalent cations (Ce^{3+}) or electron acceptor adsorbates (O,S) (88–90). Inversely, the white line area decreases upon adsorption of electron donor molecules (NH_3, C_6H_6) and upon deposition of molybdenum atoms. The lower density of charges on 1 nm platinum aggregates is also supported by the positive shift of the stretching frequencies of linearly adsorbed CO (87,42). This shift is 20 cm^{-1} with respect to 2 to 3 nm platinum particles supported on SiO_2 or on zeolite. It is related to the electron deficiency induced by electron transfer to the acidic sites of the zeolite since the neutralization of the acidity by NaOH suppresses the shift. Similar results were reported by Figueras et al. (91) and Chukin et al. (92) on PdY zeolites. The shifts of CO also confirm the electron donor effect of NH_3 and C_6H_6 and the electron acceptor effect of sulfur (87). It is remarkable that the results of the three techniques reported in Table 8.2 point to the same conclusion regarding the PtMoY catalyst; namely, the addition of molybdenum atoms by $Mo(CO)_6$ decomposition suppresses the electron deficiency of platinum aggregates. The effects of the chemical environment on the electronic structure of platinum aggregates were corroborated by a chemical method based on the competitive hydrogenation of toluene and benzene (see Section 8.5.2)

A decrease of charge density of palladium aggregates upon adsorption of unsaturated hydrocarbon has been observed by Romannikov et al. (93) in PdY zeolites. The ν CO bands at 2100 to 2140 cm^{-1} corresponding to CO adsorbed on Pd$^+$ or Pd$^{\delta+}$ species were shifted to 2080 cm^{-1} upon C_6H_6 adsorption which indicates a decrease in palladium electron deficiency. This was ascribed to a charge transfer between C_6H_6 and the electron acceptor sites of the zeolite that suppresses the electron transfer from the palladium atoms to these sites. A related interpretation was given by Bergeret and Gallezot (20) to account for the partial reduction of Pd^{2+} in sodalite cages on C_6H_6 adsorption even though C_6H_6 molecules cannot enter the cages. Similarly, the formation of TCNE$^-$ anions detected by ESR in PtY zeolites (69) was not attributed to a direct electron transfer between platinum aggregates and

TABLE 8.2 Effect of the Chemical Environment on the Electronic Properties of 1 nm Pt Aggregates

	X-ray Absorption Spectroscopy			Infrared Spectroscopy		XPS (Pt 4f 7/2)	
	Edge Shift (eV)	Normalized Peak Area[a]	Ref.	CO (cm^{-1})	Ref.	Binding Energy (eV)	Ref.
Bulk Pt	0	1.0		2070[c]	42,87	71.5	69
PtNaHY (1)[b]	0	1.2	88				
PtNaHY (2)[b]	0.6	1.3	88,89,90	2090	42,87	72.2	69
PtNaHY (3)[b]	0.6	1.6	88,90				
PtNaHY + NaOH		1.7	88	2068	42		
PtNaHY + O$_2$	0.8	1.6	90	2100	87		
PtNaHY (2) + H$_2$S		0.9	90	2025	87		
PtNaHY (2) + NH$_3$				2040	87		
PtNaHY (2) + C$_6$H$_6$							
PtCeY	0.5	1.5–2.0	88–90	2067	42		
PtMoY[d]		1.0	42			71.6	42
Pt sodalite cages	1.2	1.3	88			72.8	69

[a]With respect to Pt foil.

[b]Samples with different concentration of Brönsted sites (1) < (2) < (3).

[c]5% Pt/SiO$_2$, 2 nm Pt particles.

[d]Pt/Pt + Mo = 0.48.

TCNE but by an indirect transfer through the zeolite framework. All these results point to the importance of the role of the zeolite framework as the intermediate for charge transfer between metal clusters and electron-donor or electron-acceptor species such as cations, protons, and adsorbed molecules present in the zeolite. This role is related to the general concept of zeolites acting as solid electrolytes.

Since most of the investigations have been performed on acidic zeolites, there are very few examples where metal aggregates bear a larger charge density than in bulk metal. However, Besoukhanova et al. (94) have shown that platinum aggregates in the basic environment of L-type zeolite exchanged with alkali cations carry a higher charge density. Evidence is provided by the shift of the v CO band toward lower frequencies as the cation basicity increases (Rb > K > Li).

The electronic properties of metal aggregates of the first transition row are difficult to characterize because the samples are often heterogeneous, with variable fractions of unreduced atoms, small aggregates, and large particles. The case of the NiCaX zeolite reduced by atomic hydrogen (30,31) is of interest since nickel is totally reduced and homodispersed in the form of 1 nm aggregates. Fargues et al. (95) have shown by X-ray emission spectroscopy that a fine structure appears on the Ni $L\alpha$ line which can be attributed to a change in the density of Ni$3d$ states. Investigations on iron aggregates by Mössbauer spectroscopy and magnetic methods have been reported by Schmidt (96), Iton et al. (39), and Nazar et al. (47).

8.5   CATALYTIC PROPERTIES OF METAL CLUSTERS

In many papers dealing with nonacid catalysis on zeolites, the catalysts are not well prepared and/or characterized. Therefore, the nature of the active sites and their specificity are not known with certainty. Thus, because the metal dispersion is heterogeneous, it is difficult to establish the relative contribution of large particles and encaged aggregates in metal-catalyzed reactions. It is even more difficult to know in catalysis by metal complexes whether the active sites are mononuclear or polynuclear species. A complete review on catalysis by cations, complexes, and metal in zeolites has been published recently (97). Therefore, our attention will be focused on the few cases where encaged molecular metal clusters have been characterized in the catalysts or where encaged metal aggregates are the main active component.

8.5.1   Catalysis by Molecular Metal Clusters

Mantovani et al. (53) have shown that the treatment of Rh^{III}Y zeolite at 130°C under 80 atm of equimolar CO:H$_2$ mixtures leads to the formation of rhodium carbonyl clusters, possibly Rh$_6$(CO)$_{16}$, (see Section 8.2.2). After hexene-1 hydroformylation into aldehydes at 80°C under 80 atm of CO:H$_2$, the IR spectrum has not changed, indicating that encaged polynuclear carbonyls are still present. The small loss of rhodium in the reaction products proves that the clusters are well stabilized in zeolite cages. The catalyst is very active, showing a high selectivity to the formation of

TABLE 8.3 Activities and Selectivities of Rh Aggregates and Rh Carbonyl Clusters in CO + H₂ Reaction (Ref. 54)

Products	Rh/Y (3 nm)		Rh/Y (1 nm)		$Rh_6(CO)_{16}$[c]	
	A[a]	S[b]	A	S	A	S
CH_4	105	85.5	10.8	70.5	1.0	46.8
$C_2—C_5$	15	12.1	3.5	22.7	0.4	24.7
CH_3OH	0.4	0.3	0.1	0.6	0.5	24.2
C_2H_5OH	2.5	2.1	0.9	6.0	0.1	4.3

[a]Activity in mmole g^{-1} metal h^{-1}

[b]Selectivity in percent (%) of products formed.

[c]$Rh_6(CO)_{16}$ detected in Y zeolite after reaction.

aldehydes and a normal/isoaldehyde ratio similar to that observed in the homogeneous hydroformylation of olefins using rhodium carbonyl clusters. On the other hand, the hydroformylation of diolefins gives a higher fraction of dialdehydes than that on homogeneous catalysts.

Lefebvre et al. (54) have studied the CO + H₂ reaction (503 to 523K, 30 atm, $CO:H_2 = 1:2$) over RhY zeolites. Whatever the rhodium precursor, namely, Rh^{III}, $Rh^I(CO)_2$, $Rh_4(CO)_{12}$, and $Rh_6(CO)_{16}$ (see Section 8.2.2), the reaction data were similar. This is in agreement with the fact that all precursors were ultimately converted into $Rh_6(CO)_{16}$ under catalytic reaction conditions. Table 8.3 gives the activities and selectivities of the CO + H₂ reaction on Y zeolite containing 1 and 3 nm rhodium aggregates or when $Rh_6(CO)_{16}$ clusters are detected after reaction. The metal aggregates exhibit a significant activity at 453K, whereas the carbonyl cluster does not show any activity until the reactor temperature is raised to 503 to 523K. The selectivity pattern given in Table 8.3 indicates a contrast in the behavior of metal- and cluster-type samples, the latter producing less methane and more methanol than the rhodium aggregates. Therefore, the cluster-type catalyst can be discriminated from the metal catalyst on activity and selectivity grounds.

8.5.2 Catalysis by Metal Aggregates

Effect of Aggregate Accessibility

Metal aggregates can be completely inaccessible to reagents and therefore inactive for catalysis. Thus, Bergeret and Gallezot (20) have shown that in PdY zeolite activated at 870K in O_2 and reduced at 420K in H_2, most of the palladium atoms are in sodalite cages. The activity of the catalyst in benzene hydrogenation at 300K is low but increases as the catalyst is heated from 300 to 570K (Fig. 8.9). A parallel study by X-ray diffraction and TEM indicates that in this temperature range the palladium atoms progressively migrate out of the sodalite cages and agglomerate into 2 nm palladium particles. Thus, the higher activity is explained merely by the formation of particles that are accessible to the reactants at the expense of inaccessible atoms. The subsequent decrease of activity is due to metal sintering.

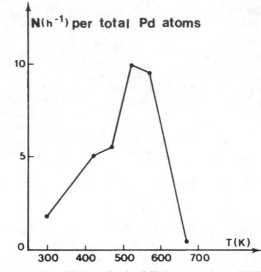

Figure 8.9   Catalytic activities of PdY zeolite for C_6H_6 hydrogenation at 300K versus reduction temperature.

In most cases only part of the aggregate surface is accessible to reactants. Thus, 1 nm platinum aggregates in Y zeolite fill the supercages (see Section 8.3.2) so that only half the platinum atoms, those in front of the four supercage windows, are available for hydrocarbon adsorption or hydrocarbon synthesis. Therefore, the kinetics and selectivities of catalytic reactions can be affected by the small size of the ensemble of atoms available. Accordingly, Jacobs et al. (98) interpreted the Fischer-Tropsch selectivity on RuY catalysts in terms of restricted ensemble sizes. Because the chain growth mechanism involves the condensation of CHOH species adsorbed on adjacent atoms, the chain length is limited to C_5 on ruthenium aggregates trapped in the supercages, whereas products up to C_{10} are formed on larger ruthenium particles occluded in zeolite holes. Chain length selectivity was also observed in the case of NaY zeolite loaded with $Fe_3(CO)_{12}$ (44). The chain length is limited to C_{10}; however, by addition of HY zeolite, there are less C_1—C_3 and C_8—C_{10} products and more C_4—C_6 hydrocarbons, which means that side reactions also play an important role in the selectivity of Fischer-Tropsch reaction on zeolites.

Effect of the Electronic Properties

Structure insensitive reactions such as hydrogenations are little or not affected by the modification of the electronic structure of noble metals in zeolite (12). Although the trend for platinum and palladium aggregates is an increase in activity with metal electron deficiency (87,91,99), the situation is more complicated in the case of nickel aggregates. Ione et al. (100) reported that with respect to Ni/SiO_2 the activity of nickel aggregates is smaller for benzene hydrogenation and larger for hexene-1 hydrogenation. In the latter case it was assumed that nickel aggregates are active because the electron acceptor sites of the zeolite are blocked by hexene which stabilizes the nickel aggregates. In contrast, benzene does not prevent the nickel aggregates from being strongly electron deficient or even disrupted into cationic form and thus inactive for benzene hydrogenation.

The selectivity of competitive hydrogenation reactions can be changed markedly because the coefficient of adsorption are sensitive to the electron-donor, electron-acceptor properties of the metal. Thus, Tri et al. (101) have determined the ratio of the adsorption coefficients of toluene and benzene b_T/b_B from a kinetic analysis of the competitive hydrogenations of these hydrocarbons on platinum aggregates with different chemical environment. Since toluene is a stronger electron donor than benzene, the ratio b_T/b_B is larger when the density of charges on platinum is smaller. Figure 8.10 gives the b_T/b_B ratios for a PtNaHY zeolite where the aggregate environment has been modified. It is noteworthy that the ratios b_T/b_B increase in presence of Ce^{3+} ions or upon H_2S adsorption favoring electron deficiency. Inversely, it decreases on neutralization of the Bronsted acidity of the zeolite or upon adsorption of electron-donor molecules (NH_3). These results are in agreement with those given by physical techniques (Table 8.2), but the modification of the electronic properties are probed with a much higher sensitivity.

The rates of hydrogenolysis reactions on platinum aggregates in acidic zeolites are enhanced because of platinum electron deficiency (86,90,99,102); details on these investigations can be found in previous reviews (11,12). On the contrary, electron deficiency can have an adverse effect on the hydrogenolysis activity of nickel. Thus, Sauvion et al. (32) have shown that 1 nm nickel aggregates on NiCeX zeolite are inactive in n-butane hydrogenolysis because hydrogen strongly adsorbed on the surface poisons the reaction. The strong binding energy of hydrogen was related to the electron deficiency of the encaged aggregates interacting with the acid sites of the zeolite.

Electron deficiency of metal aggregates can also modify the rate of $CO + H_2$

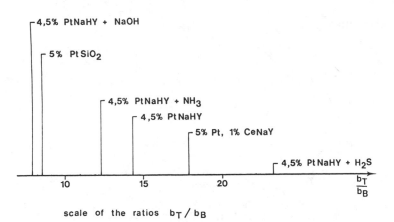

Figure 8.10 Ratios of the adsorption coefficients of toluene and benzene, b_T/b_B, for modified PtY catalysts calculated from competitive hydrogenations of benzene and toluene.

reactions. Vannice (103) interpreted the higher methanation activity of PdHY zeolite with respect to Pd/SiO$_2$ in terms of the formation of weakly bonded CO species on electron-deficient palladium. Jacobs et al. (104) reported that the decrease in methanation activity of RuY zeolites with zeolite acidity is due to a stronger interaction of CO with the electron deficient metal. Chen et al. (46) assumed that the higher selectivity of 1 to 1.5 nm ruthenium aggregates for C$_2^+$ hydrocarbon could be due to ruthenium electron deficiency.

Higher-charge density of metal aggregates in basic environment also results in modifications of the catalytic properties. The selectivity of the conversion of n-hexane into benzene is greatly improved on nonacidic X, Y, and L zeolites (94,105) where the platinum aggregates bear a higher charge density.

Effect of a Second Metal

Elliott and Lunsford (15) have studied the methanation activity of Ru—Ni and Ru—Cu bimetallic aggregates with size distribution centered around 20 Å. The dilution of ruthenium atoms by copper atoms merely results in a decrease of activity. The role of nickel atoms is more subtle; the turnover decreases but the catalysts are more stable with time. The activity for CO dissociation is decreased and the hydrogenation of carbon deposited on ruthenium is easier which limits the poisoning by carbon deposits. Tri et al. (106) have studied the catalytic properties of PtMoY catalysts prepared by decomposition of Mo(CO)$_6$ on 1 nm platinum aggregates (42). The hydrogenolysis activity in n-butane conversion is enhanced upon molybdenum addition (Fig. 8.11). The decrease in activity of molybdenum-rich samples is due to a decrease in the number of platinum sites dissociating hydrogen. From the variation of the kinetic parameters with Pt/Mo composition it was concluded that platinum and molybdenum atoms play a specific role in the reaction mechanism; molybdenum atoms act primarily as strong adsorption sites for n-C$_4$ and the hydro-

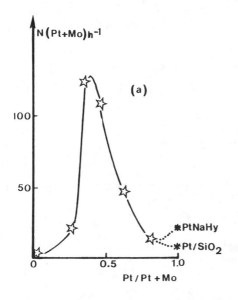

Figure 8.11 Influence of Mo content on the total hydrogenolysis rate of n-butane on Pt—Mo aggregates in Y zeolites (turnover calculated with the total number of Pt and Mo atoms).

carbon fragments are hydrogenated by hydrogen dissociated on platinum. This accounts for the positive reaction order with respect to H_2 and n-C_4 pressure since the reactants are not in competition for the same sites. This is due both to the specific adsorption properties of platinum and molybdenum and to the steric configuration of the platinum-molybdenum aggregates in zeolite cages. Thus, after molybdenum deposition on the platinum atoms facing the supercage aperture, hydrogen can still be dissociated on platinum atoms but the bulky hydrocarbon molecules can only be adsorbed on the molybdenum atoms.

8.6   CONCLUDING REMARKS

So far, preparation of metal clusters in zeolites relied mostly on ion exchange and reduction process. However, hydrogen reduction generates protonic acidity which may be detrimental in some catalytic applications. Also high silica zeolites such as ZSM-5 have a very low ion exchange capacity. Alternative methods of metal loading using adsorption of metal compounds should be developed. In situ synthesis of polynuclear metal complexes from exchanged cations, metal carbonyls, or even metal aggregates is attractive from both the academic and applied research standpoint. It can be compared to building a model ship in a bottle so that large pores zeolites are the best materials for this exercise. Metal aggregates have been prepared mainly in faujasite-type zeolites. There is also a variety of zeolites with medium-size pores that can be used to prepare aggregates of smaller sizes.

The homogeneity of the metal cluster distribution in zeolite crystals and of the bimetallic cluster composition can be studied with the analytical capabilities of the scanning transmission electron microscope which gives the chemical composition on a nanometric scale. Characterization of the atomic structure of metal aggregates and molecular metal clusters requires studies combining X-ray techniques (RED and EXAFS) and other spectroscopic techniques (NMR, UV, IR). The results obtained so far indicate that the atomic structure of metal aggregates is modified with respect to that of bulk metal and it is very sensitive with respect to adsorbates.

The electronic structure of metal clusters is even more difficult to study because the intrinsic structure might be modified by both the chemical environment and the method used to probe the electronic states. With the combined use of several techniques, it is possible to conclude that the chemical environment does modify the electronic structure of metal clusters. This is unfortunate for the physicist interested by the intrinsic structure of the cluster. But from the chemist viewpoint, the possibility of tuning the properties of the metal clusters by changing their environment adds a great deal of interest to these materials. This is especially true in the field of catalysis since parameters such as the zeolite structure (crystal field and pore size), the exchangeable cations and protons, and the electron-donor or electron-acceptor molecules hosted in cages can be chosen to tailor the metal zeolite catalysts for a specific need. Mild reaction conditions, good selectivity, and stability are expected. The few investigations conducted so far support the initial hopes—the future of catalysis on metal clusters in zeolites looks bright!

244

METAL CLUSTERS IN ZEOLITES

REFERENCES

1. D. W. Breck, *Zeolite Molecular Sieves: Structure, Chemistry and Use*, Wiley, New York, 1974.

2. J. A. Rabo, Ed., *Zeolite Chemistry and Catalysis*, American Chemical Society, Washington DC, 1976.

3. R. M. Barrer, *Zeolites and Clay Minerals as Sorbent and Molecular Sieves*, Academic Press, New York, 1978.

4. *Molecular Sieves*, Society of Chemical Industry, London, 1968.

5. R. F. Gould, Ed., *Adv. Chem. Ser.*, **102**, (1971).

6. W. M. Meier and J. B. Uytterhoeven, Eds., *Adv. Chem. Ser.*, **121**, (1973).

7. J. R. Katzer, Ed., *Molecular Sieves-II*, American Chemical Society, Washington DC, 1977.

8. L. V. C. Rees, Ed., Proc. 5th Int. Conf. Zeolites, Heyden, London, 1980.

9. A. Bisio and D. H. Olson, Eds., Proc. 6th Int. Conf. Zeolites, Butterworth, Guildford, 1984.

10. Kh. M. Minachev and Y. I. Isaakov, in J. A. Rabo, Ed., *Zeolite Chemistry and Catalysis*, American Chemical Society, Washington, DC, 1976, pp. 552–611.

11. P. Gallezot, *Catal. Rev.-Sci. Eng.*, **20**, 121 (1979).

12. P. Gallezot, in A. Bisio and D. H. Olson, Eds., Proc. 6th Int. Conf. Zeolites, Butterworth, Guildford, 1984, p. 352.

13. P. Gallezot and G. Bergeret, in P. A. Jacobs et al., Eds., *Metal Microstructure in Zeolites*, Elsevier, Amsterdam, 1982, pp. 167–177.

14. J. Verdonck, P. A. Jacobs, M. Genet, and G. Poncelet, *J. Chem. Soc. Faraday I*, **74**, 403 (1980).

15. D. J. Elliott and J. H. Lunsford, *J. Catal.*, **57**, 11 (1979).

16. L. A. Pedersen and J. H. Lunsford, *J. Catal.*, **61**, 39 (1980).

17. M. Primet, *J. Chem. Soc. Faraday I*, **74**, 2570 (1978).

18. N. Kaufherr, M. Primet, M. Dufaux, and C. Naccache, *Compt. Rend. Acad. Sci., Paris, Ser. C.*, **286**, 131 (1978).

19. G. Bergeret, P. Gallezot, and B. Imelik, *J. Chem. Phys.*, **85**, 411 (1981).

20. G. Bergeret and P. Gallezot, *J. Chem. Phys.*, **87**, 1160 (1983).

21. P. Gallezot and B. Imelik, *Adv. Chem. Ser.*, **121**, 66 (1973).

22. M. Dufaux, P. Gelin, and C. Naccache, in B. Imelik et al., Ed., *Catalysis by Zeolites*, Elsevier, Amsterdam, 1980, p. 261.

24. P. Gallezot, A. Alarcon Diaz, J. A. Dalmon, A. J. Renouprez, and B. Imelik, *J. Catal.*, **39**, 334 (1975).

25. P. Gallezot, I. Mutin, G. Dalmai-Imelik, and B. Imelik, *J. Microsc. Spectr. Electr.*, **1**, 1, 1976.

26. P. Gallezot, unpublished results.

27. P. Gallezot, in P. A. Jacobs et al., Eds., *Metal Microstructure in Zeolites*, Elsevier, Amsterdam, 1982, pp. 167–177.

28. F. Schmidt, W. Gunsser, and J. Adolph, in J. Katzer, Ed., *Zeolite Molecular Sieves-II*, American Chemical Society, Washington, DC, 1977, p. 291.

29. M. Briend-Faure, M. F. Guilleux, J. Jeanjean, D. Delafosse, G. Marriadassou, and M. Bureau-Tardy, *Acta Phys. Chem.* (Szeged Hungaria), **24**, 99 (1978).

30. M. Che, M. Richard, and D. Olivier, *J. Chem. Soc. Faraday I*, **76**, 1526 (1980).

31. D. Olivier, M. Richard, L. Bonneviot, and M. Che, in J. Bourdon, Ed., *Growth and Properties of Metal Clusters*, Elsevier, Amsterdam, 1980, p. 165.

32. G. N. Sauvion, M. F. Guilleux, J. F. Tempere, and D. Delafosse, in P. A. Jacobs et al., Eds., *Metal Microstructure in Zeolites*, Elsevier, Amsterdam, 1982, p. 229.

33. E. D. Garbowski, C. Mirodatos, M. Primet, and M. V. Mathieu, *J. Chem. Phys.*, **87**, 303, (1983).

34. Ch. Minchev. V. Kanazirev. L. Kosova, V. Penchev, W. Gunsser, and F. Schmidt, in L. V. C. Reeds, Ed., Proc. 5th Int. Conf. Zeolites, Heyden, London, 1980, p. 355.

35. P. Gallezot, Y. Ben Taarit, and B. Imelik, *J. Phys. Chem.*, **77**, 652 (1973).

36. P. A. Jacobs, H. Nijs, J. Verdonck, E. G. Derouane, J. P. Gibson, and A. J. Simoens, *J. Chem. Soc., Faraday I*, **75**, 1196 (1979).

37. A. Wiedenmann, F. Schmidt, and W. Gunsser, *Ber. Bunsen Ges. Phys. Chem.*, **81**, 525 (1977).

38. J. Scherzer and D. Fort, *J. Catal.*, **71**, 111 (1981).

39. L. E. Iton, R. B. Beal, and D. T. Hodul, *J. Mol. Catal.*, **21**, 151 (1983).

40. U.S. Patents 3.013.986, 987,988,990.

41. P. Gallezot, G. Coudurier, M. Primet, and B. Imelik, in J. R. Katzer, Ed., *Molecular Sieves-II*, American Chemical Society, Washington DC, 1977, p. 144.

42. T. M. Tri. J. P. Candy, P. Gallezot, J. Massardier, M. Primet, J. C. Vedrine, and B. Imelik, *J. Catal.*, **79**, 396 (1983).

43. D. Ballivet-Tkatchenko and G. Coudurier, *Inorg. Chem.*, **18**, 558 (1979).

44. D. Ballivet-Tkatchenko and I. Tkatchenko, *J. Mol. Catal.*, **13**, 1 (1981).

45. J. B. Nagy, M. Van Eenoo and E. G. Derouane, *J. Catal.*, **58**, 230 (1979).

46. Y. W. Chen, H. T. Wang, and J. G. Goodwin, Jr., *J. Catal.*, **83**, 415 (1983).

47. L. F. Nazar, G. A. Ozin, F. Hughes, J. Godber, and D. Rancourt, *J. Mol. Catal.*, **21**, 313 (1983).

48. J. G. Goodwin, Jr. and C. Naccache, *J. Mol. Catal.*, **14**, 259 (1982).

49. D. Ballivet-Tkatchenko, G. Coudurier, and N. D. Chau, in P. A. Jacobs et al., Eds., *Metal Microstructures in Zeolites*, Elsevier, Amsterdam, 1982, p. 123.

50. Th. Bein, P. A. Jacobs, and F. Schmidt, in P. A. Jacobs et al., Eds., *Metal Microstructures in Zeolites*, Elsevier, Amsterdam, 1982, p. 111.

51. P. Gelin, Y. Ben Taarit, and C. Naccache, *J. Catal.*, **49**, 357, (1979).

52. J. H. Lunsford, in J. R. Katzer, Ed., *Molecular Sieves-II*, American Chemical Society, Washington, DC, 1977, p. 473.

53. E. Mantovani, N. Palladino, and A. Zanobi, *J. Mol. Catal.*, **3**, 285 (1977–1978).

54. F. Lefebvre, P. Gelin, C. Naccache, and Y. Ben Taarit, in A. Bisio and D. H. Olson, Eds., Proc. 6th Int. Conf. Zeolites, Butterworth, Guildford, 1984, p. 435.

55. P. Gelin, F. Lefebvre, B. Elleuch, C. Naccache, and Y. Ben Taarit, *Intrazeolitic Chemistry*, American Chemical Society, Washington, DC, 1983, p. 445.

56. G. Bergeret, P. Gallezot, F. Lefebvre, C. Naccache, Y. Ben Taarit, and B. Shannon, to be published.

57. P. Gelin, G. Coudurier, Y. Ben Taarit, and C. Naccache, *J. Catal.*, **70**, 32 (1981).

58. J. A. Rabo, C. L. Angell, P. A. Kasai, and V. Schomaker, *Diss. Faraday Soc.*, **41**, 328 (1966).

59. P. H. Kasai and R. J. Bishop Jr., *J. Phys. Chem.*, **77**, 2308 (1973).

60. Y. Kim and K. Seff, *J. Am. Chem. Soc.*, **100**, 175 (1978).

61. Y. Kim, J. W. Gilje, and K. Seff, *J. Am. Chem. Soc.*, **99**, 7055 (1977).

62. L. R. Gellens, W. J. Mortier, R. A. Schoonheydt, and J. B. Uytterhoeven, *J. Phys. Chem.*, **85**, 2783 (1981).

63. P. A. Jacobs, J. B. Uytterhoven, and H. K. Beyer, *J. Chem. Soc. Faraday I*, **75**, 56 (1979).

64. L. R. Gellens, W. J. Mortier, and J. B. Uytterhoeven, *Zeolites*, **1**, 11 (1981); *ibid.*, **1**, 5 (1981).

65. H. K. Beyer, P. A. Jacobs, and J. B. Uytterhoeven, *J. Chem. Soc. Faraday I*, **72**, 674 (1976).

66. G. A. Ozin and F. Hughes, *J. Phys. Chem.*, **87**, 94 (1983).

67. G. A. Ozin, F. Hughes, J. M. Mattar, and D. F. McIntosh, *J. Phys. Chem.*, **87**, 3445 (1983).

68. C. Naccache, M. Primet, and M. V. Mathieu, *Adv. Chem. Ser.*, **121**, 266 (1973).

69. J. C. Vedrine, M. Dufaux, C. Naccache, and B. Imelik, *J. Chem. Soc. Faraday I*, **74**, 440 (1978).

70. U.S. Patents 3.287.257 (1966); 3.318.802 (1967); 3.835.028 (1974); 3.849.293 (1974); 3.899.441 (1975).

71. R. G. Herman, J. H. Lunsford, H. Beyer, P. A. Jacobs, and J. B. Uytterhoeven, *J. Phys. Chem.*, **79**, 2388 (1975).

72. M. Che, J. F. Dutel, P. Gallezot, and M. Primet, *J. Phys. Chem.*, **80**, 2371 (1976).

73. P. Gallezot, in J. R. Anderson and M. Boudart, Eds., *Catalysis-Science and Technology*, Vol. 5, Springer-Verlag, Berlin, 1983, Ch. 4.

74. P. Gallezot, A. Bienenstock, and M. Boudart, *Nouv. J. Chim.*, **2**, 263 (1978).

75. P. Gallezot and G. Bergeret, *J. Catal.*, **72**, 294 (1981).

76. P. Gallezot, *Zeolites*, **2**, 103 (1982).

77. P. Gallezot, *J. Chim. Phys.*, **78**, 881, 1981.

78. G. Bergeret and P. Gallezot, *J. Catal.*, **87**, 86 (1986).

79. G. Bergeret and P. Gallezot, Proc. 8th Int. Cong. Catal., Vol. 5, Dechema, Frankfurt, 1984, p. 659.

80. B. Moraweck, G. Clugnet, and A. J. Renouprez, *Surf. Sci.*, **81**, 631 (1979): *ibid.*, **106**, 35 (1981).

81. P. Gallezot, unpublished results.

82. D. M. Washecheck, E. J. Wucherer, L. F. Dahl, A. Cerioti, G. Longoni, M. Manesserao, M. Sansoni, and P. Chini, *J. Am. Chem. Soc.*, **101**, 6110 (1979).

83. M. G. Mason, *Phys. Rev. B*, **27**, 748 (1983).

84. G. V. Antoshin, E. S. Shpiro, O. P. Tkachenko, J. B. Nikishenko, M. A. Ryashentseva, V. I. Avaev, and Kh. Minachev, in T. Seiyama and K. Tanabe Eds., Proc. 7th Int. Cong. Catal., Elsevier, Amsterdam, 1981, p. 302.

85. S. L. T. Anderson and M. S. Scurrell, *J. Catal.*, **71**, 233 (1981).

86. K. Foger and J. R. Anderson, *J. Catal.*, **54**, 318 (1978).

87. P. Gallezot, J. Datka, J. Massardier, M. Primet, and B. Imelik, in G. C. Bond et al., Eds., Proc. 6th Int. Congr. Catal., The Chemical Society, London, 1977, p. 144.

88. P. Gallezot, R. S. Weber, R. A. Dalla Betta, and M. Boudart, *Z. Naturforsh.*, **34A**, 40 (1979).

89. R. S. Weber, M. Boudart, P. Gallezot, in J. Bourdon, Ed. *Growth and Properties of Metal Clusters*, Elsevier, Amsterdam, 1980, p. 415.

90. T. M. Tri, J. Massardier, P. Gallezot, and B. Imelik, in T. Seiyama and K. Tanabe, Eds., Proc. 7th Int. Cong. Catal., Elsevier, Amsterdam, 1981, p. 266.

91. F. Figueras, R. Gomez, and M. Primet, *Adv. Chem. Ser.*, **121**, 480 (1973).

92. G. D. Chukin, M. V. Landau, V. Kruglikov, D. A. Agievskii, B. V. Smirnov, A. L. Belozerov, V. D. Asrieva, N. V. Goncharova, E. D. Radchenko, O. D. Konovalcherov, and A. V. Agaforvov, in G. C. Bond et al., Eds., Proc. 6th Int. Cong. Catal., The Chemical Society, London, 1977, p. 668.

93. V. N. Romannikov, K. G. Ione, and L. A. Pedersen, *J. Catal.*, **66**, 121 (1980).

94. C. Besoukhanova, J. Guidot, D. Barthomeuf, M. Breysse, and J. R. Bernard, *J. Chem. Soc. Faraday 1*, **77**, 1595 (1981).

95. D. Fargues, F. Vergand, E. Belin, C. Bonnelle, D. Olivier, C. Bonneviot, and M. Che, *Surf. Sci.*, **106**, 239 (1981).

96. F. Schmidt, in P. A. Jacobs et al., Eds., *Metal Microstructures in Zeolites*, Elsevier, Amsterdam, 1982, p. 191.

97. I. E. Maxwell, *Adv. Catal.*, **31**, 1 (1982).

98. P. A. Jacobs, in B. Imelik et al., Eds., *Catalysis by Zeolites*, Elsevier, Amsterdam, 1980, p. 293 and references therein.

99. R. A. Dalla Betta and M. Boudart, in H. Hightower, Ed., Proc. 5th Cong. Catal., North Holland, Amsterdam, 1973., p. 1329.

100. K. G. Ione, V. N. Romannikov, A. A. Davydov, and L. B. Orlova, *J. Catal.*, **57**, 126 (1979).

101. T. M. Tri, J. Massardier, P. Gallezot, and B. Imelik, in B. Imelik et al., Eds., *Metal Support and Metal Additives Effects in Catalysis*, Elsevier, Amsterdam, 1982, p. 141.

102. C. Naccache, N. Kaufherr, M. Dufaux, J. Bandiera, and B. Imelik, in J. R.-Katzer, Ed., *Molecular Sieves-II*, American Chemical Society, Washington, DC, 1977, p. 538.

103. M. A. Vannice, *J. Catal.*, **40**, 129 (1975).

104. P. A. Jacobs, J. Verdonck, R. Nijs, and J. B. Uytterhoeven, *Adv. Chem. Ser.*, **178**, 15 (1979).

105. J. R. Bernard, in L. V. C. Rees, Ed., Proc. 6th Int. Conf. Zeolites, Heyden, London, 1980, p. 686.

106. T. M. Tri, J. Massardier, P. Gallezot, and B. Imelik, *J. Catal.*, **85**, 244 (1984).

107. J. Fraissard, T. Ito, L. C. de Menorval, and M. A. Springuel-Huet, in P. A. Jacobs et al., Eds., *Metal Microstructures in Zeolites*, Elsevier, Amsterdam, 1982, p. 179.

9

COMPARATIVE CATALYTIC
ACTIVITY OF
SUPPORTED CLUSTERS

ALAN BRENNER
Department of Chemistry
Wayne State University
Detroit, Michigan

9.1   INTRODUCTION

During the past decade there has been an explosive growth in the field of cluster chemistry. Much of this widespread enthusiasm has been centered on new catalytic applications of cluster complexes and the attempt to establish a relationship between the chemistry occurring at a metal cluster and on a metal surface. Numerous reviews (1–13) have been published dealing with the synthesis, structure, bonding, stoichiometric reactions, and surface analogy of cluster complexes. Only a perceptive account by Moskovits (10) focused on the differences between cluster chemistry and metal surfaces. However, there is much less data in the literature dealing with the chemistry of *supported* clusters and the *catalytic* activity of clusters, either in solution or supported (7,9,13). At the end of one lengthy review (3) on clusters and surfaces the authors noted that reactivity was not discussed, "simply because the comparative data are not available." Somewhat of an exception to this is the extensive recent work dealing with supported $Os_3(CO)_{12}$. The Os—Os bond is one of the strongest metal–metal bonds, and thus this cluster is particularly useful for the synthesis of well-defined cluster catalysts if the temperature is kept low. Catalysis by such well-defined materials is described by Gates in Chapter 10.

This chapter focuses on the catalytic activity of supported cluster complexes. Particular attention is given to comparing the activity of catalysts derived from carbonyl clusters to that of their mononuclear counterparts. It should be stressed that especially for reactions occurring above about 200°C (which in general is the temperature range of the industrially more challenging reactions for which special application of cluster complexes has been envisioned), most mononuclear and cluster-derived catalysts will change their nuclearity and morphology. The characterization of these materials is in itself a most demanding area at the forefront of current research. Thus, in many cases and especially for much of the activity data gathered in this laboratory, the matter of the comparative activity of cluster complexes vis-à-vis mononuclear complexes distills down to a more pragmatic question: Do catalysts *prepared* from clusters display any special activity or selectivity compared to that of catalysts *derived* from mononuclear precursors or compared to traditional catalysts? (In this chapter "traditional" catalysts is construed to be catalysts prepared in the conventional manner by the reduction of supported salts.)

9.2 HISTORICAL PERSPECTIVE

Several papers appearing in the mid-1970s did much to spur interest in the catalytic applications of cluster complexes. Particularly notable were contributions by Ugo (6,7) and Muetterties (1,2). Attention was drawn to the fact that the multiplicity of bonding modes available in cluster complexes may provide for reaction pathways that are not possible with mononuclear complexes and might mimic the chemistry of metal surfaces. Muetterties (1) described clusters as, "little pieces of metal with chemisorbed species on the periphery" and noted that, "Cluster chemistry has a great potential in modeling the chemisorption and catalytic processes at a metal surface." Particularly important was the suggestion that cluster complexes might allow the homogeneous catalysis of technologically important reactions for which there were no known homogeneous (mononuclear) catalysts but for which efficient heterogeneous catalysts existed. Specific mention was made of the reduction of nitrogen to ammonia, the hydrogenation of carbon monoxide to methanol or methane, and the reforming of normal alkanes. Equally important was the hope that clusters would provide *selective* catalysis. The energy crises was in full swing at this time, so the logical focal point was Fischer-Tropsch synthesis. Traditional heterogeneous catalysts yield a wide range of products for this reaction (14,15), reflecting the polymerization nature of the mechanism and perhaps also the inhomogeneity of the catalysts. It was argued that a molecular cluster complex, due to its well-defined structure and more limited bonding modes, might be an efficient (by virtue of being a cluster) and *selective* catalyst for such a reaction.

Although the dynamic growth in cluster chemistry and catalysis began in the mid-1970s, the study of the dependence of catalytic action on metal particle size and morphology considerably antedates it. In 1925 H. S. Taylor (16) proposed that catalysis occurs on only a small subset of "active sites" on a metal surface. These active sites were envisioned to be discontinuities such as edges and peaks. The concept of active sites quickly became a tenet of heterogeneous catalysis. In the past decade this idea has achieved elegant expression in the "kinks" and "edges" described by surface scientists such as Somorjai (17). In 1929 the Russian chemist A. A. Balandin (18) began espousing his "multiplet" theory of catalysis. Balandin believed that only those sites on a surface that were capable of proper geometric synchronization with an adsorbate could effect catalysis. Much of his work focused on explaining the patterns of activity of transition metals for the dehydrogenation of cyclohexane to benzene. Balandin argued that a sextet of sites was necessary for benzene hydrogenation: three metals atoms to interact with the reactant and three more that helped dissociate the hydrogen atoms. Only metals with a face-centered cubic or hexagonal close-packed structure possessed a sextet of atoms with the proper geometry to effect the reaction.

Although Balandin's multiplet theory has been largely discredited due to its extreme dependence on purely geometrical factors in catalysis (19), his ideas are resurfacing with more modern and careful expression. For example, Muetterties

(20) has reported that benzene undergoes rapid C—H bond scission on the (111) face of Ni due to the synchronization between the symmetry of the benzene molecule and the underlying metal atoms. Theoreticians have offered similar explanations that essentially rely on a favorable geometrical overlap between the orbitals of chemisorbed benzene and metal atoms (21).

Most reactions occurring on metal surfaces have long been believed to involve two or more metal sites (22). For example, in the case of the extensively studied reaction of olefin hydrogenation, a common mechanism requires two sites for the diadsorption of olefin and two more sites for the dissociation of H_2 (23). In some cases, as for olefin hydrogenation and isomerization, it is now well known that the reaction can occur homogeneously at a single metal center (which provides several binding sites) (24), but it is unclear how many different metal atoms are involved when the reaction occurs on a metal surface.

A signal contribution that attempted to quantize some aspects of the variation of catalytic activity with metal particle size was the introduction by Boudart and co-workers (25) of the concept of structure insensitive (or facile) and structure sensitive (or demanding) reactions in 1966. The rate of a metal catalyzed reaction under a given set of conditions can be reported as a turnover frequency, N:

$$N = \text{(no. of molecules reacting)}/\text{(no. of exposed metal atoms) (unit time)}$$

$$(9.1)$$

The number of exposed metal atoms on the surface of a catalyst can be determined by the number of techniques, but most commonly is titrated by selective chemisorption. In this technique a gas (such as H_2 or CO) adsorbs selectively on the exposed metal and allows the calculation of the fractional dispersion, D, of a catalyst,

$$D = \text{(no. of exposed metal atoms)}/\text{(total no. of metal atoms)} \qquad (9.2)$$

The total number of metal atoms in a catalyst is readily calculated from the metal loading and weight of a catalyst. Thus, the parameter N effectively normalizes the activity for catalysts of different mass or dispersion and represents the average activity of the exposed metal atoms.

The dispersion of a metal particle depends on both its size and shape. For the simple case of cubic geometry, an ensemble of less than 19 atoms will have a dispersion of 1.0. A group of 19 atoms will have a single atom in the interior and hence a dispersion of 0.95. Particles containing 44, 85, and 670 atoms will have dispersions of 0.87, 0.78, and 0.49, respectively. When the size of a metal particle changes, many parameters besides its dispersion can change. This includes the average coordination number of the surface atoms (8), physical properties (26), packing and geometry (26), electronic factors (27), and the nature of the metal-support interaction. Nonetheless, in some cases it is found that the turnover frequency is *independent* of the dispersion, termed a structure-insensitive reaction. When the turnover frequency changes (in practice, the change should be at least several fold) with dispersion, then the reaction is structure sensitive. Numerous catalyst-reaction systems have been classified. For the most part reactions involving the scission of C—H and H—H bonds are structure insensitive, whereas reactions involving C—C and often C—O bond breaking are structure sensitive.

Theoretical calculations show that the electronic properties of small ensembles of metal atoms rapidly change with size. In fact, the very concept of "metallic" properties requires the overlap of orbitals from a large number of metal atoms. Chemically speaking, a cluster containing only a few metal atoms is not metallic. The point at which a crystallite has sufficient atoms to be reasonably close to its asymptotic metallic behavior is still a subject of debate, but it appears that roughly 30 to 80 atoms are required (28–32). A crystallite containing 80 atoms has a dispersion of 0.80. In assessing the results of studies indicating the existence of structure-insensitive reactions one should note that traditional synthetic methodology often limits the range of dispersions that can be studied. Thus, in Boudart's (25) classic work the dispersion ranged from 0.73 to 4×10^{-5}. Although the range appears large, covering four orders of magnitude, the smallest particles studied contained roughly 100 atoms. Thus, the regime of ultra small ensembles—exactly the area in which special properties are most likely to be observed—was not examined. This is an area especially suited to studies of cluster complexes.

It is consequently evident that the concept that ultra small ensembles of metal atoms might have special catalytic properties was not a new idea to a catalytic chemist of the 1970s. Rather, catalysis by clusters was a natural extension of Boudart's concept of structure sensitivity which was now taken to its logical limit. Cluster catalysis focused on the hypothesis that the turnover frequency for some useful reactions would go to zero as the number of metal atoms in a particle approached one. Clusters were envisioned as having reasonable activity and good selectivity. Further, concepts and definitions of heterogeneous catalysis were recast and put into better defined and measured terms of the organometallic chemist. There was a shift in viewpoint from the macroscopic to the molecular. Instead of the vagueness of adsorption, metal particles of variable size, and active sites, one then spoke of processes such as oxidative addition on a molecular complex of known crystal structures.

9.2.1 Early Observations

The first report that cluster complexes can have special catalytic activity appeared in 1976. Thomas, Beier, and Muetterties (33) claimed that $Ir_4(CO)_{12}$ and $Os_3(CO)_{12}$ produced methane when heated in toluene at 140°C in the presence of CO and H_2 but that no mononuclear carbonyl complex had significant activity. Although this appeared to support the "cluster hypothesis" and did much to stimulate research in the area, the results are subject to question. When activated in this manner the mononuclear group 6B hexacarbonyls are inactive for the hydrogenation of ethylene ($N < 10^{-9}$ s^{-1}) (34), even though this reaction is readily catalyzed by a number of mononuclear catalysts. Thus, the inability of a material to catalyze the much more difficult methanation reaction may simply reflect the inactive form of the catalyst and not its nuclearity. Also, the rate of the methanation reaction over the two clusters was very low ($N = 1 \times 10^{-5}$ s^{-1}) and the total yield of methane was only about four molecules per complex. However, when supported on alumina both mononuclear and cluster carbonyl complexes are highly active for the hydrogenation of ethylene and produce significant amounts of methane during their *stoichiometric* decomposition when heated in a flow of H_2 (35). A more detailed analysis of the Muetterties data (34,35) suggested that the production of methane was a stoichiometric reaction and that the ability to produce methane depended on the thermal stability of a complex and not on its nuclearity.

In a related early experiment Smith et al. (36) supported various cluster complexes on *hydrated* alumina and heated them under CO in glass ampoules at about 250°C for about 10^4 s. The production of light hydrocarbons (predominately methane) was reported and it was suggested that the reaction occurred on "molecular cluster aggregates." In this system the H_2 necessary for hydrocarbon synthesis was most likely formed by the water gas shift reaction

$$CO \text{ (g)} + H_2O \text{ (g)} = H_2 \text{ (g)} + CO_2 \text{ (g)} \tag{9.3}$$

in which the H_2O (g) was provided by the dehydration and dehydroxylation of the wet alumina. Unfortunately, mononuclear complexes (which are now known to be effective precursors for methanation catalysts) (37–40) were not tested, and there is scant evidence that the catalysis occurred on "molecular" sites.

9.3 CATALYST PREPARATION AND HANDLING

There are numerous ways of supporting a complex. At one extreme there is the use of stoichiometric reactions to bind a complex to functional groups (such as hydroxyl or phosphine) on a support. In this approach considerable effort is made to define the ligand environment about the metal skeleton. Such catalysts have been made using polymers as well as functionalized and standard oxide supports. These catalysts are akin to immobilized homogeneous catalysts and are most amenable to characterization. However, their activity is often low and a well-defined structure is generally lost at elevated temperatures. As already noted, this class of catalysts is described in detail in Chapter 10. At the other extreme, complexes can be decomposed on normal oxide supports to yield catalysts of good stability (since the retention of molecular structure is not required) and activity but that are generally difficult to characterize. Although the catalysts produced in this laboratory are not highly characterized, they are carefully synthesized under rigorously clean conditions and yield materials of reproducible composition and catalytic properties.

Catalysts were prepared by physically dispersing a carbonyl complex on about 0.25 g of γ-alumina (Conoco Catapal SB previously calcined at 500°C in flowing O_2 and cooled in He; $\bar{S}$ = 203 m^2/g, average pore diameter = 7 nm, 28% hydroxylated). The alumina was contained in a Pyrex reactor and supported on a sintered glass frit of 10 mm OD. Usually the complex was added to the reactor against a backflow of He and then dispersed at roughly 70°C with agitation of the alumina to achieve an initial uniform bonding to the support. In some cases the complex was impregnated in pentane solution at 23°C followed by evaporation of the solvent. The catalyst was then activated at the desired temperature in either flowing He or H_2.

In other experiments catalyst activation was studied by temperature programmed decomposition (TPDE). In this technique the temperature of the reactor is raised at a linear rate (about 5°C/min) as He flows through the reactor and the evolution of gases (primarily CO and H_2) is continuously monitored by thermal conductivity detectors. Catalysts were also characterized by CO chemisorption. Pressure was measured with an electronic pressure transducer and the net chemisorption was determined in the standard manner as the difference between two isotherms measured at 23°C. The synthesis reaction system has a leak rate of about 1×10^{-7} cm^3/s under vacuum and the adsorption of 0.001 cm^3 of gas can be accurately measured. Details of the rigorously air, grease, and Hg-free synthetic method, and the TPDE system have been previously described (41).

Reactions were run without removing a catalyst from the reactor. Flows were controlled by Brooks electronic mass flow controllers. The reactor effluent was sampled with a motor-driven gas sampling valve interfaced to a gas chromatograph. Peaks were electronically integrated. He was purified with a Go-Getter (General Electric), CO was purified with an Oxisorb cartridge (Scientific Gas Products), and the H_2 was purified by a Matheson (catalytic) purifier followed by a 13X molecular sieve. O_2 impurity is claimed by the manufacturers to be <1 ppm with these techniques.

9.4 TPDE AND DISPERSION OF SUPPORTED CARBONYLS

The TPDE of all of the common binary carbonyls (except for the radioactive compound $Tc_2(CO)_{10}$) has been measured in the laboratory (42). However, only the TPDE of $Cr(CO)_6$ (41,43), $Mo(CO)_6$ (44), $W(CO)_6$ (45), $Fe(CO)_5$ (46), $Fe_2(CO)_9$ (46), $Fe_3(CO)_{12}$ (46), $Ni(CO)_4$ (35), and $Ir_4(CO)_{12}$ (35) have been described in journals. Others have reported the TPDE of $Ir_4(CO)_{12}$ (47), $Rh_4(CO)_{12}$, and $Rh_6(CO)_{16}$ (48), $Ru_3(CO)_{12}$ (49,50), $Os_3(CO)_{12}$ (51), $Mo(CO)_6$ (50), and $Re_2(CO)_{10}$ (50). By way of example, Figure 9.1 shows the TPDE of $Ru_3(CO)_{12}/Al_2O_3$ prepared by the dry mixing technique (42). This chromatogram is typical in that a low-temperature evolution of CO occurs, and at high temperatures the final decomposition of the complex is accompanied by the evolution of H_2. The H_2 is formed by a redox reaction between the initially zero-valent complex and hydroxyl groups on the surface of the alumina:

$$M(CO)_j + n(\sigma - OH) \xrightarrow{\Delta} (\sigma - O^-)_n M^{n+} + (n/2)H_2 + jCO \quad (9.4)$$

As a result of this reaction, after activation at high temperatures all of the complexes become oxidized (52). Small amounts of CO_2 and CH_4 (as well as tiny amounts of C_2H_6) are also detected during TPDE. In this case the gas evolutions were 8.95

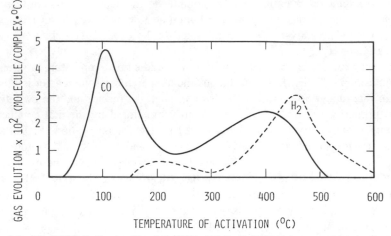

Figure 9.1 The TPDE in flowing He of $Ru_3(CO)_{12}/Al_2O_3$ (0.197% Ru) prepared by the dry mixing technique (heating rate is 5°C/min).

CO, 5.1 H_2, 0.29 CH_4, 2.56 CO_2, and 0.012 C_2H_6 molecule/complex, and the final oxidation number was about 4. Base metals yield much less CO_2 than Ru during TPDE. Although it should not be considered a definitive argument, such TPDE suggest that at temperatures below about 200°C a complex may retain some of its structure, whereas at high temperatures it clearly is destroyed. Numerous IR studies also indicate that below about 80°C most supported complexes retain their molecular nature.

The "activity spectrum" for the hydrogenation of propylene has also been determined for most of the supported carbonyls (53). Data have been published for $Fe_3(CO)_{12}$ (54) and $Mo(CO)_6$ (44). This technique is similar to TPDE except that the inert sweep gas of He is replaced with H_2 + C_3H_6 and the conversion to C_3H_8 is monitored. Figure 9.2 shows the activity spectrum for $Ru_3(CO)_{12}$. This spectrum is typical in showing that activity develops at temperatures slightly above the onset of significant CO loss as measured by TPDE. Hence, it appears that the low-temperature CO peak in TPDE roughly measures the development of coordinative unsaturation and accompanying catalytic activity. The high-temperature peaks represent far more drastic changes in the complex and its ultimate oxidation.

For most reactions the rate increases with the dispersion of a catalyst since a higher fraction of the metal atoms are exposed to reactants. The dispersion of all of the supported binary carbonyls has been measured and briefly reported (37,55). Other studies are available for $Mo(CO)_6$ (56), Fe carbonyls (46), $Ir_4(CO)_{12}$ (47,57), $Os_3(CO)_{12}$ (51,58), $Ru_3(CO)_{12}$ (59), $Co_2Rh_2(CO)_{12}$ (57), $Ni_2Cp_2(CO)_2$ and $Ni_3Cp_3(CO)_2$ (60), and Pt carbonyl cluster anions (61). The dispersion is a function of many variables such as the nature of the support, loading, synthetic method, and especially the temperature of activation. Table 9.1 shows the dispersion of carbonyl-derived catalysts after activation at 250°C for 30 min in flowing He. These values are compared to the dispersions of analogous traditional catalysts calcined for 30 min

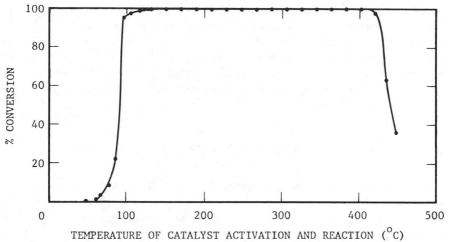

TEMPERATURE OF CATALYST ACTIVATION AND REACTION (°C)

Figure 9.2 The activity spectrum for the hydrogenation of propylene over $Ru_3(CO)_{12}/Al_2O_3$ (1.88 $\times$ $10^{-2}\%$ Ru). H_2/C_3H_6 = 4, flow = 30 cm^3/min, P = 1 atm, 0.50 g catalyst. 100% conversion corresponds to N_f = 5 s^{-1}.

TABLE 9.1   Dispersion of Supported Metals[a]

Catalyst	Dispersion %	Catalyst	Dispersion %
$Cr(CO)_6$	2.0	$Ru_3(CO)_{12}$	43
$Cr(NO_3)_3 \cdot 9H_2O$	2.0	$RuCl_3 \cdot 1\text{-}3H_2O$	11
$Mo(CO)_6$	16	$Os_3(CO)_{12}$	10
$(NH_4)_2MoO_4$	1.6	$OsCl_3$	0.8
$W(CO)_6$	9	$Co_2(CO)_8$	19
$(NH_4)_2WO_4$	1.0	$Co_4(CO)_{12}$	28
$Mn_2(CO)_{10}$	0.7	$Co(NO_3)_2 \cdot 6H_2O$	3.5
$MnCl_2 \cdot 4H_2O$	<0.05	$Rh_4(CO)_{12}$	26
$Re_2(CO)_{10}$	13	$Rh_6(CO)_{16}$	35
Re_2O_7	47	$Rh(NO_3)_3 \cdot 2H_2O$	143
$Fe_3(CO)_{12}$[b]	2.2	$Ir_4(CO)_{12}$	30
$Fe_3(NO_3)_3 \cdot 3H_2O$	2.3	$IrCl_3 \cdot 3H_2O$	124

[a]Percent (%) dispersion = 100(CO/M). Loadings are about 2% metal on alumina that had been calcined at 500°C.

[b]Catalysts of 0.5% Fe prepared from $Fe(CO)_5$, $Fe_2(CO)_9$, and $Fe_3(CO)_{12}$ had dispersions of 36, 24, and 23%, respectively, after activation at 120°C.

at 500°C in flowing O_2 and then reduced at 600°C for 30 min in flowing H_2. It is seen that in a number of cases enhanced dispersions can be obtained with carbonyl-derived catalysts (much higher enhancements are possible in some cases). The enhanced dispersion is not dependent on nuclearity. Rather, it reflects the milder activation procedures possible with carbonyl catalysts. Also, in the case of the base metal catalysts the decomposition of a carbonyl provides a route to low-valent material (which will chemisorb CO), whereas the oxides are often not fully reducible.

9.5 STABILITY OF SUPPORTED COMPLEXES

In the preceding section it has been noted that complexes can be expected to lose their molecularity at elevated temperatures. However, this can also sometimes occur at low temperatures and is a particular problem during reaction. In one of the first papers dealing with supported clusters, Watters and coworkers (62) noted that $Rh_6(CO)_{16}$, $Rh_4(CO)_{12}$, and $Rh_2(CO)_4Cl_2$ could be decomposed on alumina near room temperature in the presence of O_2. However, upon exposure to CO and extraction in refluxing $CHCl_3$ only the more stable $Rh_6(CO)_{16}$ was recovered in all three cases. This result suggests surface migration and reconstruction. It is also possible that the transformations occurred in the liquid phase. $CHCl_3$ and similar chlorinated hydrocarbons have also been used by other workers as a solvent for carbonyls that have low solubility in pentane (57,63–65). However, TPDE studies show that $CHCl_3$ can act as a reactive contaminant at elevated temperatures. For example, when prepared by impregnation from $CHCl_3$ solution, $Rh_6(CO)_{16}/Al_2O_3$ evolved 188 molecules of CO per complex during TPDE. Blank runs without complex indicated that both $CHCl_3$ and CH_2Cl_2 react with alumina to yield considerable amounts of CO and H_2 as well as small amounts of other gases (42). For this reason, care must be exercised in interpreting results in which solvents have been used.

Another early example of cluster fragmentation at mild temperatures is $Fe_3(CO)_{12}/Al_2O_3$. When He was flowed over this material at 90°C, no carbonyl was removed from the surface. However, flowing CO at 1 atm and 90°C resulted in the recovery of 14% of the Fe as $Fe(CO)_5$ (46). In this case the high volatility of $Fe(CO)_5$ relative to $Fe_3(CO)_{12}$ and $Fe_2(CO)_9$ provided the driving force for the conversion.

Other examples of changes in the structure of clusters at relatively low temperatures have been reported. Numerous workers have found that when supported on Al_2O_3, SiO_2, TiO_2, and ZnO (but not on MgO) that $Os_3(CO)_{12}$ forms mononuclear Os(II) species in the temperature range of 150 to 200°C (7,51,58,66–68). In the presence of n-hexene the reaction occurs at 120°C (69). $H_2Os_3(CO)_{10}$ and $Os_6(CO)_{18}$ also form mononuclear Os(II) fragments on Al_2O_3 between 100 and 200°C (58). There is some evidence that the fragments remain in close proximity (0.6 nm) as ensembles of three Os atoms (70,71). However, Collier et al. (64) believe that on SiO_2, Al_2O_3, and TiO_2, $Os_3(CO)_{12}$ and $Os_6(CO)_{18}$ yield species of approximate formula $Os_{12}(CO)_{30}C_2$ which contain Os—Os bonds. The transformation begins near 25°C and is complete at 250°C. $Ru_3(CO)_{12}/Al_2O_3$ is reported to begin forming oxidized structures near 100°C (65,72) and form large crystallites at 400°C (72).

$H_4Ru_4(CO)_{12}$ and $Ru_6C(CO)_{17}$ on Al_2O_3 are also reported to fragment at 80 to 200°C (72). $Fe(CO)_5$ and $Fe_3(CO)_{12}$ on Al_2O_3, MgO, and ZnO are claimed to form the anion $HFe_3(CO)_{11}^-$ at 25°C (73). When supported on SiO_2, $Fe_3(CO)_{12}$ forms some $Fe(CO)_5$ at 25°C (73), and complete formation occurs at 50°C in the presence of CO when supported on SiO_2 or SiO_2-Al_2O_3 (74). $Fe(CO)_5$, $Fe_2(CO)_9$, and $Fe_3(CO)_{12}$ supported on a dehydrated HY zeolite all form surface bonded $Fe_3(CO)_{12}$ at 60°C (75). Exposure to CO at 25°C converts this to adsorbed $Fe(CO)_5$ and $Fe(CO)_4$. Tanaka et al. (47) have reported that $Ir_4(CO)_{12}$ supported on Al_2O_3 and SiO_2 starts to form oxidized fragments above 75°C. Similar results were found by Crawford et al. (68) for $Ir_4(CO)_{12}/Al_2O_3$. $Rh_6(CO)_{16}$ has been reported to form monomeric Rh(I) species on hydrated Al_2O_3 and MgO at 25°C (63). Various authors have found that partially dehydroxylated supports afford better stability of complexes.

CO is often a reactant in catalytic systems. Consequently, a more systematic study of the mobility of supported carbonyls in the presence of CO was conducted (53). Carbonyl was deposited from the gas phase at the top of a column of alumina contained in a 4-mm-ID Pyrex tube. The rate of migration of the colored band was then measured in vacuum and in the presence of CO. The temperature of the initial deposition and the migration study was the same and approximately corresponded to the onset of significant CO loss as determined by TPDE. In vacuum no migration was observed (note that these are purely macroscopic measurements). The rate of migration increased with CO pressure. The results at 150 torr of CO are shown in Table 9.2. In general, mononuclear complexes migrate rapidly but clusters migrate slowly or not at all. The slow migration of clusters could reflect their high molecular weight as well as out-of-phase movements by two ends of a complex. A salient

TABLE 9.2 Macroscopic Migration of Supported Carbonyls in the Presence of 150 torr of CO

Complex	Temperature of Migration (°C)	Rate (mm/min)
$Cr(CO)_6$	135	0.8
$Mo(CO)_6$	100	1.0
$W(CO)_6$	135	0.8
$Mn_2(CO)_{10}$	100	0.5
$Re_2(CO)_{10}$	120	0.5
$Fe_3(CO)_{12}$	130	0.4
$Ru_3(CO)_{12}$	100	0.2
$Os_3(CO)_{12}$	120	0.2
$Co_2(CO)_8$	220	0.0[a]
$Rh_6(CO)_{16}$	180	0.0[a]
$Ir_4(CO)_{12}$	200	0.0[a]

[a]Rate $<4 \times 10^{-3}$ mm/min.

point of these studies is that even after detailed characterization, a complex could undergo transformations due to surface migration during reaction.

9.6   MEASUREMENTS OF CATALYTIC ACTIVITY

The utility of expressing catalytic activity in terms of a turnover frequency, N, has been described. However, in some cases it is difficult to obtain a good value for the dispersion of a catalyst. For the catalysts of this chapter, this primarily arises when chemisorption does not appear to titrate properly the amount of exposed metal [this may especially be a problem for base metals (38,76)] and when the loading is so low that chemisorption is inaccurate due to a small background adsorption on the support. Some workers do not report the dispersions of their catalysts. Also, a catalyst of low activity (in terms of conversion) can have a high value of N if the measured dispersion is low. Consequently, it can be useful to express activities in terms of a *formal* turnover frequency, N_f (38):

$$N_f = \text{(no. of molecules reacting)}/\text{(total no. of metal atoms) (unit time)} \quad (9.5)$$

N_f is a complementary concept to N and expresses the average activity of *all* of the metal atoms in a catalyst. N_f is always less than or equal to N and the two quantities are related by the expression

$$N_f = D \cdot N \quad (9.6)$$

The following sections describe the comparative activity of many clusters and related catalysts for a variety of reactions. Although the large majority of published data are described in these sections, this is not intended as an exhaustive review. However, there is more than adequate coverage to provide an accurate overview of catalysis by clusters.

9.6.1   Hydrogenation of Ethylene

The hydrogenation of ethylene has long been used as a model reaction to assess catalytic activity. A catalyst of low activity for this reaction is unlikely to effect more useful reactions such as the hydrogenation of hindered olefins, aromatics, or CO. The classic works in this area are the studies of Beeck (77) between 1940 and 1950 in which he determined the activities of films of many of the catalytically important transition metals. In 1958 Schuit and Van Reijen (78) did a similar study of metals supported on silica. Although there are some differences, for the most part the activities paralleled those of the metal films.

The hydrogenation of ethylene is sufficiently fast so that it can usually be run at or below ambient temperature. This makes it especially useful as a test reaction for supported carbonyls since at these mild conditions there is minimal chance of restructuring a catalyst. Also, since activity spectrum measurements indicated that hydrogenation activity set in slightly above the temperature required for moderate CO loss, catalysts can be activated at low temperatures. For these reasons the activity of most of the binary carbonyls was determined after activation at 200°C in flowing He. Examination of the TPDE spectra (42) indicated that this temperature is high enough to afford significant loss of CO in all cases but is well below the temperature for complete decomposition and significant H_2 evolution. Hence, these may be termed "subcarbonyl" catalysts. After treatment at 600°C in flowing H_2 the complexes are completely decomposed and should be similar to well-dispersed traditional catalysts. It should be noted that due to reaction (9.4), the metal will become oxidized at high temperatures. However, if the metal is reducible, then the H_2 will re-reduce the element back to the zero-valent state. Base metals will primarily remain in an oxidized form after this treatment.

Activity data for hydrogenation over a number of carbonyl catalysts have been previously reported (37). Table 9.3 contains a more complete summary. Catalysts of base metals generally show somewhat lower activity after reduction at 600°C. This is probably due to partial oxidation and lower dispersions (55). However, the more reducible metals display much higher activity and their dispersions increase after reduction (55). This is probably due to residual CO ligands acting as a poison on the subcarbonyl catalysts. Corrected to these reaction conditions, the turnover

TABLE 9.3 Hydrogenation of Ethylene

	N_f (s^{-1})[a]	
Catalyst	Activated 200°C, He	Activated 600°C, H_2
$Cr(CO)_6$	4×10^{-2}	6×10^{-3}
$Mo(CO)_6$	5×10^{-2}	3×10^{-2}
$W(CO)_6$	2×10^{-2}	7×10^{-4}
$Mn_2(CO)_{10}$	5×10^{-5}	4×10^{-5}
$Re_2(CO)_{10}$	7×10^{-2}	3.7
$Fe(CO)_5$	3×10^{-3}	—
$Fe_2(CO)_9$	3×10^{-3}	—
$Fe_3(CO)_{12}$	3×10^{-3}	3×10^{-4}
$Ru_3(CO)_{12}$	6×10^{-1}	75
$Os_3(CO)_{12}$	5×10^{-2}	3.7
$Co_2(CO)_8$	8×10^{-1}	11
$Co_4(CO)_{12}$	7×10^{-1}	11
$Rh_4(CO)_{12}$	2	35
$Ir_4(CO)_{12}$	1×10^{-2}	1.9

[a] $T = 0°C$ (or extrapolated to this T), $H_2/C_2H_4 = 4$, flow $= 30$ cm^3/min, $P = 1$ atm. N_f values for hydrogenation reported in Ref. 37 are incorrectly normalized per *complex*, rather than per metal *atom*.

frequency for metal films of Ru, Rh, and Ir are about 4×10^3, 7×10^3, and 3×10^1 s^{-1}, respectively (77) (N_f numbers have no meaning for a film). The N_f values for these metals in Table 9.3 are roughly an order of magnitude lower.

Hydrogenation of olefins has also been studied over *well-defined* clusters. Perhaps in the first such investigation, Robertson and Webb (79) found that $Ru_3(CO)_{12}/SiO_2$ was active for the hydrogenation (and isomerization) of butenes after partial decomposition at temperatures between 80 and 150°C. Insufficient data are given for an accurate calculation of N_f, but a reasonable estimate is 1×10^{-2} s^{-1} for reaction at 110°C and 0.1 atm H_2. The reaction was found to be first order in H_2, was zero order in butenes, and have an activation energy of 11 kcal/mol. This extrapolates to an N_f value of about 2×10^{-4} s^{-1} at the conditions of Table 9.3. Although the evidence is quite weak, the authors suggest that the active site is $Ru_3(CO)_5$.

Table 9.4 summarizes data on well-defined systems. It is clear that the better defined materials are several orders of magnitude lower in activity than the subcarbonyls of Table 9.3. Again, this presumably reflects the poisoning effect of the retained CO ligands. When the complexes were heated above about 120°C they began to decompose and in some cases metallic particles could be detected. This also resulted in large increases in activity. The low activity of molecular clusters compared to more decomposed materials complicates the identification of the active site since less than one part in a thousand of metallic particles could account for all of the activity.

Ichikawa (60) has also reported that di- and trinickel cyclopentadienyl carbonyl complexes supported on silica were active for the hydrogenation of ethylene and benzene at room temperature. Good activity at such a low temperature for benzene hydrogenation is unusual. Unfortunately, no rate data were given.

TABLE 9.4   Hydrogenation of Ethylene over Well-defined Supported Complexes

Catalyst	T (°C)	N_f (s^{-1})	Reference
$Ir_4(CO)_{11}PPh_3$-SIL	28	6×10^{-4}	109
$Ir_4(CO)_{11}PPh_2$-polymer	72	2×10^{-3}	110
$Ir_4(CO)_{10}(PPh_2$-polymer$)_2$	72	2×10^{-4}	110
$H_4Ru_4(CO)_{11}PPh_2$-polymer	72	2×10^{-3}	111
$H_4Ru_4(CO)_9(PPh_2$-polymer$)_3$	72	5×10^{-3}	111
$H_4Ru_4(CO)_8(PPh_2$-polymer$)_4$	72	1×10^{-2}	111
$RuPt_2(CO)_5(PPh_2$-polymer$)_3$	73	3×10^{-2}	112
$Fe_2Pt(CO)_8(PPh_2$-polymer$)_2$	75	6×10^{-3}	112
$H_2Os_3(CO)_9PPh_2C_2H_4$-SIL	75	4×10^{-5}	113
$H_2Os_3(CO)_9PPh_2$-SIL	100	0	112
$ClAuOs_3(CO)_{10}PPh_2$-polymer	73	3×10^{-4}	114
$HAuOs_3(CO)_{10}PPh_2$-polymer	73	0	114
$HAuOs_3(CO)_{10}PPh_2$-SIL	100	0	112

9.6.2 Hydrogenolysis of Ethane

The reaction of ethane in the presence of hydrogen to form methane has been widely used within the petrochemical industry as a model reaction for the scission of the C—C bond. Unlike the hydrogenation of ethylene, this reaction requires moderate temperatures (about 250°C) for reasonable activity over many catalysts. The rates and activation energy of reaction have been determined for each of the carbonyl catalysts (53). To minimize the chance that a catalyst would restructure during reaction, they were activated at 400°C for 30 min in flowing H_2. It is unlikely that any catalyst maintained its metal framework under these conditions. Thus, as noted in the Introduction, the matter being investigated is whether catalysts *derived* from clusters display unusual properties compared to their mononuclear or traditional analogs.

Table 9.5 shows the activity of carbonyl and salt-derived catalysts for hydrogenolysis at 230°C. Data for the activity at 350°C of a number of base metals and Os have been previously published (37). The metal loading was usually about 0.2% for the more active catalysts and about 1% for catalysts of low activity. In the cases of V, Mn, Fe, Co, Ni, and Os the carbonyl-derived catalysts have substantially higher activity than a corresponding traditional catalyst. With the exception of Os, these are base metals and suggests that the improved activity reflects the ability to prepare a catalyst in a low-valent state via a carbonyl. Consistent with this, when $Mo(CO)_6$ and $W(CO)_6$ were prepared on a *dehydroxylated* alumina (pretreated at 1000°C, residual hydroxyl content about 0.6%) to suppress reaction (4), their ac-

TABLE 9.5 Hydrogenolysis of Ethane

Catalyst	N_f (s^{-1})^a	Catalyst	N_f (s^{-1})^a
$V(CO)_6$	4×10^{-6}	$Ru_3(CO)_{12}$	2×10^{-2}
NH_4VO_3	3×10^{-7}	$RuCl_3 \cdot 1\text{-}3H_2)$	5×10^{-1}
$Mo(CO)_6$	3×10^{-6}	$Os_3(CO)_{12}$	8×10^{-4}
$(NH_4)_2MoO_4$	1×10^{-6}	OsO_4	4×10^{-5}
$W(CO)_6$	10^{-8}	$Co_2(CO)_3$	8×10^{-4}
$(NH_4)_2WO_4$	10^{-8}	$Co_4(CO)_{12}$	2×10^{-5}
$Mn_2(CO)_{10}$	5×10^{-5}	$Co(NO_3)_2 \cdot 6H_2)$	8×10^{-6}
$MnCl_2 \cdot 4H_2O$	2×10^{-5}	$Rh_4(CO)_{12}$	7×10^{-2}
$Re_2(CO)_{10}$	2×10^{-3}	$Rh_6(CO)_{16}$	3×10^{-2}
$HReO_4$	1×10^{-2}	$Rh(NO_3)_3 \cdot 2H_2)$	8×10^{-2}
$Fe(CO)_5$	2×10^{-5}	$Ir_4(CO)_{12}$	4×10^{-5}
$Fe_3(CO)_{12}$	2×10^{-4}	$IrCl_3 \cdot 3H_2)$	3×10^{-4}
$Fe(NO_3)_3 \cdot 9H_2O$	9×10^{-7}	$Ni(CO)_4$	2×10^{-4}
$Ru(CO)_5$	9×10^{-3}	$Ni(NO_3)_2 \cdot 6H_2O$	3×10^{-7}

$^a T = 230°C$ (or extrapolated to this T), $H_2/C_2H_6 = 2$, flow $= 15$ cm^3/min, $P = 1$ atm.

tivities increased over 100-fold (53). Reducing supported MoO_3 and WO_3 at 1000°C rather than 600°C also caused a similar increase in activity (53), again indicating that further reduction played a more dominant role in the activity than did the morphology of the catalysts.

In general, the periodic trends in activity are similar to those found by Sinfelt (80) in his comprehensive studies of hydrogenolysis over transition metals supported on silica. The order of reaction was also determined over $Ru_3(CO)_{12}/Al_2O_3$ containing 2% Ru (53). The order was 1.0 with respect to C_2H_6 and -2.2 with respect to H_2 after activation at 400°C. After activation at 1000°C the orders had a negligible shift to 0.8 and -2.5, respectively. A 2% Ru/Al_2O_3 catalyst prepared from $RuCl_3$ had orders of 0.8 and -2.2 after reduction at 400°C, and 0.8 and -2.4 after reduction at 1000°C (53). Sinfelt reported that Ru/SiO_2 had orders of 0.8 and -1.3. Orders of about 1 in H_2 and -1 to -2 in H_2 are common for this reaction (80). Thus, at the elevated temperatures of the hydrogenolysis reaction it does not appear likely that a cluster skeleton has been retained or that a cluster is a better precursor than a mononuclear complex.

9.6.3 Methanation

Methanation is a convenient model reaction to test the ability to cleave the very strong C—O triple bond. It can also be considered the prototype of the more interesting Fischer-Tropsch reaction. When run at 1 atm and 3/1 ratio of H_2/CO, methane is the dominant product (15). As with the hydrogenolysis of ethane, this reaction requires elevated temperatures and retention of a metal skeleton for prolonged periods is considered unlikely. The presence of CO will also favor surface reconstruction. In special cases, such as $Os_3(CO)_{12}/MgO$ (81), but not $Os_3(CO)_{12}$ supported on Al_2O_3, SiO_2, TiO_2, or ZnO (66), there is evidence that Os_3 clusters might be intact at reaction temperatures.

Early in the development of cluster catalysis it was suggested that mononuclear complexes do not have sufficient binding sites to activate CO to lead to bond cleavage (1,2). Although retention of nuclearity under catalytic conditions is difficult to establish, it is somewhat easier to monitor during TPDE. TPDE of supported carbonyls in He results in small amounts of CH_4 being formed, the hydrogen being derived from surface hydroxyl groups. TPDE in H_2 results in about a 30-fold increase in CH_4 formation, but it is still strictly a stoichiometric reaction involving surface species. Detailed analysis of many experiments suggests that CH_4 can be formed from coordinated CO on mononuclear complexes (34,35). Some of the data are contained in Table 9.6. In sweep gases of both He and H_2 the average amount of CH_4 formed per metal atom does not depend on the nuclearity of a complex. At the low loadings of these experiments (roughly 0.3% metal), the surface concentration of metal is very low, and it was expected that CH_4 formation would be very small for mononuclear complexes if sintering to metal clusters was a prerequisite. Also, the yield of CH_4/Mo did not vary when the loading of Mo was changed over a 270-fold range. In all cases no CH_4 was evolved during the low-temperature CO peak, but during TPDE in H_2 the CH_4 evolution commenced at roughly the position

TABLE 9.6 Methane Evolution During TPDE to 600°C

Complex	CH_4/M, He Flow	CH_4/M, H_2 Flow
$V(CO)_6$	0.10	0.48
$Cr(CO)_6$	0.11	0.43
$Mo(CO)_6$	0.12	2.1
$W(CO)_6$	0.32	1.91
$Fe(CO)_5$	0.03	1.76
$Ni(CO)_4$	0.01	0.41
Average	0.098	1.2
$Mn_2(CO)_{10}$	0.056	0.48
$Re_2(CO)_{10}$	0.050	3.26
$Fe_2(CO)_9$	0.035	—
$Co_2(CO)_8$	0.027	1.31
Average	0.042	1.7
$Fe_3(CO)_{12}$	0.021	1.10
$Ru_3(CO)_{12}$	0.091	2.69
$Os_3(CO)_{12}$	0.073	2.82
$Co_4(CO)_{12}$	0.035	1.30
$Rh_4(CO)_{12}$	0.078	1.52
$Rh_6(CO)_{16}$	0.014	1.20
$Ir_4(CO)_{12}$	0.016	2.80
Average	0.047	1.9

where the second CO peak starts (about 200°C). Thus, the time and temperature available for sintering is much less than during a typical catalytic reaction.

This group has also determined the activity for methanation of most of the carbonyls after activation at 250°C for 30 min in flowing He (37). Subsequently, the catalysts were oxidized for 30 min at 500°C and then reduced for 60 min at 500°C to yield a redox form of catalyst that should mimic the activity of a traditional catalyst. The activity data are given in Table 9.7 (37). As in the case of hydrogenolysis, there is again no apparent dependence of activity on the nuclearity of the precursor. Rather, enhanced activity is especially notable for some of the base metal catalysts that are better prepared from carbonyls than by the reduction of an oxide.

9.6.4 Fischer-Tropsch Synthesis

Fischer-Tropsch synthesis (FTS) is the reaction of CO and H_2 over transition-metal catalysts. Products can range from to C_1 chemicals to high molecular weight waxes and can be either hydrocarbons or oxygenates. The product distribution is usually well described by a polymerization mechanism (14) variously termed a Schulz-Flory

TABLE 9.7 Methanation

Catalyst	N_f (s^{-1})^a	
	Activated 250°C, He	Redox
$V(CO)_6$	5×10^{-5}	7×10^{-6}
$Cr(CO)_6$	1×10^{-6}	4×10^{-7}
$Mo(CO)_6$	2×10^{-3}	4×10^{-6}
$W(CO)_6$	1×10^{-3}	2×10^{-6}
$Mn_2(CO)_{10}$	7×10^{-8}	2×10^{-8}
$Re_2(CO)_{10}$	9×10^{-7}	6×10^{-4}
$Fe_3(CO)_{12}$	2×10^{-5}	9×10^{-5b}
$Ru_3(CO)_{12}$	2×10^{-3}	6×10^{-3b}
$Os_3(CO)_{12}$	6×10^{-5}	5×10^{-6}
$Co_2(CO)_8$	9×10^{-3}	5×10^{-3}
$Co_4(CO)_{12}$	6×10^{-3}	6×10^{-3}
$Rh_4(CO)_{12}$	9×10^{-3}	2×10^{-2}
$Rh_6(CO)_{16}$	1×10^{-2}	3×10^{-2}
$Ir_4(CO)_{12}$	8×10^{-4}	6×10^{-4}

$^aT = 250°C$ (or extrapolated to this T), $H_2/CO = 3$, flow $= 12$ cm^3/min, $P = 1$ atm.

^{b}A separate salt-derived catalyst was used.

or Anderson distribution (82) which yields the relation

$$\ln(W_n/n) = n(\ln P) + \ln[(1 - P)^2/P] \qquad (9.7)$$

where W_n is the weight fraction of the product containing n carbon atoms and P is the probability of polymerization. A corollary to this equation is that as the degree of polymerization increases [$D_p = 1/(1 - P)$], the product distribution widens. For example, the maximum yield of C_2 compounds is 30% and corresponds to $D_p = 1.4$, and the maximum yield of the gasoline fraction (C_5 through C_{11} hydrocarbons) is about 50% which occurs at $D_p = 4.1$ (82). Only for the case of $D_p = 1$, and hence $P = 0$, can a single product be formed. Both methane and methanol can be synthesized with 100% selectivity. Although originally FTS was envisioned as a source of synthetic fuels, it must also be realized that many chemical feedstocks are by-products of petroleum refining. Consequently, there is now great interest in using FTS to synthesize pure chemicals, such as ethylene. As noted in Section 9.2, mononuclear complexes are inactive for FTS and it was hoped that cluster complexes might be *selective* catalysts.

Unlike the previous cases for hydrogenation, hydrogenolysis, and methanation, no comprehensive study of FTS has been done by a single research group. However, a number of catalysts have been investigated. As already noted the product distribution is usually complex. In addition, in a complex manner it can strongly depend on the reaction conditions as well as on the conversion and the time on stream. This is especially true in the case of reactive intermediates, such as olefins. At low conversions (typical of the differential reaction conditions employed in academic studies) the yield of olefins can be high, but at practical conversions the yield can decrease to a very low value. For these reasons it is very difficult to compare the results of different workers quantitatively.

Iron Catalysts

In a series of papers Commereuc and coworkers have reported on the activity of $Fe(CO)_5$ and $Fe_3(CO)_{12}$ deposited from pentane solution on several supports. The first publication (83) is illustrative of both the potential and problems associated with the use of "well-defined" complexes to catalyze FTS. When deposited on alumina both catalysts gave a conversion of 3.3% at $P = 10$ atm, $H_2/CO = 1$, and GHSV $= 100\ h^{-1}$. The temperatures were 260°C for the $Fe(CO)_5$ catalyst (1.8% Fe) and 270°C for the $Fe_3(CO)_{12}$ catalyst (0.8% Fe). This corresponds to a somewhat low activity, $N_f = 6 \times 10^{-5}$ and $1 \times 10^{-4}\ s^{-1}$, respectively. However, the catalysts showed high selectivities to olefins (57 and 43%, respectively). In contrast, a conventional Fe catalyst of 8.1% loading prepared from $Fe(NO_3)_3$ gave a conversion of 19.4% at 270°C ($N_f = 8 \times 10^{-5}\ s^{-1}$) and yielded only 38% olefins and a much broader product distribution extending to C_{16} (but very little beyond C_{12}). By removing weakly adsorbed carbonyl, experimenters believed that both complexes could be quantitatively converted to surface-bonded $HFe_3(CO)_{11}^-$. Such catalysts also showed high selectivity to olefins and a remarkable 45% selectivity to propylene after 24 hours of reaction. Characterization by electron microscopy of a fresh catalyst revealed no Fe particles, suggesting that the particles are <1.4 nm in diameter. However, after 48 hours of reaction electron microscopy indicated an average particle size of about 50 nm and the activity and selectivity changed to that of a conventional Fe catalyst. Later papers from this group provide additional details, characterization, and a reaction mechanism (84,85).

Commercial FTS catalysts are usually promoted with alkali to increase the degree of polymerization and the olefin yield. McVicker and Vannice (86) have reported on the use of a variety of K-group VIII complexes as precursors for FTS catalysts. Decomposition of $K_2Fe(CO)_4 \cdot 2C_4H_8O$ yielded a catalyst of 5.5% K and 3.9% Fe. The dispersion, as measured by CO chemisorption on both a fresh (0.15) and used (0.063) catalyst, was about twice that for a conventionally prepared catalyst of the same composition. Near 266°C, $P = 1$ atm, $H_2/CO = 3$, and GHSV of roughly 3000 h^{-1}, this catalyst gave a CO conversion of 2.3% ($N_f = 1.4 \times 10^{-4}\ s^{-1}$), compared to 0.07% for the conventional catalyst. The carbonyl catalyst also gave much less methane and more C_3^+.

Tkatchenko and coworkers (87) have studied FTS over $Fe_3(CO)_{12}$ in zeolites. The carbonyl was sublimed onto the zeolites at 60°C, and it is claimed that the complex is localized in the supercage. Batch reactions were run at 250°C, $P = 20$ atm, and variable H_2/CO ratios. Analysis was only performed after 15 hours of reaction. $Fe_3(CO)_{12}/HY$ zeolite was variously reported to be of low activity (87), yielding

12.7% conversion of CO at $H_2/CO = 4$ ($N_f = 2 \times 10^{-4}$ s^{-1}) or inactive for FTS (88). The low activity was attributed to reaction (9.4) which resulted in the Fe being oxidized to Fe(II) and the difficulty in reducing Fe(II). Such oxidation has also been reported for Fe carbonyls decomposed on alumina (46). When supported on NaY the oxidation reaction does not occur and Fe(O) is formed. It was believed that the Fe existed in the form of very small particles. This material had an activity of $N_f = 1 \times 10^{-3}$ s^{-1} and 2×10^{-3} s^{-1} at $H_2/CO = 4$ and 1, respectively. A conventional Fe/Al_2O_3 catalyst gave 63% conversion of CO at $H_2/CO = 1$ corresponding to $N_f = 1 \times 10^{-4}$ s^{-1}.

Later papers (88,89) extended these results to include $(FeCp(CO)_2)_2$ and carbonyls of Co and Ru. The cyclopentadienyl catalyst appeared to be similar in activity to $Fe_3(CO)_{12}$, but yielded less light hydrocarbons and more C_4^+. A catalyst was also prepared from $Fe_3(CO)_{12}/NaY$ in such a manner as to cause migration of the Fe to the outside of the zeolite leading to particles 20 to 30 nm in diameter. This material gave a broader product distribution than catalysts with small particles. In general, it was found that carbonyl-derived catalysts with small particles selectively formed hydrocarbons in the C_1 through C_9 range.

The chain length limitation of C_9 could be due to either an intrinsic effect of the small particles or shape selectivity induced by the zeolite. In consideration of the well-documented shape selectivity of zeolites and the likelihood of even the small metal particles containing a substantial number of atoms (a crystallite 1.4 nm in diameter will contain about 85 atoms), the latter explanation seems more plausible. However, Nijs and colleagues (90) prepared Ru/NaY by ion exchange of $Ru(NH_3)_6Cl_3$ and found that at 252°C, $P = 14.2$ atm, and $H_2/CO = 1.5$, and that this catalyst gave a conversion of 12% ($N_f = 7 \times 10^{-3}$ s^{-1}) and showed a large drop in chain growth probability at C_9. Only 1% of the products were in the C_{11}^+ range. A Ru/SiO_2 catalyst (which is not shape selective) gave a conversion of 11% under similar conditions ($N_f = 3 \times 10^{-3}$ s^{-1}) with 60% of the products in the C_{12}^+ range.

Ru/NaY was found to be susceptible to hydrolysis by water vapor which resulted in the formation of Ru particles up to 4 nm in diameter. A stable Ru/LaY catalyst was prepared which gave smaller (exact size not specified) metal particles (91). In this case the probability of chain growth went to zero at C_5 and the authors believe that the effect is due to the small and constant size of the Ru particles and not to the shape selectivity of the zeolite.

Melson et al. (92) studied $Fe_3(CO)_{12}$ deposited by extraction from hexane solution on the acid forms of ZSM-5 and mordenite, and on the Na form of 13X. For loadings of <10% Fe it was claimed that in all cases the complex was completely decarbonylated during adsorption (the degree of air exposure was not stated) and γ-Fe_2O_3 is formed with particle sizes <5 nm. A fraction of the iron corresponding to about 1% loading was present as Fe(O). In contrast, synthesis from $Fe(NO_3)_3$ resulted in α-Fe_2O_3 of diameter >10 nm. After reduction in H_2 at 450°C and 20.7 atm. about 80% of the Fe was reduced. This then formed χ-Fe_5C_2 carbide (along with some inactive Fe_3O_4), which is the active component, during reaction. Reaction was run in a flow system at 280 and 300°C, $P = 20.7$ atm, $H_2/CO = 1$, and GHSV = 1000

h^{-1}. After eight days on stream, catalysts of about 15% Fe on each of the supports had a CO conversion of about 75% ($N_f = 2 \times 10^{-3}$ s^{-1}). For the most part, the product distributions were similar (and also changed with time). However, the ZSM-5 and mordenite catalysts (which were in the acid form and have medium-size pores) gave more unsaturates and less wax than the 13X catalyst (which is nonacidic and has larger pores).

Ruthenium Catalysts

Tkatchenko et al. (88,89) have also studied Ru$_3$(CO)$_{12}$/Y zeolites for FTS at conditions similar to those used for Fe catalysts (*vide supra*). It is claimed that after activation in vacuum at 200°C the Ru catalysts are inactive for FTS at 200°C. This is surprising and the reasons are not clear. Activation between 320 and 400°C yielded active catalysts for both HY and NaY supports (note that Ru is much more easily reduced than Fe). Both before and after reaction (200°C, H$_2$/CO = 4, P = 20 atm), electron microscopy indicated particle sizes of 1.5 to 2 nm for Ru/HY. By altering the synthetic method, particles 10 to 100 nm in diameter were produced. The former catalyst yielded products up to C$_9$ and the latter catalyst gave products up to C$_7$ and much more methane. The CO conversions (after 15 hours) were 37% ($N_f = 5 \times 10^{-4}$ s^{-1}) and 100% ($N_f = 1 \times 10^{-3}$ s^{-1}), respectively. Both catalysts exclusively yielded methane at 250°C. Ru$_3$(CO)$_{12}$ supported on SiO$_2$—Al$_2$O$_3$ yielded metallic particles 1 nm in diameter after activation at 200°C. This material had an activity of about $N_f = 1 \times 10^{-4}$ s^{-1} and yielded mostly methane. During reaction the Ru sintered to hexagonal plates 10 to 100 nm in diameter. Thus, it appears that in favorable cases the supercage of a zeolite can limit particle size growth.

In a particularly illustrative work Chen, Wang, and Goodwin (93) prepared catalysts from Ru$_3$(CO)$_{12}$, Ru(NH$_3$)$_6$Cl$_3$, and RuCl$_3 \cdot$ 1.5H$_2$O supported on NaY zeolite and then reduced in H$_2$ at 420°C for 2 hours. The particle size of the salt-derived catalysts increased with loading but was constant at about 1.3 nm for the ammine and 1.0 nm for the carbonyl catalyst. At 250°C, P = 1 atm, H$_2$/CO = 1, and GHSV = 1800 h^{-1}, the turnover frequencies (N) of the three catalysts were similar, being about 1.5×10^{-3} s^{-1}, and the activation energies were all 22 to 24 kcal/mol. However, the product distributions were markedly different and varied with loading. The salt-derived catalysts gave about 90% methane, the ammine catalysts about 80% methane, and the carbonyl catalysts about 45% methane. The order of selectivities to olefins was the reverse of this, about 80% of the C$_2$—C$_4$ hydrocarbons being olefins over the carbonyl catalysts.

Okuhara and coworkers (59) compared Ru/Al$_2$O$_3$ which was prepared from RuCl$_3$ and Ru$_3$(CO)$_{12}$ and then reduced at 450°C for 2 hours. Using a circulating system at 200°C, P = 0.7 atm, and H$_2$/CO = 2, they found that the carbonyl-derived catalyst gave $N_f = 3 \times 10^{-4}$ s^{-1} and yielded mostly C$_1$—C$_6$ hydrocarbons with 80% olefinic content. The salt-derived catalyst had an activity of $N_f = 2 \times 10^{-4}$ s^{-1} and yielded more methane and only 52% olefins. At 260°C over the carbonyl catalyst the methane yield substantially increased (from 8 to 29%) and the olefin yield dropped to 51% ($N_f = 1 \times 10^{-3}$ s^{-1}). However, the addition of K dropped the methane yield to 19% and increased the olefin yield to 74% ($N_f = 5 \times 10^{-4}$ s^{-1}).

Kellner and Bell (94) have reported a detailed kinetic study of FTS over 1% Ru/Al$_2$O$_3$ prepared from Ru$_6$C(CO)$_{17}$. The catalysts were activated at 400°C in 10 atm of H$_2$ for 10 to 12 hours. At 250°C, $P = 1$ atm, and H$_2$/CO $= 1$, the activity for methane formation was $N_f = 1 \times 10^{-3}$ s^{-1}. About 70% of the hydrocarbon products were in the C$_2$—C$_{10}$ range (small quantities of methanol were also observed) and the probability of polymerization was 0.62. This value increased with a decrease in temperature or H$_2$/CO ratio.

Ferkul and coworkers (95) have briefly reported on FTS over catalysts prepared from Ru$_3$(CO)$_{12}$, H$_2$Ru$_4$(CO)$_{13}$, and RuCl$_3 \cdot$ 3H$_2$O supported on Al$_2$O$_3$, SiO$_2$, and NaY zeolite. All catalysts gave mainly methane and the activities were similar. At 360°C, $P = 1$ atm, and H$_2$/CO $= 3$, both Ru$_3$(CO)$_{12}$ and the salt-derived catalyst gave $N_f = 4 \times 10^{-2}$ s^{-1}. High methane yield is expected at this elevated temperature, but few data were reported at lower temperatures.

Knozinger et al. (71) reported that Ru$_3$(CO)$_{12}$/Al$_2$O$_3$ had an activity of $N_f = 4.6 \times 10^{-4}$ s^{-1} at 333°C, $P = 1$ atm, and H$_2$/CO $= 3$. The product was >99% methane.

Commereuc et al. (83) briefly investigated FTS over Ru$_3$(CO)$_{12}$/Al$_2$O$_3$. At 280°C, H$_2$/CO $= 1$, $P = 10$ atm, and GHSV $= 100$ h^{-1}, the conversion was 14.7% ($N_f = 4 \times 10^{-4}$ s^{-1}). The selectivity to hydrocarbons was 100% with an olefin selectivity of 4%, and 99% of the product was C$_1$—C$_4$.

Osmium Catalysts

Commereuc (83) also reported that Os$_3$(CO)$_{12}$/SiO$_2$ gave 9% conversion of CO ($N_f = 3 \times 10^{-4}$ s^{-1}) at 200°C and otherwise the same conditions as just noted. The olefin yield was 9%, and 99% of the products were in the range C$_1$–C$_4$. Deeba et al. (81) studied Os$_3$(CO)$_{12}$/MgO. At 300°C, $P = 31.8$ atm, H$_2$/CO $= 4$, and <0.1% conversion in a flow system, $N_f = 8 \times 10^{-4}$ s^{-1} and the products were 75% methane, 23% ethane, and 2% propane. After reaction the IR spectrum was similar to that expected for H$_2$Os$_3$(CO)$_9$OMg$^-$, so the authors suggested that the active site was an Os$_3$ cluster. This assignment must be considered quite speculative at this time.

Knozinger and coworkers (71) studied H$_2$OsCl$_6$, Os$_3$(CO)$_{12}$, and H$_4$Os$_4$(CO)$_{12}$ supported on alumina for FTS. At 333°C, 1 atm, and H$_2$/CO $= 3$, the activity of H$_2$OsCl$_6$ was 2.5×10^{-4} s^{-1} and the activity of Os$_3$(CO)$_{12}$ was 1.5×10^{-4} s^{-1}. The former catalyst gave >99% methane, and Os$_3$(CO)$_{12}$ gave 86% methane, 11% C$_2$ (mostly ethylene), and 2% C$_3$. H$_4$Os$_4$(CO)$_{12}$ was tested at 32 atm and 200°C. The activity for H$_2$/CO $= 3$ was 1.3×10^{-5} s^{-1}, and the products were 63% methane and 37% C$_2$. Examination by IR and electron microscopy failed to detect particle growth in any of the Os catalysts. The authors claimed that upon decomposition Os ensembles were formed with a nuclearity corresponding to that of the precursor complex. It was also suggested that these ensembles were the active site for reaction and differences in activity and product distribution might reflect the differences in nuclearity. However, direct evidence for this was slight and, as the authors noted, one catalyst contained chlorine, the reaction conditions were not constant, and in all cases methane was the dominant product.

Psaro et al. (58) studied FTS over $Os_3(CO)_{12}$, $H_2Os_3(CO)_{12}$, and $Os_6(CO)_{16}$ supported on silica and alumina. These workers claim that above 200°C the cluster framework is broken to produce Os(II) species. The Os(II) was not reduced to Os metal below 400°C which suggested that Os particles are not the active site for FTS. Reaction was examined in a closed system at 250 and 300°C, $P = 1$ atm. and $H_2/CO = 1$. Rate data were not given. At 250°C the yield of methane varied from 70 to 92% and small amounts of $C_2–C_4$ hydrocarbons were produced. At 300°C the yield of methane was about 95%.

Cobalt Catalysts

Meyers and Hall (96) studied FTS over $Co_4(CO)_{12}$ deposited on alumina from hexane solution. Reaction was at 250°C, $P = 4.1$ atm, and $H_2/CO = 1.5$ in a closed vessel. Analysis was performed after about 30 min. A normal Anderson product distribution was observed with a probability of polymerization of 0.47 (negligible C_7^+ products). Adequate kinetic data were not given, but the activity was about $N_f = 1 \times 10^{-3}$ s^{-1}. XPS indicated that during decomposition in vacuum at 250°C most of the Co was oxidized to Co(II) with 18% residual Co(O). This amount of oxidation agrees well with earlier TPDE data (52). During FTS the amount of Co(O) increased to 22%. Laser-induced mass spectroscopy detected $Co_4(CO)_x$ and $Co_4C(CO)_x$ clusters after reaction and electron microscopy failed to detect any Co particles, indicating sizes <10 nm. It was suggested that Co_4 clusters were also the stable arrangement during FTS. Similar analyses were not applied to a traditional Co catalyst to see if such clusters were formed by the laser.

Tkatchenko et al. (88,89) briefly studied $Co_2(CO)_8/NaY$ in the same manner as previously described for other catalysts. At $H_2/CO = 1$ and 4 the products reasonably fit Anderson kinetics except for a dip at C_2. At $H_2/CO = 4$ the yield of methane was 43%, but at a ratio of 1 the yield was only 11% and $N_f = 1 \times 10^{-3} s^{-1}$. The selectivity to olefins was low.

Ferkul and colleagues (95) have also briefly examined FTS over $Co_2(CO)_8$ and $Co_4(CO)_{12}$. When supported on alumina a product distribution of saturated hydrocarbons was observed, typical of Co catalysts. However, when supported on silica the catalysts were nearly inactive below 250°C, but above 300°C ($H_2/CO = 2$) the catalysts gave 100% conversion and methane as the sole product ($N_f = 7 \times 10^{-3}$ s^{-1}).

Rhodium Catalysts

Commereuc et al. (83) reported on FTS over $Rh_4(CO)_{12}$ and $Rh_6(CO)_{16}$ supported on alumina. At 180°C, $Rh_4(CO)_{12}$ gave 4.3% conversion of CO ($N_f = 1 \times 10^{-4}$ s^{-1}), 100% hydrocarbons containing 2.5% olefins, 85% methane, and virtually no product $>C_4$. $Rh_6(CO)_{16}$ at 200°C gave 2.7% conversion ($N_f = 6 \times 10^{-5}$ s^{-1}), 80% hydrocarbons, 28% olefins, 29% methane, and virtually no product $>C_5$.

Ichikawa (97–99) has extensively studied the synthesis of methanol and ethanol by FTS over a variety of Rh carbonyls (and other complexes) deposited on a variety of supports. The data have recently been summarized (100). The complexes were impregnated from hexane, acetone, or tetrahydrofuran solution. After evaporation of the solvent, the catalysts were decomposed in vacuum at 120 to 180°C for 1 to 2 hours. Most studies were done at a loading of about 0.3% Rh using 20 g of catalyst in a circulating system. Oxygenated products were collected in a trap at −80°C. Basic supports, especially MgO and ZnO, provided high selectivity to methanol. For example, FTS over $Rh_4(CO)_{12}/ZnO$ at 220°C, $P = 0.9$ atm, and $H_2/CO = 2.3$ gave a conversion of 8.8% after 5 hours ($N_f = 4 \times 10^{-5}$ s^{-1}) and 96% selectivity to methanol. The only other product was 4.5% methane. $Rh_2Cp_2(CO)_3$, $Rh_6(CO)_{16}$, and $(NBu_4)_2Rh_{13}(CO)_{23}H_{2-3}$ supported on ZnO were also tested. These gave similar activities and methanol yields ranging from 76 to 97%. Increasing the temperature substantially lowered the selectivity in all cases. This is unfortunate since the activities are very low (an N_f value of 1.2×10^{-5} s^{-1} corresponds to reacting only 1 molecule per metal atom in a catalyst per day). A traditional catalyst (prepared by the reduction of $RhCl_3 \cdot 3H_2O/ZnO$ at 350°C) gave mainly methane (rate data were not given). $Rh_4(CO)_{12}$ supported on silica (a very weakly acidic support) gave only 0.6% methanol and 85% methane ($N_f = 3 \times 10^{-5}$ s^{-1}). When supported on alumina (which is mildly acidic), 0.2% methanol and 77% methane was formed ($N_f = 8 \times 10^{-5}$ s^{-1}).

Less basic supports such as La_2O_3, CeO_2, TiO_2, ZrO_2, and ThO_2 gave less methanol and substantial yields of ethanol. $Rh_4(CO)_{12}/La_2O_3$ gave a 36% conversion of CO in 5 hours ($N_f = 2 \times 10^{-4}$ s^{-1}) at 224°C with a 61% selectivity to ethanol, 20% methanol, and 12% methane. Again, the other Rh carbonyls gave roughly similar activities and good selectivity to ethanol. In contrast, a traditional Rh/La_2O_3 catalyst at 220°C gave only 17% ethanol, 11% methanol, and 56% methane. Ichikawa and coworkers (101) have also reported that supporting Rh on silica that has been previously coated with ZrO_2 or TiO_2 has much higher activity and selectivity to ethanol than Rh/SiO_2.

Although there are some differences in activity and selectivity based on the choice of precursor, Ichikawa does not claim that nuclearity is maintained during reaction. Direct measurements were not reported, but it was inferred that the Rh produced from the carbonyls was highly dispersed. It is not known if the chlorine in the traditional catalyst affected its selectivity. It was suggested that ethanol is produced by the hydrogenation of an acyl species formed by CO insertion into CH_x species. The latter would be formed by partial hydrogenation of C(ads) resulting from the dissociation of CO(ads).

FTS over Other Catalysts

Ichikawa has also reported on the use of $Ir_4(CO)_{12}$ and Pt carbonyl cluster anions for FTS (98,100). Even when supported on La_2O_3 or ZrO_2, methanol was the dominant product. $(NEt_4)_2Pt_{15}(CO)_{30}/ZnO$ yielded 99% methanol at 240°C, $P = 0.8$

atm, and $H_2/CO = 3$. The activity was very low, $N_f = 6 \times 10^{-6} s^{-1}$. At the same conditions, $Ir_4(CO)_{12}/ZnO$ gave 87% methanol and 13% methane ($N_f = 6 \times 10^{-6} s^{-1}$).

As part of their previously described study (83), Commereuc and colleagues also investigated FTS over $Ir_4(CO)_{12}/Al_2O_3$ at 180°C. The CO conversion was 11.8% ($N_f = 4 \times 10^{-4} s^{-1}$); 100% hydrocarbons were formed with only 0.8% selectivity to olefins, and even at this low temperature 78% of the product was methane with no C_5^+ detected.

Sivasanker et al. (38) studied FTS over $Mo(CO)_6/Al_2O_3$ which was activated at 500°C in flowing H_2. Best results were obtained when a dehydroxylated alumina (pretreated at 1000°C) was used to suppress reaction (4) (oxidized Mo is very hard to reduce). Reactions were run at 350°C, $P = 24$ atm, $H_2/CO = 1$, and GHSV = 1300 h^{-1}. Compared to a conventional catalyst (prepared by reducing MoO_3 at 500°C) the carbonyl catalyst was five-fold more active, and nine-fold more active ($N_f = 2 \times 10^{-2} s^{-1}$) if the catalysts were prepared with 4% K. The C_1–C_4 products obeyed Anderson kinetics with a probability of polymerization of 0.37 (the C_5^+ yield was only 4%). K had little effect on the reaction rate but increased the yield of olefins. It was suggested that the improved activity reflected the ability to prepare well-dispersed and low-valent Mo from the carbonyl.

Cichowlas and coworkers (39) studied $W(CO)_6/Al_2O_3$ at the same conditions. Activity was only observed if dehydroxylated alumina was used. This catalyst was 50-fold more active ($N_f = 4 \times 10^{-3} s^{-1}$) than a conventional catalyst prepared by reducing WO_3. The addition of 4% K increased the yield of olefins and the probability of polymerization was 0.26 (negligible C_4^+ products). The improved activity was attributed to the same factors as for the $Mo(CO)_6$ catalyst.

9.6.5 Catalysis over Mixed Metal Clusters

Catalysis by alloys has been extensively studied for several decades. Alloying can greatly change the activity and selectivity of a metal catalyst. However, early work is limited by the absence of surface-sensitive spectroscopies to determine the surface composition (which in general will be quite different from the bulk composition). Particularly fruitful has been the pioneering work of Sinfelt (80) who studied ethane hydrogenolysis over Cu—Ni alloys. It was demonstrated that the Cu effectively dilutes the active Ni sites and catalysis only occurs on ensembles of Ni atoms. Such observations have suggested the use of mixed metal clusters to prepare alloy catalysts of constant and well-defined composition. However, from what has already been observed with monometallic clusters, the possibility of fragmentation and secondary agglomeration of mixed metal clusters is likely at elevated temperatures.

Anderson and Mainwaring (102) reported the first study of catalysis over a mixed metal cluster. The hydrogenolysis of methylcyclopentane (MCP) was investigated over $Rh_2Co_2(CO)_{12}/SiO_2$ (deposited from hexane solution) at 260°C, 2.9 atm, and $H_2/MCP = 6.9$ in a flow system. The conversion was kept at <3%. The catalyst was first activated by reduction in flowing H_2 at 380°C for about 12 hours. H_2 chemisorption yielded a dispersion of 0.22 which was not altered if the catalyst was reduced for 5 hours at 500°C. This indicated that some sintering had occurred, although electron microscopy apparently failed to detect large particles. The products were nearly exclusively noncyclic hexane isomers. It was claimed that this is quite different from traditional catalysts of Co and Rh, each of which give much more fragmentation to lower molecular weight hydrocarbons. Unfortunately, a mixed Co—Rh catalyst prepared from salts was not evaluated by the authors. In a later report (103) these authors did compare such a traditional alloy catalyst to the mixed metal carbonyl and found virtually no difference in product distribution for the hydrogenolysis of n-hexane (hydrogenolysis of MCP was not reported). It was also found (57) that the surface of all of the Co—Rh catalysts was heavily enriched in Co. Thus, the use of a bimetallic carbonyl cluster did not result in a well-defined catalyst, and it is not clear that any special activity or selectivity existed compared to a traditional catalyst of the same composition.

The first comparative study of catalysis over a series of mixed metal clusters was by Brenner and coworkers (104,105). The hydrogenation of ethylene and the hydrogenolysis of ethane were studied over the complexes $Fe_nRu_{3-n}(CO)_{12}/Al_2O_3$, $n = 1,2,3$. Catalysts were activated at 200°C in flowing He prior to running ethylene hydrogenation at 25°C, $H_2/C_2H_4 = 4$, and P = 1 atm. As previously noted in Section 9.6.1, the relatively low temperature of activation enhances the possibility of retaining the cluster skeleton. However, olefin hydrogenation is a structure-insensitive reaction (over traditional catalysts) which likely makes it a less useful probe of catalyst morphology. Ethane hydrogenolysis was run at 230°C, $H_2/C_2H_6 = 2$, and P = 1 atm. Consequently, for this reaction the catalysts were activated at 400°C in flowing H_2. Although ethane hydrogenolysis is a structure-sensitive reaction, the more rigorous activation procedure is likely to destroy the cluster. Hence, the study again reduces to the practical matter of whether a mixed metal precursor will display different activity from the sum of its parts.

For both reactions Ru is at least 100-fold more active than Fe. In both cases it was observed that the value of N_f increased linearly with the Ru content of the cluster (or equivalently, the activity remained constant if based on the Ru content only, $N_f = 1.5$ s^{-1} at 25°C for hydrogenation, and $N_f = 2$ s^{-1} at 230°C for hydrogenolysis). Hence, in these cases no special properties can be attributed to the mixed metal clusters. However, a catalyst prepared from $RuCl_3$ was about 20-fold more active for hydrogenolysis. This difference might be due to the presence of Cl or the carbonyl-derived catalysts having a different morphology. A similar conclusion was found by this group (53) for ethane hydrogenolysis over the clusters $Co_2(CO)_8$, $Co_4(CO)_{12}$, $Co_3Rh(CO)_{12}$, $Co_2Rh_2(CO)_{12}$, $Rh_4(CO)_{12}$, and $Rh_6(CO)_{16}$.

Shapley and coworkers (106) investigated the activity of $Os_3(CO)_{12}$, $WOs_3Cp(CO)_{12}H$, and $MoOs_3Cp(CO)_{12}H$ supported on alumina for FTS between 350 and 400°C, $P = 1$ atm, and $H_2/CO = 3$. Prior to reaction the cyclohexane solvent was evaporated and the catalysts activated by slowly heating in H_2 to 500°C. Neither the activation energy (about 32 kcal/mol), activity (about $N_f = 5 \times 10^{-2}$ s^{-1} at 350°C), nor product distribution (about 83% methane) varied significantly between the catalysts. The authors concluded that the activity was solely due to the Os atoms and no special effects were derived by the use of mixed metal clusters.

Guczi et al. (49) attempted to prepare RuFe bimetallic catalysts by the codeposition of $Ru_3(CO)_{12}$ and $Fe_3(CO)_{12}$ from hexane solution on Cab-O-Sil. The catalysts were then activated by TPDE in either flowing He or H_2 up to 500°C. Total metal loading was kept constant at about 0.5% and the Ru content was 0, 14, 28, and 100 atom%. The TPDE of the mixed carbonyls was slightly different from the sum of the parts which led the authors to claim that a bimetallic alloy was formed. The catalysts were tested for FTS near 300°C, $P = 0.7$ atm, and $H_2/CO = 3$ in a flow system operated at low conversions. The pure Ru catalyst was 3×10^3 more active ($N_f = 6 \times 10^{-1}\,s^{-1}$ at 280°C) than the pure Fe catalyst, and the product distribution smoothly changed from 50% methane over Fe to 100% methane over Ru.

This group also studied butane hydrogenolysis using a circulating system at 140°C, $P = 2$ atm, and $H_2/butane = 10$. The Fe catalyst had no activity and the activity increased with Ru content up to $N_f = 3 \times 10^{-2}\,s^{-1}$ for a pure Ru catalyst decomposed in He. A similar catalyst decomposed in H_2 gave an activity of only $3 \times 10^{-4}\,s^{-1}$, and a catalyst prepared from $RuCl_3$ had an activity of $2 \times 10^{-4}\,s^{-1}$. This lower activity for a traditional Ru catalyst is opposite the behavior found by Brenner et al. (37) for ethane hydrogenolysis. The authors suggested that the reason TPDE in He resulted in higher activity than decomposition in H_2 is that He allows more C(ads) to remain on the catalyst and this helps stabilize the Ru to sintering. In another paper (107) the authors focused on the role of surface C during FTS over these catalysts.

Ferkul (95) studied FTS over $H_2FeRu_3(CO)_{13}$ and $(Ph_3P)_2NCoRu_3(CO)_{13}$ at conditions previously described for Ru catalysts. The results were very similar to that for $Ru_3(CO)_{12}$, except that the CoRu catalyst was of lower activity and inactive when supported on silica. This was probably due to poisoning by the iminium cation.

Ichikawa (108,48) has studied hydroformylation over a variety of carbonyls including the series $Rh_4(CO)_{12}$, $Rh_2Co_2(CO)_{12}$, $RhCo_3(CO)_{12}$, and $Co_4(CO)_{12}$ supported on ZnO. At 160°C, $P = 0.8$ atm, and $C_2H_4/CO/H_2 = 1/1/1$, $Rh_4(CO)_{12}$ gave 99% selectivity to propionaldehyde and $N_f = 3 \times 10^{-3}\,s^{-1}$. At the same conditions but using propylene in the feed, $Rh_4(CO)_{12}$ gave 58% selectivity to n-butyraldehyde and $N_f = 6 \times 10^{-4}\,s^{-1}$. The activity decreased and the proportion of linear aldehyde increased as the Co content of the catalysts increased. Catalysts prepared from some other complexes and from $RhCl_3$ showed poor activity, and acidic supports also gave poor hydroformylation activity (48).

The same series of complexes supported on ZrO_2 was investigated for FTS at 200°C, $P = 0.9$ atm, and $H_2/CO = 2.3$ in a closed circulating system (100). The selectivity to oxygenates decreased and the fraction of ethanol within the oxygenates increased as the Co content increased.

9.7 CONCLUSIONS

There has been tremendous interest and growth in the field of cluster chemistry during the past decade. This chapter has attempted to review much of the work relating to catalysis over supported cluster complexes and to compare these materials to related mononuclear and traditional catalysts. It is clear that the understanding of the catalytic properties of clusters considerably lags the understanding of their structure, bonding, and stoichiometric transformations. A central problem is that under reaction conditions a cluster is likely to lose its molecular integrity while undergoing a variety of fragmentation and agglomerization processes. This problem is especially severe for reactions such as FTS which are run at elevated temperatures or in the presence of gases such as CO which can encourage restructuring. Also, the analysis of surfaces is very difficult and usually yields the dominant surface species—which might not be related to the catalytically important sites. For these reasons it can be useful to take a more pragmatic approach and simply examine if cluster *precursors* provide any special catalytic properties. Even this type of investigation can be deceptive. It is important that similar mononuclear and traditional catalysts also be prepared and that they be tested at the same reaction conditions to act as a reference state for comparison. Often this has not been done.

Figure 9.3 depicts the range of states of metal in catalytic systems. At one extreme there are mononuclear complexes (which can be either homogeneous or supported), represented by Wilkinson's catalyst. At the other extreme there are large and ill-defined three-dimensional crystallites. Metal in a complex has a full (or nearly so) complement of ligands, whereas metal at the surface of a crystallite is highly coordinatively unsaturated.

The gist of the cluster hypothesis is that mononuclear complexes are inactive for some reactions readily catalyzed by metal crystallites because additional bonding sites are necessary. Consequently, by virtue of having multiple adjacent metal centers, cluster complexes should be more reactive than mononuclear complexes. Further, their well-defined structure should enhance selectivity. This argument appears reasonable. However, an extension of the same logic would suggest that cluster complexes should be *less* active than the larger metal crystallites that typically exist on traditional catalysts of supported metals. This viewpoint is supported by the comparison of activity for the hydrogenation of ethylene over well-defined complexes and over the same materials after further decomposition (Section 9.6.1). The activity of the molecular complexes is very low, but the activity increases by orders of magnitude when metal crystallites are formed. Not only should a crystallite offer more bonding possibilities than a molecular cluster, but the ligands on a stable cluster should resemble irreversibly adsorbed material on a traditional catalyst. Thus,

Figure 9.3 Metal habitats in catalysts.

a complex may be a model for a partially *poisoned* metal surface. Also, results described in Section 9.2.1 suggest that the ability of a support to stabilize coordinatively unsaturated sites is important. For this reason, supported clusters might have more application than do homogeneous systems.

The observation that molecular clusters are likely to be less active than traditional catalysts has an important corollary. Since catalysis of a reaction such as FTS typically requires a temperature of about 300°C to afford a reasonable rate, a cluster complex would need to operate at a higher temperature. However, molecular complexes generally disintegrate above about 200°C. Hence, unless a complex is able to buck the trend and be *much* more active than a traditional catalyst (or especially stable), it is likely that it would be destroyed under reaction conditions. Presumedly, any special selectivity associated with its well-defined structure would also be lost.

In summary, numerous examples have been reported in which clusters behave differently from traditional catalysts. However, it appears that the change in synthetic methodology is often the dominant feature and not the retention of a well-defined cluster framework. Cluster complexes should find application in catalysis for reactions that can be run at suitably low temperatures to prevent disintegration. Compared to traditional catalysts, clusters are unlikely to display increased activity and their value will probably be due to improved selectivity. The synthesis of complexes with improved thermal and chemical stability should increase the potential catalytic applications of clusters. Due to the ability of a support to immobilize a complex, supported clusters are likely to have a higher temperature range of application than homogeneous clusters.

ACKNOWLEDGMENT

Support by Department of Energy for the research conducted in the author's laboratory is gratefully acknowledged.

REFERENCES

1. E. L. Muetterties, *Science,* **196**, 839 (1977).

2. E. L. Muetterties, *Bull. Soc. Chim. Belg.,* **84**, 959 (1975).

3. E. L. Muetterties, T. N. Rhodin, E. Band, C. F. Brucker, and W. R. Pretzer, *Chem. Rev.,* **79**, 91 (1979).

4. E. L. Muetterties, and Stein, *J. Chem. Rev.,* **79**, 479 (1979).

5. E. L. Muetterties, *Catal. Rev.-Sci. Eng.,* **23**, 69 (1981).

6. R. Ugo, *Catal. Rev.-Sci. Eng.,* **11**, 225 (1975).

7. R. Ugo, R. Psaro, G. M. Zanderighi, J. M. Basset, A. Theolier, and A. K. Smith, *Fund. Res. Homog. Catal.,* **3**, 579 (1979).

8. J. M. Basset, and R. Ugo, *Aspects Homog. Catal.,* **3**, 137 (1977).

9. A. K. Smith, and J. M. Basset, *J. Mol. Catal.,* **2**, 229 (1977).

10. M. Moskovits, *Acc. Chem. Res.,* **12**, 229 (1979).

11. P. Chini, G. Longoni, and V. G. Albano, *Adv. Organometal. Chem.,* **14**, 285 (1976).

12. G. A. Ozin, *Catal. Rev.-Sci. Eng.,* **16**, 191 (1977).

13. D. C. Bailey, and S. H. Langer, *Chem. Rev.,* **81**, 109 (1981).

14. R. B. Anderson, *Catalysis,* **4**, 1 (1956).

15. M. A. Vannice, *J. Catal.,* **37**, 449 (1975).

16. H. S. Taylor, *Proc. R. Soc.,* **A108**, 105 (1925).

17. G. A. Somorjai, *Adv. Catal.,* **26**, 2 (1977).

18. A. A. Balandin, *Adv. Catal.,* **19**, 1 (1969).

19. B. M. W. Trapnell, *Adv. Catal.,* **3**, 1 (1951).

20. E. L. Muetterties, Symp. Metal Surf. Metal Clusters, University of Toronto, 1981.

21. L. Salem, Int. Sem. Relat. Metal Cluster Compounds, Surf. Sci., Catal., Pacific Grove, Calif. 1979.

22. G. C. Bond, *Catalysis by Metals,* Academic Press, New York, 1962.

23. M. Polanyi, and J. Hortiuti, *Trans. Faraday Soc.,* **30**, 1164 (1934).

24. J. Halpern, *Am. Chem. Soc. Adv. Chem.,* **70**, 1 (1968).

25. M. Boudart, A. Aldag, J. E. Benson, N. A. Dougherty, and C. G. Harkins, *J. Catal.,* **6**, 92 (1966).

26. J. J. Burton, *Catal. Rev.-Sci. Eng.,* **9**, 209 (1974).

27. J. R. Anderson, *Structure of Metallic Catalysts,* Academic Press, New York, 1975.

28. J. R. Schrieffer, *Preprints Div. Petrol. Chem. Am. Chem. Soc.,* **21**, 331 (1976).

29. R. P. Messmer, *Gazz. Chim. Ital.,* **109**, 241 (1979).

30. J. Demuynck, M. M. Rohmer, A. Strich, and A. Veillard, *J. Chem. Phys.,* **75**, 3443 (1981).

31. W. A. Goddard, III, Int. Sem. Relat. Metal Cluster Compounds, Surf. Sci. Catal., Pacific Grove, Calif., 1979.

32. T. H. Upton, and W. A. Goddard, III, *CRC Critical Reviews in Solid State and Materials Science,* 1981, p. 261.

33. M. G. Thomas, B. F. Beier, and E. L. Muetterties, *J. Am. Chem. Soc.,* **98**, 1296 (1976).

34. A. Brenner, and D. A. Hucul, *J. Am. Chem. Soc.,* **102**, 2484 (1980).

35. D. A. Hucul, and A. Brenner, *J. Am. Chem. Soc.,* **103**, 217 (1981).

36. A. K. Smith, A. Theolier, J. M. Basset, R. Ugo, D. Commereuc, and Y. Chauvin, *J. Am. Chem. Soc.,* **100**, 2590 (1978).

37. T. J. Thomas, D. A. Hucul, and A. Brenner, ACS Symp. Ser. Vol. 192, 1982, p. 267.

38. S. Sivasanker, E. P. Yesodharan, C. Sudhakar, A. Brenner, and C. B. Murchison, *J. Catal.,* **87**, 514 (1984).

39. A. Cichowlas, E. P. Yesodharan, and A. Brenner, *Appl. Catal.,* **11**, 353 (1984).

40. A. Brenner, and D. A. Hucul, Proc. 3rd. Int. Conf. Chem. Uses Molybdenum, 1979, p. 194.

41. A. Brenner, D. A. Hucul, and S. J. Hardwick, *Inorg. Chem.,* **18**, 1478 (1979).

42. D. A. Hucul, Ph.D. Thesis, Wayne State University, May 1980.

43. A. Brenner, and D. A. Hucul, *Prepr., Div. Petrol. Chem., ACS,* **22**, 1221 (1977).

44. A. Brenner, *J. Mol. Catal.,* **5**, 157 (1979).

45. A. Brenner, and D. A. Hucul, *J. Catal.,* **61**, 216 (1980).

46. A. Brenner, and D. A. Hucul, *Inorg. Chem.,* **18**, 2836 (1979).

47. K. Tanaka, K. L. Watters, and R. F. Howe, *J. Catal.,* **75**, 23 (1982).

48. M. Ichikawa, *J. Catal.,* **59**, 67 (1979).

49. L. Guczi, Z. Schay, K. Matusek, I. Bogyay, and G. Steffer, Proc. 7th Int. Cong. Catal., Tokyo, 1980, p. A12.

50. P. Gallezot, G. Coudurier, M. Primet, and B. Imelik, ACS, Symp. Ser. Vol. 40, 1977, p. 144.

51. H. Knozinger, and Y. Zhao, *J. Catal.,* **71**, 337 (1981).

52. D. A. Hucul, and A. Brenner, *J. Phys. Chem.,* **85**, 496 (1981).

53. A. Brenner, unpublished results.

54. A. Brenner, Proc. 2nd Int. Symp. Relat. Homog. Heterog. Catal., 1977 p. 195.

55. C. Sudhakar, E. P. Yesodharan, A. Cichowlas, M. Majer, and A. Brenner, *Prepr., Div. Pet. Chem., ACS,* **27**, 440 (1982).

56. R. G. Bowman, and R. L. Burwell, Jr., *J. Catal.,* **63**, 463 (1980).

57. J. R. Anderson, P. S. Elmes, R. F. Howe, and D. E. Mainwaring, *J. Catal.,* **50**, 508 (1977).

58. R. Psaro, R. Ugo, G. M. Zanderighi, B. Besson, and A. K. Smith, *J. Organometal. Chem.,* **213**, 215 (1981).

59. T. Okuhara, K. Kobayashi, T. Kimura, M. Misono, and Y. Yoneda, *J. Chem. Soc. Chem. Commun.,* 1114 (1981).

60. M. Ichikawa, *J. Chem. Soc. Chem. Commun.,* 26 (1976).

61. M. Ichikawa, *J. Chem. Soc. Chem. Commun.,* 11 (1976).

62. G. C. Smith, T. P. Chojnacki, K. Iwatate, and K. L. Watters, *Inorg. Chem.,* **14**, 1419 (1975).

63. A. K. Smith, F. Hugues, A. Theolier, J. M. Basset, R. Ugo, G. M. Zanderighi, J. L. Bilhou, V. B. Bougnol, and W. F. Graydon, *Inorg. Chem.,* **18**, 3104 (1979).

64. G. Collier, D. J. Hunt, S. D. Jackson, R. B. Moyes, I. Pickering, P. B. Wells, A. F. Simpson, and R. Whyman, *J. Catal.,* **80**, 154 (1983).

65. A. Zecchina, E. Guglielminotti, A. Bossi, and M. Camia, *J. Catal.,* **74**, 225 (1982).

66. M. Deeba, and B. C. Gates, *J. Catal.,* **67**, 303 (1981).

67. A. K. Smith, B. Besson, J. M. Basset, R. Psaro, A. Fusi, and R. Ugo, *J. Organometal. Chem.,* **192**, C31 (1980).

68. J. E. Crawford, G. A. Melson, L. E. Makovsky, and F. R. Brown, *J. Catal.,* **83**, 454 (1983).

69. X. J. Li, and B. C. Gates, *J. Catal.,* **84**, 55 (1983).

70. J. Schwank, L. F. Allard, M. Deeba, and B. C. Gates, *J. Catal.,* **84**, 27 (1983).

71. H. Knozinger, Y. Zhao, B. Tesche, R. Barth, R. Epstein, B. C. Gates, and J. P. Scott, *Chem. Soc. Faraday Disc.,* **72**, 53 (1981).

72. V. L. Kuznetsoz, A. T. Bell, and Y. I. Yermakov, *J. Catal.,* **65**, 374 (1980).

73. F. Hugues, J. M. Basset, Y. B. Taarit, A. Choplin, M. Primet, D. Rojas, and A. K. Smith, *J. Am. Chem. Soc.,* **104**, 7020 (1982).

74. F. Hugues, A. K. Smith, Y. B. Taarit, J. M. Basset, D. Commereuc, and Y. Chauvin, *J. Chem. Soc. Chem. Commun.*, 68 (1980).

75. D. B. Tkatchenko, and G. Coudurier, *Inorg. Chem.*, **18**, 558 (1979).

76. D. A. Hucul, and A. Brenner, *J. Chem. Soc.*, *Chem. Commun.*, 830 (1982).

77. O. Beeck, *Faraday Soc.*, **8**, 118 (1950).

78. G. C. A. Schuit, and L. L. Van Reijen, *Adv. Catal.*, **10**, 242 (1958).

79. J. Robertson, and G. Webb, *Proc. R. Soc. London. A*, **341**, 383 (1974).

80. J. H. Sinfelt, *Adv. Catal.*, **23**, 91 (1973).

81. M. Deeba, J. P. Scott, R. Barth, and B. C. Gates, *J. Catal.*, **71**, 373 (1981).

82. D. L. King, J. A. Cusumano, and R. L. Garten, *Catal. Rev.-Sci. Eng.*, **23**, 233 (1981).

83. D. Commereuc, Y. Chauvin, F. Hughes, J. M. Basset, and D. Oliver, *J. Chem. Soc., Chem. Commun.*, 154 (1980).

84. F. Hugues, B. Besson, P. Bussiere, J. Dalmon, M. Leconte, J. M. Basset, Y. Chauvin, and D. Commereuc, ACS Symp. Ser. Vol. 192, 1982, p. 255.

85. F. Hugues, P. Bussiere, J. M. Basset, D. Commereuc, Y. Chauvin, L. Bonnevoit, and D. Oliver, *Proc. 7th Int. Cong. Catal.*, Tokyo, 1980, p. A28.

86. G. B. McVicker, and M. A. Vannice, *J. Catal.*, **63**, 25 (1980).

87. D. B. Tkatchenko, G. Coudurier, H. Mozzanega, and I. Tkatchenko, *J. Mol. Catal.*, **6**, 293 (1979).

88. D. B. Tkatchenko, and I. Tkatchenko, *J. Mol. Catal.*, **13**, 1 (1981).

89. D. B. Tkatchenko, N. D. Chau, H. Mozzanega, M. C. Roux and I. Tkatchenko, ACS Symp. Ser. Vol. 152, 1981, p. 187.

90. H. H. Nijs, P. A. Jacobs, and J. B. Uytterhoeven, *J. Chem. Soc. Chem. Commun.*, 180 (1979).

91. H. H. Nijs, P. A. Jacobs, and J. B. Uytterhoven, *J. Chem. Soc. Chem. Commun.*, 1095 (1979).

92. G. A. Melson, J. E. Crawford, J. W. Crites, K. J. Mbadcam, J. M. Stencel, and V. U. S. Rao, ACS Symp. Ser. Vol. 218, 1983, p. 397.

93. Y. W. Chen, H. T. Wang, and J. G. Goodwin, Jr., *J. Catal.*, **83**, 415 (1983).

94. C. S. Kellner, and A. T. Bell, *J. Catal.*, **70**, 418 (1981).

95. H. E. Ferkul, J. M. Berlie, D. J. Stanton, J. D. McCowan and M. C. Baird, *Can. J. Chem.*, **61**, 1306 (1983).

96. G. F. Meyers, and M. B. Hall, *Inorg. Chem.*, **23**, 124 (1984).

97. M. Ichikawa, *J. Chem. Soc. Chem. Commun.*, 566 (1978).

98. M. Ichikawa, *Bull. Chem. Soc. Jpn.*, **51**, 2268 (1978).

99. M. Ichikawa, *Bull. Chem. Soc. Jpn.*, **51**, 2273 (1978).

100. M. Ichikawa, *Chemtech.*, 674 (1982).

101. M. Ichikawa, K. Sekizawa, and K. Shikakura, *J. Mol. Catal.*, **11**, 167 (1981).

102. J. R. Anderson, and D. E. Mainwaring, *J. Catal.*, **35**, 162 (1974).

103. J. R. Anderson, and D. E. Mainwaring, *Ind. Eng. Chem. Prod. Res. Dev.* **17**, 202 (1978).

104. A. Brenner, S. Hardwick, D. A. Hucul, and G. D. Stucky, *Abstract 234, Div. Inorg. Chem., ACS*, 178th Nat. Mtg., 1979.

105. T. J. Thomas, G. Mistalski, D. A. Hucul, and A. Brenner, *Abstract 11, Div. Coll. Surf. Chem., ACS,* 179th Nat. Mtg., Houston, 1980.

106. J. R. Shapley, S. J. Hardwick, D. S. Foose, and G. Stucky, *Prepr., Div. Petrol. Chem., ACS* **25**, 780 (1980).

107. K. Lazar, Z. Schay, and L. Guczi, *J. Mol. Catal.,* **17**, 205 (1982).

108. M. Ichikawa, *J. Catal.,* **56**, 127 (1979).

109. D. W. Studer, and G. L. Schrader, *J. Mol. Catal.,* **9**, 169 (1980).

110. J. Lieto, J. Rafalko, and B. C. Gates, *J. Catal.,* **62**, 149 (1980).

111. Z. O. Schipper, J. Lieto, and B. C. Gates, *J. Catal.,* **63**, 175 (1980).

112. R. Pierantozzi, K. J. McQuade, B. C. Gates, M. Wolf, H. Knozinger, and W. Ruhmann, *J. Am. Chem. Soc.* **101**, 5436 (1979).

113. M. B. Freeman, M. A. Patrick, and B. C. Gates, *J. Catal.,* **73**, 82 (1982).

114. R. Pierantozzi, K. J. McQuade, and B. C. Gates, *Proc. 7th Int. Cong. Catal.,* Tokyo, 1981, p. 941.

10

SUPPORTED METAL CLUSTER CATALYSTS

BRUCE C. GATES
Center for Catalytic Science and Technology
Department of Chemical Engineering
University of Delaware
Newark, Delaware

10.1   INTRODUCTION

Supported metals, especially of the platinum group, are among the most important catalysts used in chemical technology, finding application in processes ranging from reforming of petroleum distillates to conversion of automobile exhaust. These catalysts consist of aggregates (or crystallites) of metal dispersed on a support, typically a metal oxide such as γ-Al_2O_3 with a microporous structure and a surface area of as much as several hundred square meters per gram. The metal aggregates are nonuniform, having a variety of sizes, shapes, interactions with the support, and catalytic activities. Usually these aggregates are small ($\sim$10 to 100 Å in diameter) so that a large fraction of the metal atoms are exposed at a surface, where they can be catalytically engaged. These catalysts, like almost all those used in chemical technology, are subject to severe conditions and lose activity during operation, for example, as a result of accumulation of carbonaceous deposits (coke) covering the surface or as a result of loss of metal surface area by aggregation (sintering) of the metal particles.

Some supported metal catalysts, including those used commercially for α-olefin polymerization, are simpler in structure, consisting of mononuclear (single metal atom) complexes on the surfaces of the oxide supports. The metals in these catalysts are usually present in positive oxidation states. Typical metals are Cr, Zr, Ti, and Mo, and typical supports are SiO_2 and Al_2O_3. The activities of these catalysts are restricted to structure-insensitive reactions (those requiring single metal centers)—unless the support itself is part of the catalytically active site.

An intermediate class of supported metals consists of small, nearly uniform metal aggregates held in the molecular-scale cages of molecular sieve zeolites. An important catalyst in this group is Pd/H-mordenite, used in hydroisomerization of gasoline-range paraffins to provide increased branching and, therefore, increased octane number. The size of the metal aggregate is restricted as long as the metal is retained inside the pore structure of the support. The zeolite-encaged metals are often not stable, however, since the metal migrates out of the pores, forming larger crystallites.

Metal clusters* have been used to prepare catalysts in all the categories mentioned previously. Supported analogs of the molecular clusters themselves have also been prepared. These are a new class of supported metal, being unique because of the simplicity of their structures; they are the focus of the present review.

Supported metal clusters have been prepared by routes either inferred from known organometallic chemistry (combined with the functional group chemistry of the support surface) or determined empirically. Most attempts at synthesis have led to

---

*The term *metal cluster* is used here to mean a molecule with more than one metal atom or a surface-bound species that is a molecular analog. The term *aggregate* is used to denote the small (nonuniform) metal structures found in typical supported metal catalysts. In the literature of surface catalysis the latter are often referred to as metal clusters.

structurally complex materials; we consider only those that appear to be structurally unique.

The motivation for investigation of this new class of supported metal derives from the importance of metals in industrial catalysis. The compelling goal is the preparation of supported metals having simple, uniform structures with neighboring metal centers to allow determination of relations between structure and catalytic properties. (With conventional supported metals, the best that one can do is to determine relations between catalytic properties and some average structural properties.) Another goal is to determine the nature of the interactions between the metal and the oxide support; specific chemical bonds are important in the surface-bound organometallics, whereas in typical supported metals the nature of the metal support interactions is poorly understood.

There is also a more practical goal motivating research with metal clusters: catalysts with new activities and, especially, selectivities. Metal clusters offer new combinations of metal centers and surrounding ligands, and the most common ligands (e.g., H, CO, small hydrocarbons) are important building blocks in the chemical industry. The initial high hopes for metal clusters as catalysts have not been realized, however; the primary difficulty is the lack of stability of the clusters.

The practical interest in attaching metal clusters to supports is related to the obvious advantages of having the catalytic species isolated in a separate phase— ease of separation from products and minimized corrosion—and also the prospects of stabilizing metal clusters (or simple structures related to them) by immobilizing them. The most successful industrial processes involving organometallic catalysis [e.g., olefin hydroformylation (1) and methanol carbonylation (2)] are notable for their high selectivities; a high selectivity is indicative of a catalytic cycle with one predominant pathway; metal clusters, if they can be stabilized, are good candidates as catalysts offering high selectivities. There is an extensive patent literature suggesting that rhodium carbonyl clusters are catalytically active for formation of ethylene glycol from $CO + H_2$ (synthesis gas), but the selectivity (methanol and other products are also formed) and activity (high pressures and temperatures are required) are not sufficient to justify commercialization.

10.2 SYNTHESIS OF SUPPORTED METAL CLUSTERS

10.2.1 Clusters on Polymers

The first attempt to prepare a polymer-supported metal cluster was reported by Collman et al. (3) in 1972; these authors brought $[Rh_6(CO)_{16}]$ in contact with phosphine-functionalized poly(styrene-divinylbenzene), forming structures that are still not well understood (4). Perhaps the first supported clusters with well-defined structures, described in 1978 (5), were tetrairidium carbonyls supported on phosphine-functionalized polymers. The majority of the known supported metal clusters have been prepared with functionalized poly(styrene-divinylbenzene) (Table 10.1).

TABLE 10.1 Polymer-supported Metal Clusters with Well-defined Structures

Metal Framework Composition	Support	Structures	Principal Characterization Method	References
Ir_4	Poly(styrene-divinylbenzene)	$Ir_4(CO)_{12-y}(Ph_2P$—℗$)_y$ $y = 1$ or 2	IR	5,6
Ru_4	Poly(styrene-divinylbenzene)	$H_4Ru_4(CO)_{12-x}(Ph_2P$—℗$)_x$ $x = 1, 3,$ or 4	IR	7
Fe_2Pt	Poly(styrene-divinylbenzene)	$Fe_2Pt(CO)_8(Ph_2P$—℗$)_2$	IR	8
$RuPt_2$	Poly(styrene-divinylbenzene)	$RuPt_2(CO)_5(Ph_2P$—℗$)_3$	IR	8
$AuOs_3$	Poly(styrene-divinylbenzene)	$HAuOs_3(CO)_{10}(Ph_2P$—℗$)$	IR	9
$AuOs_3$	Poly(styrene-divinylbenzene)	$ClAuOs_3(CO)_{10}(Ph_2P$—℗$)$	IR	9
$PtOs_3$	Poly(styrene-divinylbenzene)	$H_2PtOs_3(CO)_{10}(Ph_2P$—℗$)_2$	IR	9
Os_3	Poly(styrene-divinylbenzene)	$H_2Os_3(CO)_9Ph_2P$—℗, $H_2Os_3(CO)_{10}Ph_2P$—℗	IR	10,11
Os_3	Poly(styrene-divinylbenzene)	$Os_3(CO)_8(\mu\text{-}Cl)_2(Ph_2P$—℗$)_2$	IR	12
Os_3	Poly(styrene-divinylbenzene)	$[HOs_3(CO)_{12}]^-\,Et_3\overset{+}{N}$—$CH_2$—℗	IR	13
$RhOs_3$	Poly(styrene-divinylbenzene)	$H_2RhOs_3(CO)_{10}(acac)Ph_2P$—℗	IR	14
Fe_3	Poly(styrene-divinylbenzene)	$[HFe_3(CO)_{11}]^-\,Et_3\overset{+}{N}$—$CH_2$—℗	IR	15
Fe_4S_4	Poly(styrene-divinylbenzene)	$[Fe_4S_4(St\text{-}Bu)_4]^-\,Et_3\overset{+}{N}$—$CH_2$—℗	far IR	16
Co_3C	Poly(styrene-divinylbenzene)	$Co_3C(CO)_9$—CH_2—℗	(far) IR	17

286

Hydrocarbon polymers offer the advantages of easy functionalization, for example, with phosphine, amine, and thiol groups, and virtual inertness of the remainder of the support (18). Consequently, a number of syntheses of supported metal clusters have been carried out in direct analogy to known solution chemistry.

Cluster attachment has been effected by simple ligand association involving a coordinatively unsaturated cluster, for example, as follows (10,11):

$$H_2Os_3(CO)_{10} + Ph_2P–Ⓟ \longrightarrow H_2Os_3(CO)_{10}(Ph_2P–Ⓟ) \qquad (10.1)$$

A number of supported clusters have also been prepared by ligand exchange, for example (9),

$$ClAuOs_3(CO)_{10}(PPh_3) + Ph_2P–Ⓟ \longrightarrow ClAuOs_3(CO)_{10}(Ph_2P–Ⓟ) + PPh_3$$

$$(10.2)$$

Others have been prepared by in situ cluster synthesis from a mononuclear complex, the chemistry being analogous to that occuring in solution (6):

$$Ph_2P–Ⓟ + Ir(CO)_2Cl(p\text{-toluidine}) \xrightarrow{CO,Zn} Ir_4(CO)_{11}Ph_2P–Ⓟ \qquad (10.3)$$

It is crucial in these syntheses that appropriate solvents be used; they must dissolve the reagents without reacting with them and at the same time swell the polymer to allow entry of the reagents.

It is expected that many functionalized polymers in addition to poly(styrene-divinylbenzene) will be suitable supports. The reason so much work has been done with poly(styrene-divinylbenzene) is that it is easily prepared with a wide range of physical properties and concentrations of functional groups (18). Even high surface area macroporous polymers are available (18), and these are the most stable and, in prospect, the best candidates for application. A major disadvantage of the polymers is their lack of stability at high temperatures [≳150°C for poly(styrene-divinylbenzene)].

Both block and random copolymers of phosphine-functionalized poly(styrene-divinylbenzene) have been prepared; these offer contrasting reactivities in the synthesis of supported metal clusters (6,18). Random copolymers with low concentrations of functional groups are best for synthesis of monosubstituted clusters (e.g., $Ir_4(CO)_{11}PPh_2—Ⓟ$), whereas block copolymers (having regions with high concentrations of functional groups) are best for multiple substitution of the clusters. The flexible polymer networks have the ability to chelate the metal cluster; even four phosphine groups can be bonded to a single tetraruthenium cluster in the polymer (7). The degree of substitution can be further regulated with synthesis conditions such as time, temperature, and solvent; variously substituted clusters of tetrairidium and tetraruthenium have been prepared in high yields with phosphine-functionalized polymers (Table 10.1).

In summary, the syntheses of polymer-supported metal clusters are often straight-

forward extensions of organometallic chemistry in solution, and manipulation of
the synthesis conditions and physical properties of the support allows systematic
variations in the ligand surroundings of the cluster. Extension of this class of supported metal clusters will be straightforward.

10.2.2 Clusters on Functionalized Oxides

Metal oxides can be functionalized with pendant groups such as those mentioned
in connection with the polymers (19), and routes for attachment of metal clusters
are analogous to those previously mentioned. A list of the metal clusters supported
on functionalized oxides is given in Table 10.2. Complications sometimes ensue in
these syntheses, since the surfaces of the oxides (unlike the hydrocarbon polymers)
are not inert; the methods are therefore less general than with the polymers.

10.2.3 Clusters on Unfunctionalized Oxides

The unmodified oxides themselves are the supports most commonly used in industrial
catalysts, being stable, inexpensive, and available with high surface areas. Attachment of metal clusters to oxide surfaces proceeds through reactions with the groups
present on the (partially hydrated) surfaces. The native functional groups include
the ubiquitous -OH groups in various surroundings. For example, on Al_2O_3 there

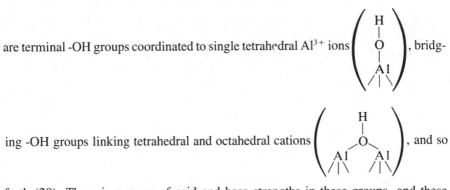

are terminal -OH groups coordinated to single tetrahedral Al^{3+} ions ( ), bridging -OH groups linking tetrahedral and octahedral cations ( ), and so

forth (29). There is a range of acid and base strengths in these groups, and these
strengths vary from oxide to oxide, MgO being generally basic, for example, and
SiO_2 almost neutral. There are also coordinatively unsaturated (Lewis acid) centers
such as Al^{3+}. The concentrations of these groups can be adjusted by controlled
dehydration of the surface; for example, calcining of a γ-Al_2O_3 sample at 300°C
gives a sample with about half the maximum concentrations of -OH groups, namely,
about 7×10^{14} groups/nm^2 (29).

There are only several examples of oxide-supported clusters with well-defined
structures (Table 10.3), the most thoroughly investigated by far being triosmium
clusters on SiO_2 and Al_2O_3, first described in the seminal work of Ugo et al. (30)
published in 1979. A more thorough account of this work has also appeared (31).
This synthesis proceeds via oxidative addition involving the triosmium cluster and

TABLE 10.2 Supported Metal Clusters with Well-defined Structures on Functionalized Oxides

Metal Framework Composition	Support	Structures	Principal Characterization Method	References
Ru_4	Ph_2P—SIL^a	$H_4Ru_4(CO)_{12-x}(Ph_2P$—$SIL)_x$ $x = 1, 2,$ or 3	IR	8
Os_3	Ph_2P—SIL	$H_2Os_3(CO)_9(Ph_2P$—$SIL)$, among others	IR	8,20–22
Os_3	$Ph_2P(CH_2)_2$—SIL	$Os_3(CO)_9(\mu$-$Cl)_2PPh_2(CH_2)_2$—SIL	IR	12
Ir_4	Ph_2P—SIL	$Ir_4(CO)_9(PPh_3)_2(Ph_2P(CH_2)_3$—$SIL)$ $Ir_4(CO)_{10}(PPh_3)(Ph_2P(CH_2)_3$-$SIL)$		23,24
$AuOs_3$	$Ph_2P(CH_2)_2$—SIL	$HAuOs_3(CO)_{10}Ph_2P(CH_2)_2$—$SIL$	IR	25
	$Ph_2P(CH_2)_2$—SIL	$ClAuOs_3(CO)_{10}Ph_2P(CH_2)_2$—$SIL$	IR	25
Os_3	HS—SIL	$HOs_3(CO)_{10}S(CH_3)_2$—SIL	IR	27
Ru_3	HS—SIL	$HRu_3(CO)_{10}S(CH_3)_2$—SIL	IR	27
$FeCo_3$	$H_2N(CH_2)_3$—SIL	$[FeCo_3(CO)_{12}]^- \overset{+}{H_3}N(CH_2)_3$—$SIL$	IR, uv-visible	28

^{a}SIL is silica.

TABLE 10.3 Oxide-supported Metal Clusters with Well-defined Structures

Metal Framework Composition	Support	Structures	Principal Characterization Methods	References
Os_3	SiO_2, Al_2O_3, TiO_2,[a] ZnO[a]	$HOs_3(CO)_{10}$—O—M	IR Raman XPS EXAFS TEM uv-visible	30–42
Os_3	MgO	Exact structures undetermined	IR	43
Ru_3	SiO_2	$HRu_3(CO)_{10}$—O—Si	IR	44
Fe_3	MgO, Al_2O_3, ZnO	$[Fe_3(CO)_{11}]^-$ / support	IR	45
Os_4	γ-Al_2O_3	$[H_3Os_4(CO)_{12}]^{-\,+}Al$	IR	46
$RuOs_3$	γ-Al_2O_3	$[HRuOs_3(CO)_{13}]^{-\,+}Al$, $[H_3RuOs_3(CO)_{12}]^{-\,+}Al$	IR	47

[a]Structure based on IR spectrum only.

a surface -OH group; the stoichiometry of the synthesis was established by measurement of CO evolution (31). The reaction can be reversed in the presence of CO

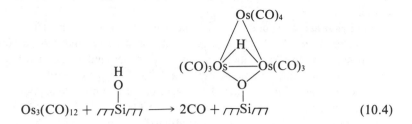

$$Os_3(CO)_{12} + \text{/77}Si\text{/77} \longrightarrow 2CO + \text{/77}Si\text{/77} \qquad (10.4)$$

(31). The synthesis seems to be rather general with surface -OH groups; it proceeds not only with SiO_2 and η-Al_2O_3 (31), but also with γ-Al_2O_3, TiO_2, and ZnO (32). The syntheses have been carried out with vapor-phase $[Os_3(CO)_{12}]$ (33), with $[Os_3(CO)_{12}]$ in refluxing octane (or heptane) (32,34), or by heating the physisorbed cluster on the support (31). The key requirement appears to be the temperature of the reaction; the role of water is significant and not fully understood (31), and the subtle differences between the oxides are not well understood.

Alternatively, the Al_2O_3-supported triosmium cluster has been prepared with a starting compound having a labile cyclohexadiene ligand (35)

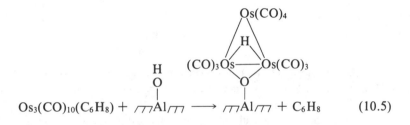

$$Os_3(CO)_{10}(C_6H_8) + \text{/77}Al\text{/77} \longrightarrow \text{/77}Al\text{/77} + C_6H_8 \qquad (10.5)$$

A similar synthesis on SiO_2 has been carried out with $[Os_3(CO)_{10}(CH_3CN)_2]$ (31). These preparations are analogous to many carried out with mononuclear metal complexes and surface -OH groups (48); they can be expected to have more general applicability than that illustrated by equation (10.3), especially with less robust clusters than triosmium.

The anchoring of triruthenium clusters to SiO_2 proceeds similarly to the anchoring of triosmium clusters (27,44)

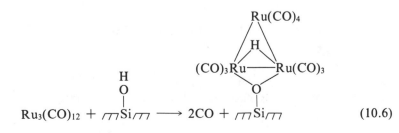

$$Ru_3(CO)_{12} + \text{/77}Si\text{/77} \longrightarrow 2CO + \text{/77}Si\text{/77} \qquad (10.6)$$

The synthesis with $[Ru_3(CO)_{12}]$, in contrast to that with $[Os_3(CO)_{12}]$, must be carried out in the absence of air. The anchored ruthenium cluster is much less stable than the osmium analog, decomposing into particles of metal and oxidized ruthenium complexes (44).

Similar organoruthenium surface chemistry has been elucidated with tetraruthenium clusters on Al_2O_3 (49,50). A variety of structures have been characterized by infrared spectroscopy, including metal particles and mononuclear complexes; there is evidence for the anionic cluster $[H_3Ru_4(CO)_{12}]^-$ (49,50).

Several other anionic clusters have been prepared on oxide supports. Hugues et al. (45) investigated the reactivity of $[Fe_3(CO)_{12}]$ with Al_2O_3, MgO, and ZnO, determining infrared, ultraviolet, and ^1H NMR spectra of the surface-bound organometallics. The spectra of the Al_2O_3-supported species are in fair aggreement with those expected for the molecular anion cluster $[HFe_3(CO)_{11}]^-$. The following surface structure was suggested (45):

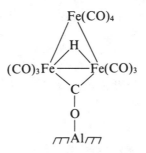

The anion cluster could be extracted from the surface with NEt_4Cl, giving $[NEt_4]^+$ $[HFe_3(CO)_{11}]^-$. Similar surface chemistry was observed with MgO and ZnO (45).

Al_2O_3-supported anion clusters $[H_3Os_4(CO)_{12}]^-$ ^+Al have been prepared by refluxing an n-octane solution of $[H_4Os_4(CO)_{12}]$ in the presence of Al_2O_3 (46). The chemistry is similar to that observed for the neutral cluster in methanolic KOH, evidently involving the basic groups on Al_2O_3. The cluster could be extracted from the surface as the tetraphenylarsonium salt.

Al_2O_3-supported anion clusters have also been prepared from $[H_2RuOs_3(CO)_{13}]$, as inferred from infrared spectra (48). The surface reaction was suggested to be

$$Al\!-\!OH + H_2RuOs_3(CO)_{13} \xrightarrow{-H_2O} Al^+[HRuOs_3(CO)_{13}]^- \qquad (10.7)$$

Heating the sample in $CO + H_2$ gave a new surface species, inferred to be $[H_3RuOs_3(CO)_{12}]^-$. This cluster was identified by its infrared spectrum; it could be extracted from the surface with Ph_4AsCl, giving $[Ph_4As]^+[H_3RuOs_3(CO)_{12}]^-$ (47).

This supported cluster anion appears to be relatively robust, being stable in H_2 + CO at temperatures up to about 200°C (47).

10.3 CHARACTERIZATION

Supported metal clusters, being simpler in structure than typical supported metal catalysts, are much more susceptible to exact structural characterization; some striking results have been obtained, helping to pave the way in the new field of surface organometallic chemistry (51). It is expected that the results obtained with the simple supported metals will also help open the way to improved characterization of traditional supported metals.

Infrared spectroscopy is the most useful and easily applied technique for characterization of supported metal clusters. The intense stretching vibrations in the carbonyl region have been used to fingerprint all of the reported materials (Tables 10.1 to 10.3), and most supported metal clusters have been characterized by this technique alone. The infrared results have provided clear indications of the structures of supported clusters which are simple analogs of molecular clusters. Far infrared spectra are useful for characterizing metal–metal bonds (16,17).

For example, the carbonyl spectra of $[Ir_4(CO)_{12-x}(PPh_3)_x]$ (x = 1, 2, or 3) are compared with those of their poly(styrene-divinylbenzene)-supported analogs in Figure 10.1; the carbonyl spectra of each molecular cluster and its supported analog match almost exactly, and the results provide a sharp distinction among the several species with different ligand environments. Not all samples, however, lend themselves to such direct identification, especially those on oxide supports, which often provide less intense spectra than polymer membranes. The reported results for oxide-supported clusters have been obtained with semitransparent wafers pressed from powders; some of the best data have been obtained by multiple scanning with Fourier transform infrared spectrophotometers.

The other vibrational spectroscopies, although less easily applied, provide valuable, complementary structural information. Raman spectroscopy has been used to detect the Os—Os bonds in Al_2O_3-supported triosmium clusters (Fig. 10.2) and Ru—Ru bonds in similar silica-supported triruthenium clusters (44). The method can be expected to find much wider application for supported metal clusters. Perhaps it will be of value for characterization of small metal aggregates as well, but it is still not routine; samples are subject to destruction by laser beams, and fluorescence often prevents measurement of useful spectra.

Inelastic electron tunneling spectroscopy is potentially of broad applicability for characterization of surface-bound organometallics (52). It offers the advantages of (1) a much wider range of wavelengths ($\sim$250 to 4000 cm^{-1}) than is accessible with infrared spectroscopy (absorption by the support causes cutoffs in the neighborhood of 1500 cm^{-1}); (2) high sensitivity ($\sim$1% of a monolayer can be detected); and (3) accessibility of all vibrational modes, some of which are excluded from the infrared and Raman spectra by the selection rules. The disadvantages of the technique are (1) the requirements for deposition of a metal such as lead over a significant

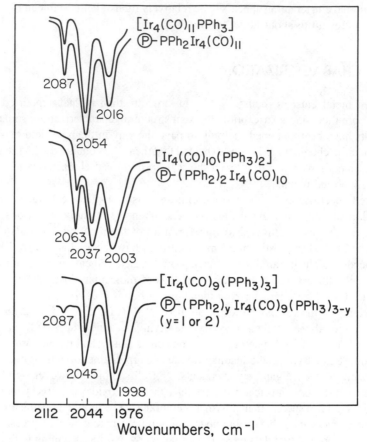

Figure 10.1 Identification of polymer-supported metal clusters by infrared spectroscopy. Spectra of membranes of polymer-supported tetrairidium carbonyl clusters and solutions of their molecular analogs (6).

fraction of the sample surface; (2) somewhat complicated sample preparation; and (3) the need to measure spectra at temperatures near absolute zero (typically 4K) to ensure superconductivity of the deposited metal.

Inelastic electron tunneling spectroscopy has been used to characterize several surface-bound organometallics, including physisorbed $[Ru_3(CO)_{12}]$ on Al_2O_3 (53) and triosmium clusters on Al_2O_3 (likely a mixture including $HOs_3(CO)_{10}$—O—Al)

(Fig. 10.3) (38). The spectra of the latter species provided the first direct evidence of the bridging hydride ligand in the surface-bound cluster.

A fourth type of vibrational spectroscopy (requiring ultra high vacuum)—electron energy loss spectroscopy (EELS)—has not yet been applied to supported metals, but the technique could be valuable, although difficult, and these materials might

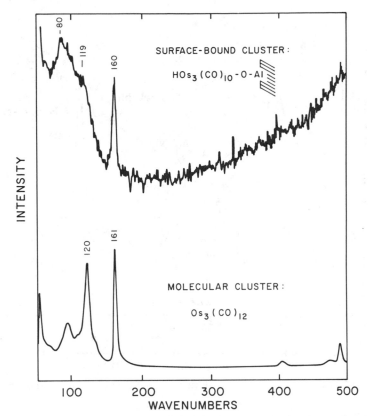

Figure 10.2 Raman spectra of a powder of the alumina-supported triosmium carbonyl cluster $HOs_3(CO)_{10}$—O-Al and of crystals of the molecular cluster $[Os_3(CO)_{12}]$ (37).

be good candidates for demonstrating the applicability of EELS to characterization of dispersed catalysts.

Electron spectroscopies have also provided some useful characterizations of supported metal clusters. X-ray photoelectron spectroscopy (XPS) [also known as electron spectroscopy for chemical analysis (ESCA)], an ultra high vacuum technique, has been used to characterize the oxide-supported triosmium clusters mentioned earlier (36). The results are indicative of the oxidation state of the metal, but alone they are far from definitive. The technique is most valuable in characterizing changes in metal oxidation states. Samples must often be handled as air sensitive.

Ultraviolet/visible reflectance spectroscopy of powder samples has provided complementary information about oxide-supported osmium clusters (37,42). The results are easily obtained but do not provide a detailed fingerprint of the surface structures. The spectra have been suggested to be useful for detecting the presence of metal–metal bonds, even in small metal aggregates derived from the original clusters (54).

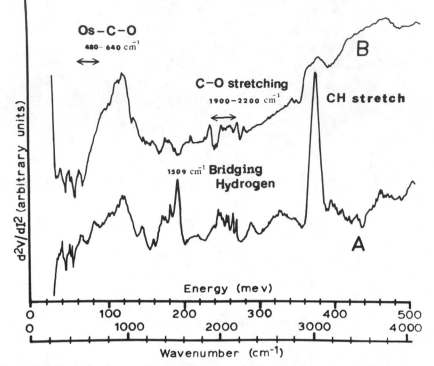

Figure 10.3 Inelastic electron tunneling spectrum of triosmium clusters on γ-Al₂O₃ obtained at 4K
(38).

Other spectroscopic techniques have been applied to characterize supported metal
clusters. The method offering the best prospects of providing exact structural information is X-ray absorption spectroscopy, including extended X-ray absorption
fine structure spectroscopy (EXAFS). EXAFS data have been reported

for $HOs_3(CO)_{10}$—O—Al, confirming the structural inferences derived from

infrared, Raman, and XP spectroscopies, and the stoichiometry of the synthesis
(39,40). Recent results establish the Os—Os distance (2.88 Å, the same as in
$[Os_3(CO)_{12}]$) and confirm that the average coordination number of osmium is 2 (39).

Only few EXAFS results have been reported because of the need for synchrotron
radiation to do the experiments [although in-house instruments have now become
available (55)] and the difficulty of data interpretation; experiments with standard
samples of known structure are essential to provide a good basis for interpretation
of data. The method is powerful and is expected to find wide application with
supported metal catatsts of many structures.

Another set of spectroscopic techniques is referred to as temperature-programmed
desorption and temperature-programmed decomposition. These are described by

Brenner in Chapter 9 of this book. Temperature-programmed decomposition provides quantitative information about the composition of the ligands surrounding a supported organometallic and information about the surface reactivity.

X-ray fluorescence is a recommended method of analysis for the amount of metal on a surface. A set of standard samples is required for calibration of the instrument, and the physical form of the sample should be similar to that of the standards. Most analyses of supported metal clusters for metal content have been carried out by dissolution of the sample and elemental analysis of the product; results obtained by commercial laboratories are sometimes erratic, and repeat analyses are strongly recommended.

Other surface spectroscopic methods offering good prospects for supported metal clusters include the following:

1. ^{1}H NMR spectroscopy, which has recently been applied to characterize hydrocarbon ligands of mononuclear rhodium complexes on SiO_2 (56). The method is similarly expected to be powerful for supported clusters. ^{13}C NMR and metal NMR are also expected to be valuable, but the methods are more difficult because of the requirement of magic angle spinning of the samples— usually in a controlled atmosphere cell.

2. Mössbauer spectroscopy, which can be expected to be of value, especially for iron-containing samples.

3. Ultraviolet photoelectron spectroscopy, which might ultimately be expected to offer some prospects as techniques are developed for supported species; ultra high vacuum is required.

4. Far infrared spectroscopy, which offers prospects of characterization of metal–metal and metal–support bonds.

5. Laser-induced mass spectrometry, which has been used recently to identify cobalt clusters on a catalyst used for CO hydrogenation (57).

Other techniques expected to be of value with the appropriate metals are electron spin resonance and magnetic susceptibility measurements.

Some of the most striking results characterizing metal clusters on supports have been obtained by high-resolution transmission electron microscopy (36,41). Figure 10.4 is a micrograph of triosmium clusters on γ-Al_2O_3. The layers in the Al_2O_3 structure are evident in the micrograph. The scattering centers indicative of the triosmium clusters are all of the same size (~ 7 Å); similar results have been obtained with tetraosmium, hexaosmium, and ruthenium-triosmium clusters (58). Even better resolution is attainable when films of the Al_2O_3 support are specially prepared by a process described as reactive evaporation of aluminum onto freshly cleaved sodium chloride crystals in the presence of oxygen (59).

In summary, supported metal clusters are among the best characterized supported metals and among the best characterized molecular species on oxide surfaces. They have provided the opportunities for demonstrating the power of some of the newer surface characterization techniques, including laser Raman spectroscopy, EXAFS,

Figure 10.4 High-resolution electron micrograph of triosmium clusters HOs$_3$(CO)$_{10}$—O-Al on

γ-Al$_2$O$_3$ (58).

inelastic electron tunneling spectroscopy, and high-resolution transmission electron microscopy. What sets the supported clusters apart is their discrete, uniform structures; many of the surface science techniques are well suited to these materials but difficult to apply incisively to the more traditional supported metals. There is a gap to be bridged here: As the simple surface structures are perturbed, they may be transformed into structures more representative of typical supported metal catalysts; as the changes in structure are following with the battery of techniques mentioned earlier, opportunities will be recognized for broader application of the techniques to dispersed catalysts.

10.4   REACTIVITY

Reactivities of supported metal clusters are for the most part poorly defined. It has been difficult to observe changes in the ligand environments of supported clusters without there being significant changes in the cluster frameworks; the most commonly observed reactions involve cluster breakup and/or aggregation of the metals on the surfaces.

A thorough set of reactivity data has been determined by infrared spectroscopy with triosmium clusters anchored to poly(styrene-divinylbenzene) and to silica through pendant phosphine ligands (10,11,20,22). The supported cluster $H_2Os_3(CO)_{10}Ph_2P$—Ⓟ, upon refluxing in hexane, loses a CO ligand to give the coordinatively unsaturated cluster $H_2Os_3(CO)_9Ph_2P$—Ⓟ (10,11,20); the silica-supported analog has also been formed, for example, from $[H_2Os_3(CO)_{10}]$ and $Ph_2PC_2H_4$-SIL (20). The unsaturated clusters react as expected from the known solution chemistry, for example, undergoing ligand association with alkynes and alkenes (10). These triosmium clusters are the only supported clusters known to give stable coordinatively unsaturated species by simple ligand dissociation. The

silica-supported clusters $Os_3(CO)_9(\mu\text{-Cl})_2PPh_2CH_2CH_2\text{-}\overset{\mathsf{E}}{\underset{\mathsf{E}}{Si}}$ have even more labile

CO ligands, and one of them is easily and reversibly exchanged with 1-butene (12).

There is evidence from gas uptake measurements and infrared spectra that SiO_2-supported triosmium clusters, $HOs_3(CO)_{10}\text{—O-}\overset{\mathsf{E}}{\underset{\mathsf{E}}{Si}}$, can reversibly coordinate eth-

ylene and hydrogen (60). Structures of the surface species have been suggested (60). The results appear to be in disagreement with results of experiments carried

out with 1-butene and $HOs_3(CO)_{10}\text{—O-}\overset{\mathsf{E}}{\underset{\mathsf{E}}{Si}}$; further experiments would be appropriate

(42) to clarify this chemistry.

The thermal stabilities of the silica- and alumina-supported triosmium clusters have been investigated in depth by Psaro et al. (31). At elevated temperatures, the clusters break apart, giving mononuclear Os^{2+} carbonyl complexes; several research groups have confirmed these observations (31,32,34). The stoichiometry of the surface reaction has been determined by measurements of CO and H_2 evolution (31). The mononuclear complex, having 2 (or 3) carbonyl ligands, has been characterized in detail by infrared spectroscopy (34) and EXAFS (39); Raman spectra of the Al_2O_3-supported sample with the broken-up cluster no longer gave evidence of Os—Os bonds, as expected (37).

The mononuclear complexes on Al_2O_3 are resistant to reduction—temperatures of 300 to 400°C in H_2 are required for formation of osmium metal, present in the form of crystallites (31,32,34). Infrared spectra have provided evidence of reconstruction of the supported cluster from the complex in the presence of CO [the reverse of the cluster breakup (43)].

All the known supported metal clusters are subject to degradation processes at elevated temperatures; products include mononuclear metal complexes and/or aggregates of metal. These processes are complex and for the most part poorly characterized. It is important to be aware of the complications that these reactions can cause; if the new structures go undetected, then incorrect conclusions about cluster reactivity (and catalytic activity) can easily be drawn.

In summary, there have been only a few systematic investigations of reactivities of surface-bound clusters. Often the reactions are complex; even if a structurally

well-defined cluster can be prepared on a surface, it will often react to give structurally complex mixtures. Understanding of surface reactivity can be expected to
develop most rapidly for clusters with stable metal frameworks (such as triosmium);
characterization by the concerted application of a variety of spectroscopic techniques
(plus determination of stoichiometries of reactions involving gain or loss of ligands)
can be expected to lead to the most rapid progress.

10.5 CATALYTIC ACTIVITY

Metal complexes, to be catalytically active, require sites of coordinative unsaturation
for binding of reactants. The best documented examples of catalysis by supported
metal clusters involve clusters that are easily stabilized in a state of coordinative
unsaturation, for example the previously mentioned triosmium clusters bonded to
phosphine-functionalized supports. The catalytic reactions involve alkenes and are
analogous to those occurring in solution in the presence of the coordinatively unsaturated cluster $[H_2Os_3(CO)_{10}]$ (Fig. 10.5) (9–11,22). Evidently, coordination of
an alkene to $H_2Os_3(CO)_9Ph_2P$—Ⓟ or $H_2Os_3(CO)_9Ph_2PC_2H_4$-$\overset{|}{\underset{|}{Si}}$ initiates a cycle
similar to that shown in Figure 10.5. The kinetics of the reaction of 1-pentene to
give 2-pentene at 75°C catalyzed by the supported cluster $H_2Os_3(CO)_9Ph_2PC_2H_4$-$\overset{|}{\underset{|}{Si}}$
in toluene has been found to be

$$r = kC_{1\text{-pentene}} \qquad (10.8)$$

where k is 0.08 ± 0.02 L/(mol of $Os_3 \cdot s$) and r has dimensions of molecules/Os_3
cluster $\cdot$ s) (10). For comparison, rates of the same reaction at the same temperature
in the presence of $[H_2Os_3(CO)_{10}]$ in toluene at 75°C were measured, the results
determining the kinetics

$$\frac{r}{C_{Os_3 \text{ cluster}}} = kC_{1\text{-pentene}} \qquad (10.9)$$

where k is 0.026 L/(mol $\cdot$ s) (10). The approximate agreement of the kinetics
supports the conclusion of a close analogy between the soluble cluster $[H_2Os_3(CO)_{10}]$
and its supported analog (10).

The supported triosmium clusters, like the soluble cluster $[H_2Os_3(CO)_{10}]$ (Fig.
10.5), are active for alkene hydrogenation, although higher pressures are required,
and again there are close similarities between the soluble and supported clusters
(10,22). The supported clusters, like their soluble analog, are unstable in the alkene
conversions, ultimately forming saturated (and inactive) triosmium species with
bridging vinyl ligands (10,22).

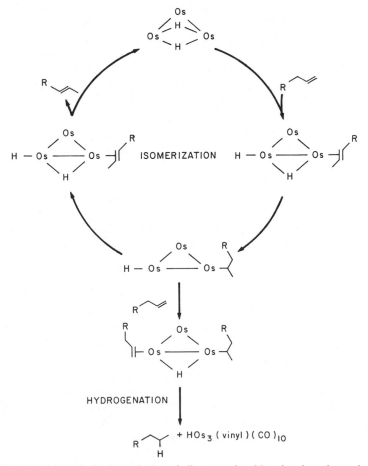

Figure 10.5 Catalytic cycle for isomerization of alkenes catalyzed by triosmium clusters in solution (61,62). The carbonyl ligands are omitted for simplicity.

Other coordinatively unsaturated clusters on supports have been prepared, including the butterfly cluster [ClAuOs₃(CO)₁₀(Ph₂P—Ⓟ)], which is an active and stable catalyst for ethylene hydrogenation at 73 to 92°C, the rate equation being

$$ r = \frac{kP_{H_2}P_{C_2H_4}}{1 + KP_{C_2H_4}} \qquad (10.10) $$

Rates were found to be in the range 10^{-3} to 10^{-2} molecules/cluster · s (9). Infrared spectra of the functioning catalyst indicated the preceding cluster as the only detectable metal species. The catalyst was stable during steady-state operation in a flow reactor, and its synthesis was reproducible, suggesting that the cluster itself provided the catalytic sites. Confirmation of this suggestion is provided by the contrast with the similar—but coordinatively saturated—supported metal cluster $HAuOs_3(CO)_{10}(Ph_2P$—Ⓟ$)$, which has a closed tetrahedral framework structure and immeasurably low catalytic activity under the same conditions (9). One cannot, however, exclude the possibility that the alkene hydrogenation was catalyzed by small amounts of metal formed from the clusters; undetectably small amounts of metal would be sufficiently active to account for the catalysis. However, with metal-catalyzed ethylene hydrogenation, a reaction order in ethylene is usually near zero, in contrast to the observations.

A number of coordinatively saturated supported metal clusters apparently are catalytically active for alkene hydrogenation. These include the polymer-supported tetrairidium and tetraruthenium clusters listed in Table 10.1. The nature of the catalytic species remains obscure, and at higher temperatures than those applied ($<100°C$), metal aggregates formed in all the samples; it is likely that in some instances small metal aggregates were the true catalytic species. This possibility seems to be fairly well established for catalysts containing tetrairidium clusters on phosphine-functionalized SiO_2 (24), and it is also considered likely for catalysts containing tetraruthenium clusters on polymers (7), which were apparently not so stable as the polymer-supported catalysts containing tetrairidium clusters.

The likelihood of decomposition of a cluster in the hydrocarbon matrix of the polymer is clearly less than that of the analogous cluster on an oxide surface. The stability of the supported tetrairidium clusters, the reproducibility of their syntheses, the indications from infrared spectroscopy that the clusters were the only detectable metal species under catalytic reaction conditions, the form of the kinetics (different from what would be expected for metal catalysis), and the influence of changes in the ligand environment on the catalytic activity (Fig. 10.6) all suggest that the clusters themselves provided the catalytic sites (6). If they did, then some undetermined mechanism provided the sites of coordinative unsaturation in a small fraction of the clusters, possibly metal–metal, metal–phosphine, or metal–CO bond dissociation (6).

It has been speculated that the mechanism involves metal–metal bond dissociation, for example, that converting a closed tetrahedral cluster framework into an open butterfly framework (9).

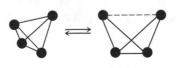

There are literature reports (e.g., Ref. 63) attesting to the importance of metal–metal bond cleavage in the activation of organic reactants by metal clusters.

Triosmium clusters anchored directly to SiO_2 (42) and $\gamma-Al_2O_3$ (64) have been shown to be catalytically active for alkene isomerization. Infrared spectra were measured during catalysis in a flow reactor/infrared cell. The spectra of the SiO_2-

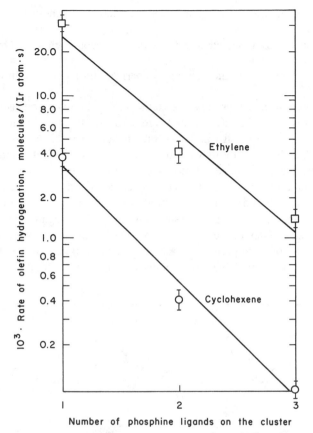

Figure 10.6 Effect of ligand environment of polymer-supported tetrairidium clusters on the catalytic activity for alkene hydrogenation (6). Ethylene hydrogenation at 70°C with partial pressures of ethylene and hydrogen of 0.2 and 0.8 atm, respectively. Cyclohexene hydrogenation at 80°C with partial pressures of cyclohexane and hydrogen of 0.15 and 0.72 atm, respectively.

supported catalyst indicated that the predominant form was the previously mentioned
$$HOs_3(CO)_{10}-O-Si,$$ which is coordinatively saturated. As flow of 1-butene was
initiated over the sample at temperatures in the range 50 to 90°C, the catalytic activity slowly increased to a maximum and then remained constant; simultaneously, there were small changes in the infrared spectra in the carbonyl region: Two new bands indicated the formation of small amounts of a cluster of different symmetry, inferred to have been formed by coordination of reactants (1-butene and/or hydrogen) (42). When CO was introduced into the feed stream, the changes in the spectrum were reversed and the catalytic activity ceased. All the results point to catalysis by

a small fraction of the surface-bound clusters—intact—at these low temperatures (42).

Similar results were observed with the γ-Al_2O_3-supported cluster $HOs_3(CO)_{10}$—O-Al (64). Catalytic reaction rate data for isomerization of 1-hexene in the presence of H_2 are shown in the upper curve of Figure 10.7. The reaction was zero order in alkene and in H_2, and the rate constants are shown in the plot. Again, all the data point to catalysis by a small fraction of the clusters themselves at these low temperatures. When the temperature was raised, however, to 120°C the catalytic activity declined, reaching a new steady state in about 2 hours; during this period the cluster broke up into the mononuclear Os^{2+} complexes described earlier, as shown by the infrared spectrum (64,65). Catalytic activity data for the broken-up clusters (i.e., the mononuclear complexes) are also shown in Figure 10.7 (again, the isomerization reaction was zero order in the reactant and in H_2). The complexes are less active than the clusters; this comparison is important in confirming that the clusters themselves were catalytically active—we can discount the possibility that small amounts

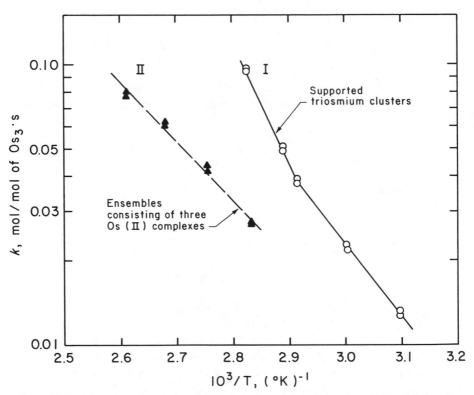

Figure 10.7 Comparison of catalytic activities of alumina-supported triosmium clusters and of ensembles formed by cluster breakup on the surface (64). The catalytic reaction was isomerization of 1-hexene taking place in a flow reactor at atmospheric pressure.

of cluster degradation products were actually responsible for the activity attributed to the clusters, since the degradation products are significantly less active than the clusters themselves.

It is not possible to determine exactly the activity of each cluster engaged in the catalysis; evidently the fraction of clusters engaged is small in each sample. The data do indicate that these oxide-bound clusters are more active for alkene isomerization than the aforementioned clusters anchored through phosphine ligands.

The silica-supported clusters have also been reported to be active catalysts for ethylene hydrogenation at 70 to 100°C, but quantitative results are not yet reported (60). Infrared spectra, mentioned previously, indicated that ethylene and hydrogen coordinated to the cluster, and structures have been suggested for the surface intermediates (60). The reaction orders in ethylene and hydrogen were found to be 0 and 1, respectively—these are the values expected for metal catalysis. These results are intriguing, and further experiments are warranted.

There are several reports of CO hydrogenation in the presence of supported metal clusters. When $HO_3(CO)_{10}$—O-Al was introduced into a CO hydrogenation reactor, catalytic conversion to methane and other hydrocarbons was observed at low pressures (1 atm) (31,36); infrared spectra of the used catalysts indicated that the clusters had broken up and that osmium (II) carbonyl complexes were the predominant—and possibly the catalytically active—species. When a triosmium carbonyl cluster (of undetermined structure) supported on MgO was introduced into a CO hydrogenation reactor, methane and other hydrocarbon products were observed at 32 atm (66). In this case, however, infrared spectra of the used catalyst indicated the presence of osmium carbonyl clusters as the only detectable osmium species and presumably those responsible for the catalytic activity. The surface clusters have been identified recently by infrared and ultraviolet/visible reflectance spectroscopy as the very stable anions $[H_3Os_4(CO)_{12}]$ and $[Os_{10}C(CO)_{24}]^{2-}$ (67); these are stable even at 275°C in CO + H_2. The catalysis is tentatively attributed to the tetraosmium clusters (67).

In summary, there are still only few data documenting catalysis by supported metal clusters—and all but one of these for low-temperature reactions. The lack of stability of the clusters is a major limitation to their application as catalysts. The available data suggest several working hypotheses for maximizing stability of clusters on supports:

1. Use of stable metal frameworks, for example, tetraosmium.
2. Use of optimum support/metal combinations, for example, the basic MgO to stabilize cluster anions such as $[H_3Os_4(CO)_{12}]^-$.
3. Use of high CO partial pressures to stabilize the clusters. High CO partial pressures may sometimes stabilize metal carbonyl clusters, but at equilibrium one might expect mononuclear complexes, which would maximize the CO to metal ratio.

10.6 SUPPORTED METALS WITH SIMPLE STRUCTURES DERIVED FROM SUPPORTED METAL CLUSTERS

Reactions leading to changes in the metal framework structure of a supported cluster usually give complex mixtures of surface structures. There is an intriguing exception, however, involving triosmium clusters on oxides. When $HOs_3(CO)_{10}$—O-

$\vdash$

Al is heated to roughly 100 to 200°C, cluster breakup into supported mononuclear

$\vdash$

$Os(CO)_2$ and/or $Os(CO)_3$ complexes takes place, as previously discussed. The changes in the infrared spectrum indicating this change are shown in Figure 10.8. Knözinger et al. (36) concluded that the mononuclear complexes were present in the form of ensembles, each consisting of three osmium ions. These ensembles have been observed by electron microscopy (36,41)—the micrographs exhibit uniform scattering centers, all the same size (about 7 Å), on the Al_2O_3 surface. Evidence for the formation of the ensembles on MgO was inferred by Deeba et al. (66), who observed that cluster reconstruction from the complexes occurred in the presence of CO; a similar observation was made for the SiO_2-supported species (42).

The ensembles are unique among supported metals with respect to their uniformity of structure; the ensembles on γ-Al_2O_3 have been observed to be stable in the high-energy electron beam of the electron microscope (41), consistent with the strong ionic bonding of the osmium to the oxide surface. When the sample was treated in hydrogen at 400°C, reduced osmium ensembles of the same size were observed—these perhaps constitute the first example of a zero-valent metal supported on an amorphous oxide and having a single size of metal aggregate. These aggregates, in contrast to the ions, were not stable in the electron beam, an indication of the weaker metal-support interaction.

Triosmium clusters have been brought in contact with a wide range of oxide supports (SiO_2, γ-Al_2O_3, MgO, TiO_2, ZrO_2, Cr_2O_3, and ThO_2) and broken up into complexes, which likely took the form of three-ion ensembles (68). These were tested as catalysts for isomerization of 1-hexene (68). The activities of the osmium supported on the transition-metal oxides were less (by an order of magnitude or more) than the activities of the osmium supported on the nontransition-metal oxides. Infrared spectra indicated that coordinatively unsaturated mononuclear osmium complexes were the catalytically active species. The activity may be related to the ease of electron transfer between the osmium and the support.

To summarize, metal clusters offer new opportunities for preparation of supported metals with unique, well-defined structures. These materials offer opportunities for determination of the effects of various metal-support interactions, ensemble size, and ensemble composition on reactivity and catalytic activity. The opportunities parallel those provided by intact metal clusters on supports—but the complications associated with lack of stability of the samples are expected to be less with some of the cluster-derived materials than with the supported clusters themselves.

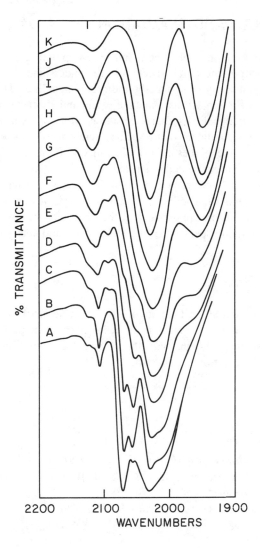

Figure 10.8 Breakup of the alumina-supported triosmium clusters $HOs_3(CO)_{10}$—O—Al indicated by changes in the infrared spectrum (62). (A) sample at 25°C; (B) sample heated to 120°C in flowing helium; samples held in flowing helium for various times: (C) 0.5 h; (D) 1 h; (E) 1.5 h; (F) 2 h; (G) 3 h; (H) 4 h; (I) 5 h; (J) 10 h; (K) after use as a catalyst in 1-hexene isomerization at 120°C for 30 h [under these conditions, the broken-up clusters (ensembles) were the catalytically active species].

10.7 SUMMARY AND EVALUATION

Supported metal clusters are still a new and little explored class of materials, but the available results portend rapid advances in the understanding of structure, reactivity, and catalytic activity of surface-bound organometallics. The key is the simplicity of structure of the materials—metal clusters and species derived from them. The structural simplicity allows characterization with a powerful and expanding arsenal of techniques available to the surface scientist. The most rapid advances will result when these techniques are used in concert. There is a need for discovery of more stable surface organometallic structures, not only to allow characterization of more materials over wider ranges of conditions, but to provide good candidate catalysts. The prospects of wholly new catalytic properties associated with structurally unique metal clusters are unrealized; the prospects are valid, but much work remains to develop stable species that are structurally and catalytically novel.

ACKNOWLEDGMENTS

This review was written at the Institute of Physical Chemistry of the University of Munich; thanks are offered to Professor H. Knözinger for helpful comments and to the Deutsche Forschungsgemeinschaft, the Fulbright Commission in Bonn, and the National Science Foundation for support.

REFERENCES

1. R. L. Pruett, *Adv. Organometal. Chem.*, **17**, 1 (1979).
2. D. Forster, *Adv. Organometal. Chem.*, **17**, 225 (1979).
3. J. P. Collman, L. S. Hegedus, M. P. Cooke, J. R. Norton, G. Dolcetti, and D. N. Marquardt, *J. Am. Chem. Soc.*, **94**, 1789 (1972).
4. M. S. Jarrell and B. C. Gates, *J. Catal.*, **54**, 81 (1978).
5. J. J. Rafalko, J. Lieto, B. C. Gates, and G. L. Schrader, Jr., *J. Chem. Soc. Chem. Commun.*, 1063 (1978).
6. J. Lieto, J. J. Rafalko, and B. C. Gates, *J. Catal.*, **62**, 149 (1980).
7. Z. Otero-Schipper, J. Lieto, and B. C. Gates, *J. Catal.*, **63**, 175 (1980).
8. R. Pierantozzi, K. J. McQuade, B. C. Gates, M. Wolf, H. Knözinger, and W. Ruhmann, *J. Am. Chem. Soc.*, **101**, 5436 (1979).
9. R. Pierantozzi, K. J. McQuade, and B. C. Gates, in *Proc. 7th Int. Cong. Catal.*, Kodansha, Tokyo, Elsevier, New York, 1981, p. 941.
10. M. B. Freeman, M. A. Patrick, and B. C. Gates, *J. Catal.*, **73**, 82 (1982).
11. J.-B. N'Guini Effa, J. Lieto, and J.-P. Aune, *J. Mol. Catal.*, **15**, 367 (1982).
12. M. Wolf, J. Lieto, B. A. Matrana, D. B. Arnold, B. C. Gates, and H. Knözinger, *J. Catal.*, **89**, 100 (1984).

13. H. Marrakchi, J.-B. N'Guini Effa, J. Lieto, M. Haimeur, and J.-P. Aune, *J. Mol. Catal.*, **30**, 101 (1985).

14. J. Lieto, M. Wolf, B. A. Matrana, M. Prochazka, B. Tesche, H. Knözinger, and B. C. Gates, *J. Phys. Chem.*, **89**, 991 (1985).

15. J.-B. N'Guini Effa, J. Lieto, and J.-P. Aune, *Inorg. Chim. Acta*, **65**, L105 (1982).

16. M. D. Monteil, J.-B. N'Guini Effa, J. Lieto, P. Verlaque, D. Benlian, *Inorg. Chim. Acta*, **76**, L309 (1983).

17. H. Atif, J.-B. N'Guini Effa, J. Lieto, and J.-P. Aune, to be published.

18. J. Lieto, D. Milstein, R. L. Albright, J. V. Minkiewicz, and B. C. Gates, *Chemtech*, **13**, 46 (1983).

19. H.-P. Boehm and H. Knozinger, in J. R. Anderson and M. Boudart, Eds., *Catalysis— Science and Technology*, Vol. 4, Springer, Berlin, 1983.

20. S. C. Brown and J. Evans, *J. Chem. Soc. Chem. Commun.*, 1063 (1978).

21. J. Evans and B. P. Gracey, *J. Chem. Soc. Chem. Commun.*, 852 (1980).

22. S. C. Brown and J. Evans, *J. Mol. Catal.*, **11**, 143 (1981).

23. T. Castrillo, H. Knozinger, and M. Wolf, *Inorg. Chim. Acta*, **45**, L235 (1980).

24. T. Castrillo, H. Knozinger, M. Wolf, and B. Tesche, *J. Mol. Catal.*, **11**, 151 (1981).

25. M. Wolf, H. Knozinger, and B. Tesche, *J. Mol. Catal.*, **25**, 273 (1984).

26. T. Catrillo, K. Knozinger, J. Lieto, and M. Wolf, *Inorg. Chim. Acta*, **44**, L239 (1980).

27. T. Castrillo de Castro, Dissertation, University of Munich, 1982; T. Castrillo and H. Knozinger, to be published.

28. R. Hemmerich, W. Keim, and M. Röper, *J. Chem. Soc. Chem. Commun.*, 428 (1983).

29. H. Knozinger and P. Ratnasamy, *Catal. Rev.-Sci. Eng.*, **17**, 31 (1978).

30. R. Ugo, R. Psaro, G. M. Zanderighi, J. M. Basset, A. Theolier, and A. K. Smith, in M. Tsutsui, Ed., *Fundamental Research in Homogeneous Catalysis*, Vol. 3, Plenum, New York, 1979, p. 579.

31. R. Psaro, R. Ugo, G. M. Zanderighi, B. Besson, A. K. Smith, and J. M. Basset, *J. Organometal. Chem.*, **213**, 215 (1981).

32. M. Deeba and B. C. Gates, *J. Catal.*, **67**, 303 (1981).

33. M. Deeba and B. C. Gates, unpublished results.

34. H. Knozinger and Y. Zhao, *J. Catal.*, **71**, 337 (1981).

35. P. L. Watson and G. L. Schrader, *J. Mol. Catal.*, **9**, 129 (1980).

36. H. Knozinger, Y. Zhao, B. Tesche, R. Barth, R. Epstein, B. C. Gates, and J. P. Scott, *Faraday Disc. Chem. Soc.*, **72**, 54 (1981).

37. M. Deeba, B. J. Streusand, G. L. Schrader, and B. C. Gates, *J. Catal.*, **69**, 218 (1981).

38. L. J. Hilliard and H. S. Gold, *Appl. Spectr.*, **39**, 124 (1985).

39. F. B. M. Duivenvoordan, D. C. Koningsberger, Y. S. Uh, and B. C. Gates, to be published.

40. S. L. Cook, J. Evans, and G. N. Graves, *J. Chem. Soc. Chem. Commun.*, 1287 (1983).

41. J. Schwank, L. F. Allard, M. Deeba, and B. C. Gates, *J. Catal.*, **84**, 27 (1983).

42. R. Barth, B. C. Gates, Y. Zhao, H. Knozinger, and J. Hulse, *J. Catal.*, **82**, 147 (1983).

43. R. Psaro, C. Dossi, and R. Ugo, *J. Mol. Catal.*, **21**, 331 (1983).

44. A. Theolier, A. Choplin, L. D'Ornelas, J. M. Basset, G. Zanderighi, R. Ugo, R. Psaro, and C. Sourisseau, *Polyhedron,* **2**, 119 (1983).

45. F. Hugues, J. M. Basset, Y. Ben Taarit, A. Choplin, M. Primet, D. Rojas, and A. K. Smith, *J. Am. Chem. Soc.,* **104**, 7020 (1982).

46. T. R. Krause, M. E. Davies, J. Lieto, and B. C. Gates, *J. Catal.,* **94**, 195 (1985).

47. J. R. Budge, J. P. Scott, and B. C. Gates, *J. Chem. Soc. Chem. Commun.,* 342 (1983).

48. Y. I. Yermakov, B. N. Kuznetsov, and V. A. Zakharov, *Catalysis by Supported Complexes,* Elsevier, Amsterdam, 1981.

49. V. L. Kuznetsov, A. T. Bell, and Y. I. Yermakov, *J. Catal.,* **65**, 374 (1980).

50. R. Pierantozzi, E. G. Valagene, A. F. Nordquist, and P. N. Dyer, *J. Mol. Catal.,* **21**, 189 (1983).

51. J. M. Basset and A. Choplin, *J. Mol. Catal.,* **21**, 95 (1983).

52. W. H. Weinberg, in J. R. Durig, Ed.,, *Vibrational Spectra and Structure,* Vol. 11, Elsevier, New York, 1982.

53. W. M. Bowser and W. H. Weinberg, *J. Am. Chem. Soc.,* **102**, 4720 (1980).

54. G. Collier, D. J. Hunt, S. D. Jackson, R. B. Moyes, I. A. Pickering, and P. B. Wells, *J. Catal.,* **80**, 154 (1983).

55. S. Khalid, R. Emrich, R. Dujari, J. Shultz, and J. R. Katzer, *Rev. Sci. Instr.,* **53**, 22 (1982).

56. H. C. Foley, S. J. DeCanio, K. D. Tau, K. J. Chao, J. H. Onuferko, C. Dybowski, and B. C. Gates, *J. Am. Chem. Soc.,* **105**, 3074 (1983).

57. G. F. Meyers and M. B. Hall, *Inorg. Chem.,* **23**, 124 (1984).

58. E. O. Odebunmi, B. A. Matrana, A. K. Datye, L. F. Allard, Jr., J. Schwank, W. H. Manogue, A. Hayman, J. H. Onuferko, H. Knözinger, and B. C. Gates, *J. Catal.,* **95**, 370 (1985).

59. B. Tesche, E. Zeitler, E. Alizo Delgado, and H. Knozinger, Proc. 40th Ann. Electr. Microsc. Soc. Am. Mtg., Washington, DC, 1982, p. 658.

60. J. M. Basset, B. Besson, A. Choplin, and A. Theolier, *Phil. Trans. R. Soc. London A,* **308**, 115 (1982).

61. A. J. Deeming and S. Hasso, *J. Organometal. Chem.,* **114**, 313 (1976).

62. J. B. Keister and J. R. Shapley, *J. Am. Chem. Soc.,* **98**, 1056 (1976).

63. G. Huttner, J. Schneider, H. D. Muller, G. Mohr, J. von Seyrl, and L. Wohlfahrt, *Angew. Chem. Int. Ed. Engl.,* **18**, 76 (1979).

64. X.-J. Li, J. H. Onuferko, and B. C. Gates, *J. Catal.,* **85**, 176 (1984).

65. X.-J. Li and B. C. Gates, *J. Catal.,* **84**, 55 (1983).

66. M. Deeba, J. P. Scott, R. Barth, and B. C. Gates, *J. Catal.,* **71**, 373 (1981).

67. H. H. Lamb and B. C. Gates, *J. Am. Chem. Soc.,* **108**, 81 (1986).

68. X.-J. Li, B. C. Gates, H. Knozinger, and E. Alizo Delgado, *J. Catal.,* **88**, 355 (1984).

INDEX